INDUSTRIAL, MECHANICAL AND MANUFACTURING SCIENCE

International Research Association of Information and Computer Science

IRAICS Proceedings Series

ISSN: 2334-0495

VOLUME 1

PROCEEDINGS OF THE INTERNATIONAL CONFERENCE ON INDUSTRIAL, MECHANICAL AND MANUFACTURING SCIENCE (ICIMMS 2014), TIANJIN, CHINA, 12–13 JUNE 2014

Industrial, Mechanical and Manufacturing Science

Editor

Dawei Zheng

International Research Association of Information and Computer Science, Beijing, China

CRC Press
Taylor & Francis Group
Boca Raton   London   New York   Leiden

CRC Press is an imprint of the
Taylor & Francis Group, an **informa** business

A BALKEMA BOOK

CRC Press/Balkema is an imprint of the Taylor & Francis Group, an informa business

© 2015 Taylor & Francis Group, London, UK

Typeset by MPS Limited, Chennai, India
Printed and bound in Great Britain by CPI Group (UK) Ltd, Croydon, CR0 4YY

All rights reserved. No part of this publication or the information contained herein may be reproduced, stored in a retrieval system, or transmitted in any form or by any means, electronic, mechanical, by photocopying, recording or otherwise, without written prior permission from the publishers.

Although all care is taken to ensure integrity and the quality of this publication and the information herein, no responsibility is assumed by the publishers nor the author for any damage to the property or persons as a result of operation or use of this publication and/or the information contained herein.

Published by: CRC Press/Balkema
P.O. Box 11320, 2301 EH Leiden, The Netherlands
e-mail: Pub.NL@taylorandfrancis.com
www.crcpress.com – www.taylorandfrancis.com

ISBN: 978-1-138-02656-8 (Hardback)
ISBN: 978-1-315-75225-9 (Ebook PDF)

Industrial, Mechanical and Manufacturing Science – Zheng (Ed.)
© *2015 Taylor & Francis Group, London, ISBN: 978-1-138-02656-8*

Table of contents

Preface	ix
Organizing committee	xi
An analysis of the influence of affordable housing system on price Y. Li, Q. Li, Y.F. Wang & Z.H. Kang	1
The research of in-pipeline robot with double driver and variable helix angle H. Zhou, L. Xu, X.Y. Cai & H.C. Zhuo	5
Simulation of coal mine trackless rubber tyre vehicle regenerative braking control strategy based on ADVISOR C. Zhang & C. Li	9
Project risk-based reliability control S. Sorooshian	13
Using adhesive instead of welds to improvement torsion stiffness of body-in-white S. Han & X. Cui	17
Application research on vinegar culture in vinegar packaging design L. Zhang & S.Y. Wang	21
Analysis of articulated chassis kinematics H. Chi, B. Gong, X. Li, H. Zang & F. Tian	25
Simulation and analysis of chatter in six-roll tandem cold rolling mills Y.L. Gong, S.J. Wang, H.R. Liu & Y. Zhang	29
Design and verification of the current transducer based on the TMR effect L.Y. Xue, K.J. Yi, W. Huang & D.Y. Song	33
Adaptive genetic algorithm for mixed-model assembly line balancing problem X. He & J. Hao	39
Research on the erosion of valve of positive pulse measurement while drilling and structural improvement C. Zhao, K. Liu & X. Chai	45
Research on the application of storm model in fatigue life prediction for offshore platform X. Yan, X. Huang & F. Liu	49
Reliability analysis of machine interference problem with vacations and impatience behavior M.J. Ma, D.Q. Yue & B. Zhao	53
Digital simulation of sensitivity in the planar mechanism motion accuracy G. Hu & S. Liu	59
Using the modified zeolites for treatment of industrial wastewater and synthesis of chemicals on the stage of recovery from aqueous solutions A.S. Konovalov, D.I. Stom, M.V. Butyrin, M.N. Saksonov & V.V. Tyutyunin	63
Catalysts for purification of sulfur-containing industrial wastewater K.A. Yurevich, K.R. Prohorovna & S.D. Iosifovich	67

Using microbial fuel cells for utilization of industrial wastewater *E.Yu. Konovalova, D.I. Stom, A.E. Balayan, E.S. Protasov, M.Yu. Tolstoy & V.V. Tyutyunin*	71
A new type of tilting mechanism for tilting tray sorter *M. Du, X. Yang, Y. Qian & L. Huang*	75
Research on baggage modeling of airport baggage handling system *Y. Chen, X. Yang, Y. Qian & C. Tang*	79
Design of automotive networking gateway based on telematics *Z.H. Dong & Z.L. Zong*	83
Processing and simulation of thin-walled shell with inner ribs formed by inner spinning technology *Z. Wang, B. Guo & S. Ma*	87
Analysis of process parameters influence on electrical discharge grinding PDC cutting tool processing *Y.H. Jia & C.Z. Guan*	93
Research on the wear and failure mechanism of tools in machining CFRP with CVD diamond film coated tools *J.D. Liu, D.M. Yu, J.W. Jin, X.Z. Ye & X.F. Yang*	99
Study on the Human-Machine interaction interface of the elderly scooter *X.Y. Fu, Y.Z. Zhu & J.F. Wu*	103
Experimental study on the fabrication of metal fibers to render micro-fin structures *M. Guo, Z.P. Wan, Z.G. Yan, W.C. Tang & X. Zheng*	107
Determination of the optimum initial operation pressure of a steam turbine unit based on a SVM and GA *G. Wu, X. Peng, Z. Yu & X. Ma*	113
Research vehicle collision avoidance warning system based on CompactRIO and LabVIEW *B. Peng, S. Li & F. Bai*	117
An improved algorithm based on LTE downlink channel modeling *Y. Jin & Q. Xu*	121
Study on the benefit of ethnic culture tourism industry in Yunnan based on method of SSA *H. Xie, J. Yi, J. Gan & Z. Zhao*	125
Numerical simulation of the flow field and cavitation in centrifugal pump *J. Wang*	131
Development of a full-wave underground MRS receiver system based on LabVIEW *G.X. Cao, J. Lin, X.F. Yi, Q.M. Duan & L.B. Feng*	135
The cooling effect and cooling energy savings potential in Beijing for metamerically color-matched cool colored coatings *J. Song, J. Qin, J. Qu, W. Zhang, Z. Song, Y. Shi, L. Jiang, J. Li, X. Xue & T. Zhang*	141
Improved K-means algorithm with better clustering centers based on density and variance *G.C. Deng, J.C. Tao, M.J. Zhou & Y.C. Xu*	147
Kind of image defect detection algorithm based on wavelet packet transform and Blob Analysis *L.X. Ao & X.B. Zhou*	151
A resolution test method for high precision accelerometer based on two-axis turntable *R. Ma, G. Yang, K. Zhang & W. Zheng*	157
Pressure fluctuation model of hydraulic turbine based on LS-SVM *X.L. An & F. Zhang*	161
The application of wireless network on quadrotor temperature observe based on NRF905 *W. Wang, Y. Zhu, T. Yu & M. Wang*	165
Hypterball batch key update method based on members' behavior *S. Liu, Y. Xu & F. Yang*	169
Waterproof time-dependent analysis of polymer modified cement mortar *D.X. Ma, Y. Liu, Y. Lai & Z.G. Luo*	175

Comparative study of highway coarse aggregate gradation used in airport pavement concrete *D.X. Ma, Y. Liu, Y. Lai & Z.G. Luo*	179
Design of rapid pesticide residue detection system based on embedded technology *S. Gong, Q. Liu, Z. He & P. Bian*	183
Application of water electrolysis oxy-hydrogen generator in the continuous cast products cutting *Z. Nie, Ying Wang, Z. Gao, C. Wang, D. Zhang, S. Liu, Y. Ren & Y. Wang*	187
Piezoelectric energy harvesting from vibration induced by jet-resonator *H. Zou, H. Chen & X. Zhu*	191
Load characteristic and strategy of power quality improvement of railways in Jiangsu *Y. Ning*	195
Experimental research on apparent viscosity behavior of different wormlike micelles *N. Li & R. Zhang*	201
Control of Lithium Battery/Supercapacitor hybrid power sources *J. Chen & S. Wu*	205
Study on compensation characteristic of ICPT system *W. Tong & S. Wu*	211
Fiber pullout with stress transfer and fracture propagation *J.Y. Wu, H. Yuan & H.W. Gu*	217
Problems and suggestions on Physical Education and Training methods in the new era *P. Cao*	223
Impact of Roman Mythology on English literature *H. Wang*	227
Practical research on mode of combining learning with working in higher vocational English education *Z. Zhang*	231
Improvement of computer multimedia technology of English education *N. Shi & N. Li*	235
Industrialization management of sports economy under market economy *A. Qu*	239
Analysis of sports news transmission and its significance *W. Zhan*	243
Analysis on the role of ideological and political education in the cultivation of sense of worth *H. Ma*	247
Development strategy of internet platform based on bilateral market theory *S. Ju*	251
Concept of art and design education *H. Sun*	255
Points of the Party's mass line education and purity construction *B. Du*	259
Dynamical and biomechanical model of side kick in free combat *S. Zhang*	263
Application of engineering technology in chemical production *X. Jiang*	267
Study on ideological and political education evaluation system of universities *Y. Qian*	271
Technical analysis of anti-seismic design for construction engineering *B. Yang*	275
Channel transmission technology of computer network environment *F. Wang*	279

Influence of digital technology on urban public art and its interactivity—a case study of cities in Jiangxi *J. Li*	283
Future transformation and development of physical education in colleges and universities *L. Wang*	287
Problems and policies of financial products innovation *Y. Shen*	291
Analysis on visual field and skill training in table tennis teaching *H. Li*	295
Native language transfer on English listening and speaking skills *H. Wei*	299
Ideological resources of humanistic care of ideological and political education *X. Wang*	303
Visualization of numerical methods for ordinary differential equations based on computer network *L. Li*	307
Study on quality control model of CPA firms under the current situation *H. Xiang*	311
Author index	315

Preface

We cordially invite you to attend the ICIMMS conference in Tianjin, China during June 12–13, 2014. The main objective of the conference is to provide a platform for researchers, engineers and academics as well as industry professionals from all over the world to present their research results and development activities in Industrial, Mechanical and Manufacturing Science. This conference provides opportunities for the delegates to exchange new ideas and experiences face to face, to establish business or research relations and to find global partners for future collaboration.

The conference received over 220 submissions which were all reviewed by at least two reviewers. As a result of our highly selective review process about 70 papers have been retained for inclusion in the proceedings, less than 40% of the submitted papers. The program of the conference consists of invited sessions, technical workshops and discussions covering a wide range of topics. This rich program provides all attendees with the opportunity to meet and interact with one another. We hope your experience is a fruitful and long-lasting one. With your support and participation, the conference will continue its success for a long time.

The conference is supported by many universities and research institutes. Many professors play an important role in the successful holding of the conference, so we would like to take this opportunity to express our sincere gratitude and highest respects to them. They have worked very hard in reviewing papers and making valuable suggestions for the authors to improve their work. We also would like to express our gratitude to the external reviewers, for providing extra help in the review process, and to the authors for contributing their research results to the conference. Special thanks go to our publisher. At the same time, we also express our sincere thanks for the understanding and support of every author. Owing to time constraints, imperfection is inevitable, and any constructive criticism is welcome.

We hope you will have a technically rewarding experience, and use this occasion to meet old friends and make many new ones. We wish all attendees an enjoyable scientific gathering in Tianjin, China. We look forward to seeing all of you next year at the conference.

The Conference Organizing Committee
June 12–13, 2014
Tianjin, China

Organizing committee

General Chair

Prof. E. Ariwa, *London Metropolitan University, UK*

Technical Committee

Prof. M. N. B. Mansor, *University Malaysia Perlis, Malaysia*
Prof. A. M. Leman, *Universiti Tun Hussein Onn Malaysia, Malaysia*
Dr. F. J. Shang, *Chongqing University of Posts and Telecommunications, China*
Prof. A. Srinivasulu, *Vignan University, India*
Prof. M. D.H, *Mangalore University, India*
Prof. C. Kumar, *Siddhant College of Engineering, India*
Prof. R. Latif, *Ibn Zohr University, Morocco*
Dr. L. T. JayPrakash, *International Institute of Information Technology-Bangalore (IIIT-B), India*
Prof. S. H. Ali, *R.M.K. Engineering College, India*
Dr K. Arshad, *University of Greenwich, School of Engineering, UK*
Prof. M. F. S. Ferreira, *University of Aveiro, Portugal*
Dr. B. S. Ahmed, *Salahaddin University, Kurdistan*
M. M. Siddiqui, *Integral University, India*
P. C. C. Anyachukwu, *University of Nigeria, Nigeria*

An analysis of the influence of affordable housing system on price

Y. Li & Q. Li
Department of Management Engineering, University of Si Chuan Jin Cheng, Chengdu, China

Y.F. Wang
Urban Planning Research Center, Planning Administration of WenJiang District, Chengdu, China

Z.H. Kang
Department of Management Engineering, University of Si Chuan Jin Cheng, Chengdu, China

ABSTRACT: In order to solve the housing problem, the government is implementing the housing affordable system, which provides the limited standards, check price or rental housing for the low income family only.

Research on the impacts of affordable house on housing price has been always a hot topic. In this paper, by constructing the effect mechanism about the affordable housing system and the housing price, analyze the interdependent and interactive relationship between them. The influence mechanism includes four factors, which are affordable housing investment, demand, land supply and the government policy. The paper comes to the conclusion that the affordable housing system has a certain inhibitory effect on the housing price and restrains the price up, however, it does not achieve the desired effect finally and existed some problems. At last, this paper put forward improvement measures to solve the difficulties and is expected to realize the goal of affordable housing to reduce the price.

Keywords: Affordable housing; The property sector; The influence on housing price; the effect mechanism

1 THE AFFORDABLE HOUSING SYSTEM

1.1 Affordable housing content

Affordable housing refers to the housing constructed by the government, provided to the housing difficulties of low-income families at a limited price. It has the characteristic of both safeguard and commodity, and includes affordable housing, low rent housing, public rental housing, limited prices housing and shanty towns transformation.

1.2 Difference between affordable housing and commercial housing

As can be seen from table 1, the difference between affordable housing and commercial housing is mainly reflected in land supplying, the price, property, dwelling unit type, economic conditions of buyers and construction funds sources [1].

2 INFLUENCE OF AFFORDABLE HOUSING ON HOUSING PRICES

2.1 Establish the effect mechanism

Since the central government has paid more attention to the construction of affordable housing, affordable

Table 1. Difference between affordable housing and commodity housing.

	affordable housing	commercial housing
land supplying	administrative transfer	To obtain the right to use land through auction or tender
price	government-guided or government-set	The parties negotiated pricing
property	Partially ownership	Full ownership
dwelling unit type	Strictly prescribed by the government	Decided by the developer
economic conditions of	The directional supply, from the current situation to see	Supply is operated according to the
construction funds sources	The central government housing subsidies, local	no preferential policies, built by

housing will occupy more market share in the future. With the increase of house supply the low and middle income families' pressure from affording a house will decrease, and the average price of flats will come down because of the increased turnover.

But the housing market is affected by the land market, prices fell from the housing will transfer to the land, land supply will reduce as land sales motivation of the local government go down. At the same time, housing prices fell also hit developers' enthusiasm, new housing construction area reduced, leading to the housing market could hardly keep up with demand in

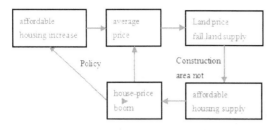

Figure 1. Ring shaped logic operation of the housing market.

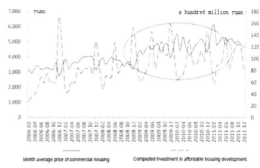

Figure 2. The reverse relationship exists between affordable housing development investment and average price of commercial housing price.

the next one or two years, housing prices will be in a new round of rising. This time the government will start the macro-control by policy intervention, but it is difficult to achieve the expected effect, while reducing the supply of land will hinder the construction of affordable housing, it is difficult to form a large-scale affordable housing market, housing market will be into a vicious spiral. The mechanism of the effect of affordable housing to housing prices can be seen in figure 1.

2.2 Analysis of the effect mechanism

2.2.1 Affordable housing supply status

From the beginning of 2007, all levels of government shift their focus of work to the affordable housing, low rent housing, affordable housing, two limited housing construction is progressing rapidly for two years. According to statistics, by the end of 2006, 512 cities had established low rent housing system in total 657 cities of China, establish rate reached 77.9%; in the first three quarters of 2007, the national affordable housing investment rose 30.5%, higher 21.6 percentage points than the same period in 2006 [2]; in 2008, "the general office of the State Council opinions on promoting the healthy development of the housing market" (GuoBanFa [2008] No. 131), the opinion request to try multiple channels to raise funds for construction, and increase affordable housing construction, the construction of low rent housing got a total of 630000 sets; in 2009, China completed 2000000 sets new and expansion of affordable housing, the various shantytowns reconstruction reached 1300000 households, 769000 households in the rural reconstruction; in 2010, the national urban affordable housing construction projects as well as various shantytowns reconstruction projects were 4100000 sets in total, accounting for 70% of the plan, dilapidated houses rebuilt in rural areas got a total of 680000 households, accounting for 56%.

The "Eleventh Five Year Plan" period of China's affordable housing construction has made the preliminary results, During the "eleven five" period, affordable housing construction appears "sprint" trend with the increasing emphasis of the central government. In the "Twelfth Five Year Plan" period, the construction of affordable housing projects got 36000000 sets, of which the scheme started in 2011 10000000 sets, the national urban affordable housing coverage rate from the current 7% to 8% increased to more than 20%, basically solved the difficulties of urban low-income families' housing problem [3].

In 2010, the State Council "on Resolutely Curbing the parts of the city house prices rise rapidly in the notice" mentioned, supply of public rental housing, affordable housing and limited prices housing must be increased in the area where the prices are too high and rising too fast. By increasing the affordable housing to lower price policy is very obvious.

The national average price of commercial housing sales from the beginning of the 08 year kept stable, and affordable housing development investment from that time began to exist reverse changes in the relationship with the commodity housing fold obviously (see Figure 2). Because the supply of affordable housing is positively related to the investment, increasing the supply can indeed inhibit the commodity housing prices.

2.2.2 Demanders of affordable housing

The current commercial housing market mainly rely on strong demand to pull up the turnover, affordable housing market and commercial housing market has some coincidence, there is a part of strong demand buyers meet the conditions of affordable housing or limited prices housing in the market. With the government to increase efforts to promote affordable housing, affordable housing will occupy more market share, which would make this part of strong demand buyers have more choice with less pressure, further more, they may turn into the affordable housing market from the commercial housing market. Meanwhile, a part of strong demand buyer who cannot afford to buy a house may withdraw from the commercial housing rental market and turn to low rent housing or public rental housing market.

Strong demand buyer of commercial housing is mainly composed of three parts: the purchase demand of newly married family, rural Chinese migrated to cities, demand caused by demolition [4]. If the rural population find works in the city and settled in, they will enter the housing guarantee system, and some of the newly married family also meet the affordable

Figure 3. Commercial housing strong demand have some coincidence with affordable housing.

housing application conditions, they may also turn to affordable housing market, which will have a certain impact on housing prices, see figure 3.

2.2.3 *Affordable housing land supply*

Land supply is fixed, expanding the construction of affordable housing will inhibit the enthusiasm of commercial housing construction. Affordable housing will reduce the competitiveness of commercial housing in the small-sized apartment and in the low-end housing developments, so these enterprises will transfer to the high-end housing developments competition. That is to say, commercial housing will be only for high-income groups, this will cause the supply and demand is relatively stable, followed will be the inevitable price fall.

2.2.4 *Affordable housing policy*

The central government has paid more attention to affordable housing construction, in the "Twelfth Five Year Plan" period, and put forward to construct 36000000 sets of affordable housing. Large amount of affordable housing construction or is a key link in the housing market's repositioning, chairman of SOHO Pan Shiyi said, 36000000 sets of affordable housing will completely change the Chinese housing market pattern, from commercial housing ruling all the land to affordable housing occupying for more than half of the market, there will be more than half of the housing market will replaced by the affordable housing.

Conclusion: apparently using a large number of affordable housing supply could adjust the housing market structure, curb the housing prices.

3 AFFORDABLE HOUSING DEVELOPMENT POLICY RECOMMENDATIONS

Whether such a large-scale affordable housing could be successfully completed is still problematic. Where to get land for affordable housing construction? How to get construction funds? Construction ability is enough or not? How to ensure the construction quality? How to allocate the house? Is the purchase family willing to live in the completed house? All these constitute the determinant of whether affordable housing construction could solve the problem of excessively high prices,

now the prices reflects the imbalance of supply and demand, therefore we need to strengthen the responsibility of local government, improve the housing distribution system, establish reasonable affordable housing supply structure, to achieve a balance and stable housing market through the housing guarantee system.

3.1 *To strengthen the responsibility of local government*

In the "Twelfth Five Year Plan", China will build 36000000 sets of affordable housing, the coverage rate will be 20%, then there will be 2830000000 square meters of affordable housing. If each set of affordable housing is 50 square meters, 2830000000 square meters of construction area will be made up of 56600000 sets of affordable housing. But 19530000 sets were started in 2009–2011, there are 2570 sets in 36000000 sets of the original plan have not yet been started, there are 11370000 sets of gap, conservatively. Moreover, the new project in 2012 was cut down to 7000000 sets for the reason of funding constraints and preferred speed to quality. In view of this, 5 years to get the coverage rate to achieve 20% coverage is very difficult.

Therefore, we need to strengthen the responsibility of local government, improve the understanding of local governments, to mobilize the initiative of local government to affordable housing construction. To establish long-term and stable investment mechanism of government finance. Based on the fully aware of the government in affordable housing construction responsibility, [4] use land grant revenue and general budget to establish long-term and stable affordable housing construction financing mechanism.

Meanwhile, affordable housing construction tasks are arduous, and need a huge demand for funds, it is difficult to guarantee the sustainability of affordable housing just relying on government financial input. So it is necessary to establish an affordable housing investment and financing platform based on government finance investment. Establishing institutions of the government to manage affordable housing construction investment, and play a role as investment and financing platform, make overall planning of resources, unified security standards, provide better service for the housing difficulties of low-income families.

3.2 *To improve the housing distribution system*

Affordable housing policy is difficult to implement, lax scrutiny in housing applicant, low transparency. It's not uncommon that rich people live in affordable housing. Therefore, we need to strengthen the regulatory measures of housing, strengthen the affordable housing management responsibilities, adhere to open, fair, just principles, establish an orderly and effective management mechanism. City Hall need to strengthen the affordable housing fund management, strictly regulate the use and management of housing, stop and deal with irregularities in time, to ensure that affordable housing resource is preferential for the most difficult families. At the same time, to improve the affordable housing application, approval, publicity, waiting, review, withdrawal system, create a sound system of three level audit public community, streets and housing security departments. Establish a standard income, property and housing review system, improve the affordable housing access and exit management mechanism, form a scientific and orderly, behavior standard, information sharing, open and transparent housing security system, strictly regulate the standard of housing for low-income people, implement dynamic tracking management, take comprehensive measures to resolutely corrected the sublet, resell behavior.

3.3 *To establish reasonable affordable housing supply structure*

It is important to adjust the housing structure, gradually reduce the affordable housing, focus on the development of public rental housing, because the paying ability of the new employment workers and resident foreign staff is weak. Encouraging the local government to increase the supply of public rental housing and to establish a sound system of low rent housing by paying rental subsidies, purchasing the house, guaranteeing public rental housing in the long-term rental market where the market of rental housing and the vacant housing is ample. Give priority to ensuring that low-cost housing, small set of ordinary housing and low-cost housing land supply. Reduce the supply of land in villa housing development projects, strictly limit low-density, land supply of large suites housing, to build the diversified market.

4 CONCLUSION

The affordable housing can effectively curb the excessive growth of house prices, if the affordable housing policy can get strong execution, then it will cause the inhibition of housing prices. But there are still some problems in our affordable housing development. Firstly, we need the government to expand the major premise of the construction of affordable housing and keep the housing regulation. Secondly, it is necessary to solve the problem of local finance, let the funds of building affordable housing to be implemented, mobilize the enthusiasm of the local government of building affordable housing. Once more, we need to strengthen the construction of affordable housing, perfect the affordable housing distribution system, make the purchase of housing more transparent. Moreover, in order to ensure the healthy development of the housing market and achieve the purpose of curb rising prices, we should considering the needs of people who need the affordable housing, avoiding blind building houses surveying the actual demand of the housing difficulties, to ensure that the affordable housing will real benefit the low incomes.

REFERENCES

[1] Chen xiaobo, improving the system arrangement, low-income housing can not poor money. First capital, 2011, (06)
[2] Zhang Rong. China's affordable housing situation analysis. Southland today (THEORY EDITION), 2009 (07).
[3] Wang Chunyan, On the supply of affordable housing curbing house prices. Contemporary economy, 2011, 10
[4] Wang guo, The impact on prices of low rent housing. Chinese construction enterprises, 2012, 2

The research of in-pipeline robot with double driver and variable helix angle

H. Zhou, L. Xu, X.Y. Cai & H.C. Zhuo
College of Mechanics Engineering, Donghua University, Shanghai, China

ABSTRACT: Most in-pipeline robots provide insufficient driving force and are hard to go through bending pipelines. Also, it is inconvenient to install detection sensors and other devices in the front of the robot body in practical application. Aimed at this problem, a research of in-pipeline robot using double driver and variable helix angle is put forwarded. The driving force is increased by the variable helix angle in rotating parts and the reduction gear in joint part. Using joint part which consists of coupling in parallel and hollow polyester rods, the in-pipeline robot can go through bend pipelines with certain curvature radius. CCD camera and ultrasonic detection devices are installed in the rotating part and the data are transmitted to computer by wires which go through the hollow rod. Helix angle is able to be changed to adapt to the change of load torque while crawling pipeline of different tilt. Finally, the effectiveness of the robotic system is proved in experiments.

Keywords: In-pipeline robot; Variable helix angle; Double driver

1 INTRODUCTION

In-pipeline crawling robots can crawl and work inside the pipeline. They have the function of detection and repairing in different environment which have been widely used in petrochemical and aqueduct industry to transport gas and oil. There is a great demand that a robot which can go through the pipeline and perform ventiduct cleaning and other detection tasks be developed.

With the development of technology, various robots were successfully developed to adapt to different environment in many countries around the world, made a great improvement in working efficiency and working quality. But most robots has its disadvantages such as the lack of driving force, the difficulty to go through bending pipelines and inconvenience to take detection devices. To solve these problems, we have made an improved designing scheme in our project and designed a robot with the functions of variable helix angle and multiple working mode based on the hybrid-driving robot prototype.

2 THE INNOVATION OF DESIGN

As displayed in Figure 1 and Figure 2, the pipeline consists of rotating parts, support components, detecting components and joint components. Rotating parts consists of screw driving wheels, variable helix angle mechanism; support components include helix drive motor, linear drive motor, straight line driving wheels and support wheels; joint parts include a pair of gears, flexible coupling, cardan, camera cable.

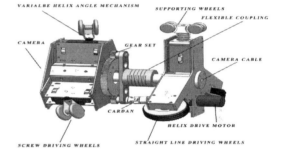

Figure 1. In-pipeline robot model.

Figure 2. Object of In-pipeline robot.

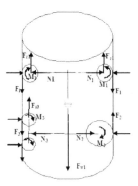

Figure 3. Linear driving force analysis.

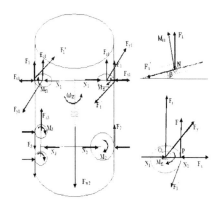

Figure 4. Helix driving force analysis.

3 THE PRINCIPLE OF INCREASING DRIVING FORCE

3.1 Force analysis model

Robot needs a great driving force to move forward or backward. By the force of adjusting instrument, the screw driving wheels can change its angle from 0 to 90. Rotating part rotates overall and drives the robot forward. In-pipeline robot gets bigger driving force at the same power because of the screw driving motor and joint part.

Figure 3 and Figure 4 are the simplified model for force analysis. Here, wheel 1 denotes the helical driving wheel, wheel 2 denotes the straight driving wheel and wheel 3 represents the supporting wheel. The diameter of wheel 1 is d_1, retightening force of spring is N_1, the frictional force of antifriction bearing is σ_{11}, the frictional force between the wheel and pipeline is σ_{12}, coefficient of sliding friction is f_1; The diameter of wheel 2 is d_2, retightening force of spring is N_2, the frictional force of antifriction bearing is σ_2, coefficient of sliding friction is f_2; The diameter of wheel 3 is d_3, retightening of spring is N_3, the frictional force of antifriction bearing is σ_{31}, the frictional force between the wheel and pipeline is σ_{32}, coefficient of sliding friction is f_3. Helical driving motor supports the maximum torsion of M_1, straight driving motor supports the maximum torsion of M_2. The maximum driving force is Fw_1 when drive straightly, the maximum driving force is Fw_2 when drive helically. The force analysis of robot is as follows.

3.2 Mathematical Model of force analysis

Linear driving: suppose F1, F2 and F3 are friction forces between wheels and pipe wall respectively.
Force analysis in Z-axis:

$$\sum F_Z = 0 \Rightarrow F_{W1} + 6F_1 + 4F_3 = 2F_2 \qquad (1)$$

Center O1 moment balance:

$$\sum M_{O1} = F_1 \frac{d_1}{2} - M_1 = 0 \quad M_1 = \sigma_{11}(\frac{N_1}{2})$$
$$\Rightarrow F_1 = \frac{2M_1}{d_1} = \sigma_{11}\frac{N_1}{d_1} \qquad (2)$$

Center O2 moment balance:

$$\sum M_{O2} = F_2\frac{d_2}{2} - M_2 = 0$$
$$F_2 = f_2 N_2 \Rightarrow M_2 = f_2 N_2 \frac{d_2}{2} \qquad (3)$$

Center O3 moment balance:

$$\sum M_{O3} = F_3 \frac{d_3}{2} - M_3 = 0 \quad , \quad M_3 = \sigma_{31}(\frac{N_3}{4})$$
$$\Rightarrow F_3 = \frac{2M_3}{d_3} = 2\frac{\sigma_{31}N_3}{4}\frac{1}{d_3} = \sigma_{31}(\frac{N_3}{2d_3}) \qquad (4)$$

From (1), (2), (3) and (4), we got

$$F_{W1} = 2F_2 - 4F_3 - 6F_1 = 2f_2 N_2 - \frac{2\sigma_{31}N_3}{d_3} - \frac{6\sigma_{11}N_1}{d_1} \qquad (5)$$

Helix driving: Suppose F_1 is sliding friction force of wheel 1 along the pipeline axis; F_1' is rolling friction force of wheel 1 along the spiral direction; Fx_1, Fy_1, Fz_1 is the force which wheel 1 bear in center location.
Force analysis:

$$\sum F_Z = F_{W2} + 4F_3 - 2F_2 - 6(F_1 - F_1'\sin\beta) = 0 \qquad (6)$$

$$\because F_1 = f_1 N_1 \quad F_2 = f_2 N_2 \Rightarrow F_3 = \frac{\sigma_{31}N_3}{2d_3} \qquad (7)$$

The force balance of wheel 1 in Y-axis:

$$\sum F_{PY} = 0 \Rightarrow F_Y - F_1'\cos\beta = 0 \qquad (8)$$

The moment balance between wheel 1 and wall contact point P:

$$\sum M_{PZ} = M_{f1}\cos\beta + M_{f12} \qquad (9)$$

$$M_{f1} = M_{f11} + M_{f12} = \sigma_{11}(\frac{N_1}{2}) + \sigma_{12}(\frac{N_1}{2}) \quad (10)$$

From (8) (9) (10) we got

$$F_1' = \frac{(\sigma_{11} + \sigma_{12})N_1}{d_1} \quad (11)$$

From (7) and (11), we have:

$$F_{W2} = 6(F_1 - F_1' \sin\beta) + 2F_2 - 4F_3$$
$$= 6(f_1 - \frac{\sigma_{11} + \sigma_{12}}{d_1}\sin\beta)N_1 + 2(f_2 - \frac{\sigma_{31}}{d_3})N_2 \quad (12)$$

From (5) and (12), the extra force provided by double driver is:

$$F_W = F_{W2} - F_{W1} = 6(f_1 - \frac{(\sigma_{11} + \sigma_{12})}{d_1}\sin\beta + \frac{\sigma_{11}}{d_1})N_1 \quad (13)$$

From the force analysis above, we came to a conclusion: The driving force in spiral is bigger than in a straight line, the size of the F_w is associated with the helix angle (β), a driving force can be changed by adjusting the helix angle.

3.3 Torque increase by level reduction gear

The level reduction gear reduces the speed to increase torque in the rotating part, the torque of big and small gears are T_2 and T_1, rotate speed is n_1 and n_2, number of teeth is $Z_1 = 88$ and $Z_2 = 17$. We have

$$T = 9550\frac{P}{n}N \bullet m \quad (14)$$

Where $n = 2\pi\omega$, $\omega = \frac{v}{R}$ and $R = \frac{1}{2}Zm$
$\frac{T_2}{T_1} = \frac{Z_1}{Z_2} = \frac{86}{17} = 5.06$, it is clear that torque increased five times.

4 THE ANALYSIS OF GOING THROUGH BENDING PIPELINE

The components of flexible parallel coupling and telescopic universal joints make the robots to move in bent pipeline smoothly, improve the crawling performance to go through the bending pipeline significantly. Figure 5 demonstrates the simplified principle. Here, arc AB represents the bent State of flexible coupling. AE, BF are the robot's rigid shaft and our design can ensure $L_{AE} = L_{BF}$. We selected to use flexible coupling which have an axial dimensions of Lo and maximum bending angle of θ. The performance analysis is as follows [7]:

Robot goes through a minimum bend radius of curvature:

$$R = \frac{L_{GF}}{\tan(\frac{\gamma}{2})} \quad (15)$$

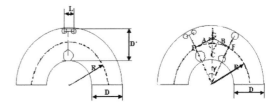

Figure 5. Performance analysis of robot in bent pipe.

From fundamental geometric relations, we know that:

$$\gamma = \theta \quad L_{GF} = L_{BF} + L_{BG} \quad L_{BG} = L_{BC}\tan\frac{\theta}{2}$$

$$L_{BC} = r$$

and

$$Lo = L_{AB} = \frac{\pi\theta}{180°}r \Rightarrow r = \frac{180° Lo}{\theta\pi}$$

From the formula above we got

$$R = \frac{L_{BC}\tan\frac{\theta}{2} + L_{BF}}{\tan\frac{\theta}{2}} = L_{BC} + \frac{L_{BF}}{\tan\frac{\theta}{2}} = \frac{180° L_O}{\theta\pi} + \frac{L_{BF}}{\tan\frac{\theta}{2}} \quad (16)$$

The performance of the robot to pass through bending pipes is related to the flexible length of the couplings, the maximum bending angle and the axial dimensions of the front and rear rigid connections. To have a better performance of passing through the bending pipelines, the robot must have a longer flexible part of the couplings, a larger bending angle, and a shorter rigid connection.

5 DETECT PART INSTALLATION

Detecting parts consist of a CCD camera and is composed of three ultrasonic testing devices. A CCD camera is used to check the larger obstacles in the pipes while ultrasonic detection devices acquire the erode and crack information of the pipe. Because the connecting component is a hollow polyester rods and coupled in parallel, the signal cables which connected the detection parts and the computers outside can go through the robot smoothly even when the rotating parts are rotating with a high speed.

6 EXPERIMENTAL VERIFICATION

6.1 Piping robot experiments

When climbing in the vertical pipelines, the robot need to overcome its' own gravity as well as the detection equipment carried. The maximum load capacity of

Figure 6.

Figure 7.

robot when climbing through the vertical tube can be found out in experiment. Figure 6 shows the crawling robot without any load while Figure 7 demonstrates the robot's performance when driving upward with a heavy load. Experiments showed that the screw-driven robot we designed can drag a weight of 2 kg load with a speed of 2.2 m/min.

7 CONCLUDING REMARKS

In this paper, the double driver in-pipeline robot whose helix angle can be adjusted is designed to meet the needs of increasing the propel forces. Also, a level reduction gears is introduced to further increase the driving torque, and facilitate transmitting of data cables together with the help of hollow rod connections. The bearing forces when crawling in the pipeline are analyzed in detail. Finally, Experiments are carried out which verified their effectiveness.

REFERENCES

[1] Jiang W.Q., Miranda, Nan Y. Pipeline construction crossover. China oil news. http://News.CNPC.com.CN/System/2013/01/25/001410444.shtml
[2] Gan X.M., Xu B.S., Dong S.Y., Zhang X.M. Current development of pipeline robot. Robotics and applications. 2003, 6:5–10.
[3] Maki K.Habib. Bioinspiration and Robotics: Walking and Climbing Robots. Hyouk Ryeol Choi, Se-gon Roh.-pipe Robot with Active Steering Capability for Moving Inside of Pipelines. ISBN 978-3-902613-15-8, pp. 544, I-Tech, Vienna, Austria, EU, September 2007:376–402.
[4] Hirose S., Ohno H., Mitsui T., et al. Design of in-pipe inspection vehicles for Φ25, Φ50, Φ150 pipes. J. of Robotics and Mechatronics, 2000, 12(3): 310–317.
[5] Zhu J.D., Zhou M., Precious. Machinery design of active oil pipeline inspection robot. Journal of Shanghai University of technology (natural science Edition), 2001, 7 (1): 57–59.
[6] Xie W.B., Yang J.G., Li Beizhi, et al. Study on pipeline inspection robot [j]. Mechanical engineer, 2005, 1:14–16.
[7] Xu F.P., Deng Z.Q. of pipeline robot in elbow by. Robots, 2004, 26 (2):155–160.
[8] Chiang Kai-shek sent June, Zhao M.M., Wang B. Design of a driving circuit for DC motor based on PWM

Simulation of coal mine trackless rubber tyre vehicle regenerative braking control strategy based on ADVISOR

Chuanwei Zhang & Chenxi Li
Department of Mechanical Engineering, Xi'an University of Science and Technology, Xi'an, Shannxi, China

ABSTRACT: The idea of EV was introduced into the coal mine trackless rubber tyre vehicle. The author aimed at the shortage of the original regenerative braking control strategy in the vehicle simulation software ADVISOR, set up regenerative braking model based on fuzzy control strategy in terms of fuzzy control strategy in order to improve regenerative braking energy recycling effect, and redeveloped ADVISOR, inserted the new model of regenerative braking into original model base of ADVISOR. Simulated the original and the new and compared the results, verified the conclusion of new strategy superior the original one by comparing SOC (state of charge) pattern of the battery load condition.

Keywords: vehicle simulation software ADVISOR; trackless rubber tyre vehicle; regenerative braking energy

1 INTRODUCTION

Mine auxiliary transportation is the indispensable part of the whole coal mine transportation system [1]. It introduces the concept of electric vehicles to the coal mine trackless rubber tyre vehicle, which has lots of advantage such as high efficiency, low environmental pollution, mechanical structure diversification and control performance, etc. [2]. It is very suitable for the terrible environment filled with dust, noise, high temperature, air humidity in deep coal mines. However, the electric vehicle's mileage of the single charge is too short due to the intrinsic rigid structure, the charging time needs too long, the charging station is not popularized; these deficiencies seriously affect and restrict the promotion and popularization of the electric vehicle industry. Using the regenerative braking system to recycle the braking energy in electric vehicles is an effect measure to increase the mileage of a single charge in an electric car at present [3]. So far, one method to improve the efficiency of vehicle regenerative braking is how to realize the optimal allocated proportion between friction braking and regenerative braking during the process of braking.

The author used MATLAB fuzzy control module to establish a fuzzy controller, and SIMULINK software to establish the model of regenerative braking strategy based on fuzzy control, developed the ADVISOR, compared the new strategy simulation and the original tactics of the ADVISOR.

2 GETTING STARTED

The regenerative braking control strategy in ADVISOR has some defects and limitations. ADVISOR

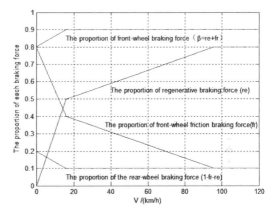

Figure 1. The proportion of each braking force.

default vehicle is front-wheel drive, the regenerative braking is implemented on the front wheels. The proportion of braking force is distributed according to the actual vehicle speed. In the total braking force, providing the proportion of front-wheel friction braking force is *fr*, the proportion of regenerative braking force is *re*, the proportion of front-wheel braking force is ($\beta = re + fr$), there is no regenerative braking force in the rear-wheel and all of braking effect depends on the friction braking force. The proportion of the rear-wheel braking force is (*1-fr-re*). The proportion of each braking force varies with the speed of Figure 1 [4].

We can see in Figure 1, the lower the speed is, the smaller the proportion of regenerative braking force is, and vice versa[5]. During braking, when the vehicle speed $V > 96$ km/h, the proportion of the front-wheel regenerative braking force remains 80%. As the vehicle brakes continually, the vehicle speed decreases gradually. When the vehicle speed $V = 16$ km/h, the proportion of the front-wheel regenerative braking force gradually decreased to 50%. Then to zero, the role of front-wheel regenerative braking force is no longer.

This control strategy is simply determining the change of the speed to distribute the proportion of the friction braking force and the proportion of regenerative braking force, while ignoring the motor's braking torque and rechargeable battery's real time power, making the ability of recovering of regenerative braking system is not fully shown, result in a huge loss of energy of the brake system. The regenerative braking system is effectless.

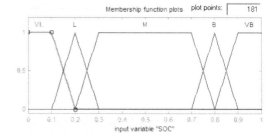

Figure 2. The membership function of SOC.

3 FUZZY CONTROLLER DESIGN

3.1 Selecting the fuzzy input variables

Adopt SUGENO fuzzy model, the proportion of the actual regenerative braking torque to the total braking torque request is represented as fuzzy output variable K.

Select battery's SOC (state of charge) as an input variable. When the SOC value is too high, it indicates that there is enough power in the battery at the time, the vehicle doesn't need excessive regenerative braking energy to charge the battery during braking; when the SOC value is too low, it indicates that the power is low and needs to be charged, in this case, the full effect of the regenerative braking system needs to charge the battery.

Select T as an input variable. $T = T_{req}$ (the actual regenerative braking torque)$-T_{max}$ (the maximum of motor's regenerative braking torque motor), T is closely related to motor's regenerative braking torque, select T as one of the fuzzy control system's input variables[6].

Select speed V as an input variable. The conclusion in Figure 1, we can see that during braking, as the vehicle speed changes, the proportion of regenerative braking is gradually changed, the vehicle's speed affects the size of regenerative braking force. When we design the fuzzy input variables, the vehicle speed V should also be taken into account

In summary, select the battery SOC, T and V as fuzzy input variables of the fuzzy controller. We develop fuzzy rules according to these three fuzzy input variables and output variables, enable vehicle system to maximize the recovery of regenerative braking energy.

3.2 Selecting the membership functions

The membership function of SOC is shown in Figure 2.

The range of SOC is [0 1], the params of SOC is {VL L M B VB}.

The membership function of T is shown in Figure 3.

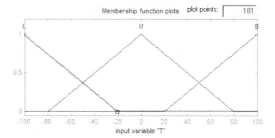

Figure 3. The membership function of T.

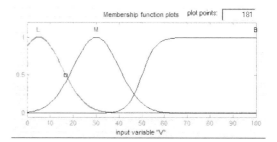

Figure 4. The membership function of V.

The range of T is [−100 100], the params of T is {L M B}.

The membership function of V is shown in Figure 4.
The range of V is [0 100], the params of V is {L M B}.

Output variable K is {0,0.1,0.2,0.3,0.4,0.5,0.6, 0.7,0.8,0.9,1}

3.3 Fuzzy Rule

The form of Fuzzy logic rule is "*if x1 is A1 and x2 is A2 and x3 is A3, then u=K*", set up fuzzy logic rule table (table 1), generate fuzzy logic rules controller. The fuzzy logic rules controller is shown in Figure 5.

Each of the control rules can work out the corresponding fuzzy relationship. We can see that the calculation of fuzzy inference is quite complicated from the established fuzzy inference system. We design and calculate it by MATLAB, get the fuzzy result by inference SUGENO. We can get the curve and the table of relationship between input and output by the assistant designing of MATLAB.

Table 1. The table of fuzzy logic rules.

NAME	SOC	V	T	K
1	VB	B	B	0
2	VB	B	M	0.1
3	VB	B	L	0.1
4	VB	M	B	0.1
5	VB	M	M	0.1
6	VB	M	L	0.2
7	VB	L	B	0.1
8	VB	L	M	0.2
9	VB	L	L	0.2
10	B	B	B	0.5
11	B	B	M	0.8
12	B	B	L	1
13	B	M	B	0.5
14	B	M	M	0.7
15	B	M	L	0.9
16	B	L	B	0.1
17	B	L	M	0.2
18	B	L	L	0.3
19	M	B	B	0.6
20	M	B	M	0.8
21	M	B	L	1
22	M	M	B	0.5
23	M	M	M	0.7
24	M	M	L	0.9
25	M	L	B	0.1
26	M	L	M	0.2
27	M	L	L	0.3
28	L	B	B	0.6
29	L	B	M	0.8
30	L	B	L	1
31	L	M	B	0.5
32	L	M	M	0.7
33	L	M	L	0.9
34	L	L	B	0.1
35	L	L	M	0.2
36	L	L	L	0.3

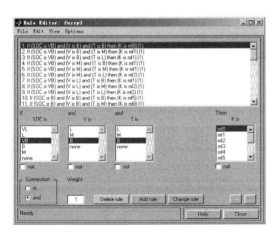

Figure 5. Fuzzy logic rules controller.

4 CREATE A NEW REGENERATIVE BRAKING CONTROL STRATEGY MODULE

ADVISOR's regenerative braking control strategy model is located in EV vehicle simulation module's vehicle control module "<vc>".The proportion of regenerative braking force and the proportion of friction braking force are located in "braking strategy" and "braking force supplied by front friction brakes", depend on the fuzzy logic controller's input variables to establish a new regenerative braking control module, change the original braking force supplied by front friction brakes. The new regenerative braking control module is shown in Figure 6. The new braking force supplied by front friction brakes is shown in Figure 7.

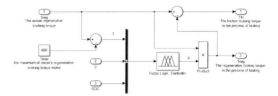

Figure 6. The new regenerative braking control module.

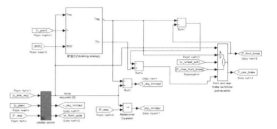

Figure 7. The new braking force supplied by front friction brakes.

Table 2. The parameters of electric vehicle.

Parameter	Value	Parameter	Value
Body mass/Kg	1144	Wheel radius/m	0.282
Height of center of gravity/m	0.5	Coefficient of rolling resistance/C_D	0.009
Frontal area/m^2	2.0	Coefficient of drag/f	0.335
Wheelbase/m	2.6		

5 THE SIMULATION AND ANALYSIS

We compare the new regenerative braking control with the original regenerative braking control of ADVISOR, select the parameters in Table 2 as the experimental parameters of the coal mine trackless rubber tyre vehicle, select driving cycle "UDDS", which was developed by U.S. Environmental Protection Agency (The parameters in Table 3), select AC induction motor as the drive motor (The parameters in Table 4). Battery SOC curve the simulation of the two strategies is shown in Figure 8.

As is shown in Figure 8, under the model of the vehicle driving conditions "CYC_UDDS", regenerative braking strategy design based on fuzzy logic control

Table 3. The parameters of driving cycle "CYC_UDDS".

Parameter	Value	Parameter	Value
Time/s	1369	Max accel/(m/s^2)	1.48
Distance/km	11.99	Max decal/(m/s^2)	−1.48
Max speed/(km/h)	91.25	Avg accel/(m/s^2)	0.5
Avg speed/(km/h)	31.51	Max decal/(m/s^2)	−0.58

Table 4. The parameters of the electric motor.

Parameter	Value	Parameter	Value
Rated power/kw	75	Base speed/(r/min)	2650
Continuous maximum torque/Nm	271	Max speed/(r/min)	10000
Peak torque/Nm	288	Max current/A	480
Min voltage/V	120	Mass/Kg	91

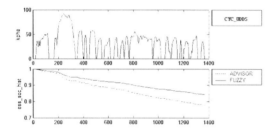

Figure 8. Change the battery SOC comparison chart.

can recover more energy than regenerative braking strategy in ADVISOR. It can improve the coal mine trackless rubber tyre vehicle's endurance mileage in mine effectively, reduce the number of charge cycles and improve the efficiency of the vehicle.

6 CONCLUSION

The power that is recovered by regenerative braking system can improve the mileage of the single battery charge of the coal mine trackless rubber tyre vehicle, conserve the petroleum energy, reduce the emission of harmful gas, improve the quality of the gas in the mine, prevent the mine being polluted secondly by the off-gas from the diesel rubber. As the regenerative braking system is added to the electric vehicle. It reduces the rate of utilization of the friction braking system and the wear of brake pads and brake disks, and also decreases the noise pollution generated by the friction of brake disks at the same time.

The design of the SIMULINK model of the regenerative braking control strategy based on fuzzy control, allocate the proportion of the vehicle regenerative braking torque to the friction braking torque reasonably through real-time monitoring three input variables of SOC, T, V, bring the regenerative braking capability into play maximumly. Thereby recycle the regenerative braking energy as much as possible. Under the road conditions model CYC_UDDS, simulate two different strategies and compare the results (battery SOC simulation graph). We can clearly see that the fuzzy logic regenerative braking strategy model SIMULINK can make electric cars recover more regenerative braking energy.

ACKNOWLEDGEMENT

The author is grateful to the school of Mechanical Engineering, Xi'an University of Science and Technology. The author is also grateful to Dr Zhang-Chuanwei for his constant encouragement and active support. The author is also grateful to Natural Science Foundation of Shannxi Province (Grant No.2012JM7021) and Ministry of Education of University Discipline Specialized Research Fund (Grant No.20126121120005).

REFERENCES

[1] Huanying Chen, Hanjun Jiang. Accelerate the Development of Several Problems of Coal Mine Auxiliary Transportation Modernization. China Coal. 2005(02):12–16.
[2] Chuanwei Zhang, Wei Guo. Trend of Coal Mine Trackless Rubber Tyre Electric Vehicle. Coal Mine Machinery. 2012(06):7–9.
[3] Donghai Xu. Regenerative Braking Control Strategy for Energy Efficient Recovery. Shandong University. 2007:5–20
[4] Pingze Wang, Overall Design and Performance Simulation of Electric Vehicle. Hefei: Hefei University of Technology. 2007:10–56.
[5] Xun Jiang, Miaohua Huang, Simulation of Electric Vehicle Regenerative Braking Control Based on ADVISOR. Beijing Automotive Engineering, 2008(01): 28–31.
[6] Danhong Zhang, Zhou Jiayang, Su Yixin. Research of Regenerative Braking Energy Recovery of HEV Based on Fuzzy Logic. Journalofwut (Information & Management Engineering) 2011-10-15:41–44.

Project risk-based reliability control

Shahryar Sorooshian
Faculty of Industrial Management, University Malaysia Pahang, Malaysia

ABSTRACT: Reliability, being an inseparable part of any industry, also is a major issue in constructing industry. It is responsible for huge losses in capital investments and progress in the industry. This study was chosen to investigate the reliability of the building construction industry. To achieve this goal, questionnaires were distributed to thirty respondents based on literature and expert recommendation after which analysis was carried out on the Risk Priority Number (RPN) of delay sources. Based on the findings, this article introduces and tests a new index to analyse the risk-based reliability of the industries. The Risk-based Reliability Index (RRI) introduced and tested using found RPNs. The calculated RRI has the ability to play a benchmark tool role for industry reliability control.

Keywords: Reliability; Project; Risk; Risk-based Reliability Index

1 INTRODUCTION

Machinery, staffing as well as capital investment are one of the areas that requires considerable level of funding in a construction project. However, delays in the expected completion time will lead to huge losses. Just like any other project, a construction project should have a time limit within which it is completed which should be stated before the physical project take off. Some of the major effects that causes delays include time overrun, settlements total abandonment, cost overrun, lawsuits and disputes between parties (Aibinu & Jagboro, 2002; Sambasivan & Soon, 2007). For effective delay prevention or decrease, parties in the construction should find out the major causes of the delays as well as determine proper strategies to solving them and decrease their effects. Literature and other research conducted throughout the world have shown that contractors, project financing, clients and designers form major associated setbacks. Whereas problems and issues such as multinational workforce, multicultural environment, using different parties and participants in a project as well as employing international designers are not seen as a major setback. The majority of the issues suggested above are consistent with the study (Assaf & Al-Hejji, 2006) where it was stressed that the undesired condition of delay occurs due mainly to regular hindrances while making decisions and owner approvals, limitations in obtaining work permits and lack of harmony and understanding between various parties in the construction project (Assaf & Al-Hejji, 2006). Baldwin (1971) have tested the delay sources and found that construction projects encounter huge financial losses when delay occurs. A survey conducted a few years ago (Assaf & Al-Hejji, 2006) in the construction industry about the source of delay have been assessed. Failures as mentioned above in building construction in this study can be grouped by owners, engineers and contractors in relation to most factors are delays. Salleh (2009) has stated the major causes of delay in building construction such as: inadequate planning; subcontractor performance; slow decision making; lack of communication between parties; change orders; lack of contractor experience; labour surplus. The major causes perceived by contractors include: site management; lack of communication; shortage of materials; sluggishness in making decision; disagreement and mistakes in the contract; payment and finance for work completed; subcontractor performance.

The most important seven causes considered by engineers are: management of site; shortage of materials; sluggishness in making decisions; lack of communication; payment and finance for work completed; disagreement and mistakes in contracts; and below contract performance. Engineers perceived seven major important causes include: insufficient contract plan; material shortage; loss of communication; change orders; payment and finance for work completed; sluggishness in making decision; below performance of contract. The number of priority risks in each of the seven stated delay causes identified in this study will be estimated of the industry building construction as well as owners, engineers and contractors viewpoints' presented calculation for the case study.

Owners: the owners are the future possessors of the building however, they take part in projects' construction stage. The owners were chosen from the list of the building completed.

Contractors: based on the system of grouping by the Government, contractors are listed and graded as building contractors.

Engineer: Any project manager or designer that assisted the owner meet his/her requirement and create a reasonable blueprint of the project is regarded as an engineer. Another objective of the study is to apply the findings from this research on real-world construction projects as well as improving the effectiveness of the project. A proper study case will also assist in determining the theoretical findings of this report. Five floors residential building construction sited in Iran were chosen as the case study. Based on interview with Dr. S. Rashidi (architecture), construction of five-floor buildings have shown greater improvement in growth compared to the four floor buildings or less. This is associated with the country's boom in production, recently and the need for more houses compared to shortage of lands in the urban areas. The people are willing to destroy their old-fashioned single floor houses and in place of modern five-floor or higher apartments.

2 METHODOLOGY

The aim of this study is to recognize and index the delay risk-based reliability of projects from owners', engineers' and contractors' perspective. To identify the sample population, this study investigates the building construction so as to grade the causes of design for specific delay critical factors assessing RPN for each of the group as included. According to Salleh (2009), the causes of delay factors are listed in the pilot survey asked. The aim of the study is to verify the completeness of the survey questionnaire to capture the focus needed in the study in Shiraz, Iran. From the pilot study it can be seen that all the respondents' questionnaire is sufficient to capture the delay causes. Therefore modifications on the causes of delays are needed as reported by Salleh (2009). To determine the cause of failure in project construction, causes obtained from the literature were distributed among the participants who were asked to choose the points from 1–5 of the three factors that may result to the causes. The number that represents the highest risk or seriousness is ranked 5. The three factors include severity (S), occurrence (O) and detection (D). Risk priority number (RPN) was used to analyse the risk associated with the potential problems that are found during failure mode and effect analysis. As shown in equation 1 RPN include severity, occurrence and detection. The major differences of RPN are a function of the three variants.

Severity (S) – this shows the level of extremity the next user (customer) or end user understands the failure EFFECT, it is however a subjective number that the factor estimates.

Occurrence (O) – this is an estimate of the subject number of possible cause. Occurrence is sometimes called LIKELIHOOD in place of occurrence.

Detection (D) – this is an estimate of subject number of the efficiency of control to decimate or differentiate cause or failure mode before the failure gets to the customer. In this study it is assumed that the cause has occurred.

$$RPN = S*O*D \tag{1}$$

Data collected was performed using a questionnaire. A balanced total of 30 respondents were chosen for data collection using a random sampling for each group selected. Based on the collected data, RPNs calculated. Using the found RPNs, this study introduces a Risk-based Reliability index (RRI). The RRI is the main contribution of this article to the knowledge of project management, which is defined as the average number of averaged multi-approach RPNs.

3 RESULT AND DISCUSSION

RPN calculated, using mean of data obtained from the owners, engineers, and contractors and indication that each issue presented is the average of all their response. Therefore RPN obtained can separately be regarded for the owners, engineers, and contractors. Figure 1 part A presents the owners' view and it shows that the most important critical delay in construction of project delays are lack of communication among the parties, the highest RPN whereas the least was labour supply.

The received from the engineers is shown in figure 1 part B. It shows the major significant factor in contractors' project construction delays. From here subcontractors' performance had the highest RPN received from the engineers. The second cause being lack of communication while the least of the RPN operation research is finance and payment for completed work. This shows that finance and payment for completed work according to engineers' opinion had the lowest priority for delay in financing a project.

The obtained from contractors is shown in figure 1 part C. It shows the significant factor that delays construction projects. From the view point of contractors, mistakes and discrepancies in the contract had the highest RPN research operation received from contractors. The next significant cause is material shortage whereas RPN is for site management; this means that site management according to contractor's opinion had the lowest priority delay.

Based on the resulted RPNs, average of RPNs for owner, engineer and contractor approaches calculated as 16.49, 18.33 and 20.12. Referring to the risk-based reliability index definition by this study, the RRI is calculated as 18.31.

4 CONCLUSION

Conclusions from this study can be drawn as follows. The most significant cause of delay based on the owners' perspective is lack of communication among the

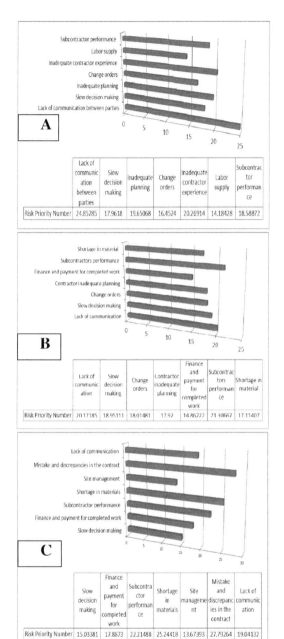

Figure 1. Risk priority numbers.

parties in the project. Mistakes and disagreement in material shortage and contract are caused according to the contractors' view point. Engineers believe that the major significant factors in project delays are subcontractor performance and lack of communication. The found RRI can be a useful benchmark tool for the construction industry advisors to be able to compare the multi-approach risk-based reliability of the industry with their benchmark and stated strategic mission. Based on the study limitations more studies are needed to fill in the gap where the study encountered lapses especially finding the RRI in other countries to compare with building construction to develop a RRI benchmark for construction industries as the study focused on the causes of delays in building construction projects. Researches may choose to adopt the effect analysis technique (Sorooshian et al, 2007) to test the findings of the study. Moreover further studies are needed to look at the risk factors integration in schedule analysis as well as the outcome of effect and common cause of failure modes and the relationship between delay risks.

ACKNOWLEDGEMENT

This study is supported by University Malaysia Pahang research grant (RDU130375).

REFERENCES

[1] Aibinu. A and Jagboro. G, "The effects of construction delays on project delivery in Nigerian construction industry". International Journal of Project Management, 2002, 20(8), pp. 593–599.
[2] Assaf. A and Al-Hejji. S, "Causes of delay in large construction projects". International Journal of Project Management, 2006, 24(4), pp. 349–357.
[3] Baldwin J. R, "Causes of delay in the construction industry'. Journal of the Construction Energy. Division, 1971, 97(2), pp. 177–187.
[4] Salleh. R, "Critical success factors of project management for Brunei construction projects: improving project performance", Ph.D. Thesis, Queensland University of Technology, 2009.
[5] Sambasivan. M and Soon. Y.W, "Causes and effects of delays in Malaysian construction industry". International Journal of Project Management, 2007, 25(5), pp. 517–526.
[6] Sorooshian. S, Norzima. Z, Yusof. I, Rosnah. M.Y, "Effect analysis on strategy implementation drivers", World Applied Sciences Journal, 2010, 11(10), pp. 1255–1261.

Using adhesive instead of welds to improvement torsion stiffness of body-in-white

Shujie Han & Xiangdong Cui
Qingdao Ocean Shipping Mariners College, Qingdao, Shandong, China

ABSTRACT: The finite element model of body-in-white was built in ANSYS. The torsion stiffness was analyzed using the finite element method. The results were compared to the test results, which proved the reliability of the finite element model. The adhesives were applied to the car body instead of welds; the torsion stiffness has been studied using the finite element method, the results showed that the body's torsion stiffness was improved, and the longer the adhesives were used, the torsion stiffness was improved greater. While the research results have an important reference value for the lightweight design of automotive.

Keywords: Torsion stiffness; Finite Element; Body-in-white; Adhesive

1 INTRODUCTION

The light weight of a car body is the key to improve car power, economy and reducing manufacturing costs. To meet the requirements of lightweight, all-bearing structure was adopted in most modern body structure designs. All-bearing structure make the car body bears almost all loads, including torsion, bending and impact loads, etc., which makes the body's stiffness of utmost importance [1]. Body stiffness will directly affect the vehicle's reliability, security, power, NVH performance and fuel economy and other key indicators. Therefore, the body's stiffness is one of the important evaluation indicators to car performances. Reasonable body structure can achieve light weight under the premise of meeting the strength and stiffness of the car body [2].

Domestic and foreign research institutions are carrying out researches on improved body stiffness and light weight, and have made some achievements. But most studies were structural optimization, improving the local structure and processes or reduce the material of the body [3-10]. In this paper, used adhesives instead of welds in some positions of the BIW. The torsion stiffness of load-bearing body–in-white was analyzed using finite element methods. The results shows that used Dow BETAMATE instead of welds really can improve the torsion stiffness, while also can achieve the lightweight body.

2 INTRODUCTION THE TORSION STIFFNESS TEST OF BIW

Torsion stiffness test of body-in-white (BIW) included the parts as follow: GLASS ASSY-W/SCRGRN LMND, GLASS ASSY-BKLG GRN, ARMATURE ASSY-FRTBPR, ARMATURE ASSY-RRBPR, SUBFRAME ASSY-FRT SUSP, SUBFRAME ASSY-RR SUSP, BODY SHELLS THROUGH THE PRINT PROCESS, TANK-ASSY MEMBER-TUNL, MEMBER-TUNL, RAIL ASSY-FAC, REINFORCEMENT, DASH ASSY ASSEMBLED.

Locate and attach simulated suspension mountings to body. Adjust the positions of simulate suspension base on the front and rear axle positions. Adjust the front beam section, ensuring that the beam is parallel to datum plane. Adjust the front suspension section, ensuring that they are perpendicular to test beam and are symmetric to the test beam axis. Position body shell on test beam and suspensions so that it is centered relative to mechanical assemblies' longitudinal axis. Ensure that body shell is parallel to datum plane. Position body over test rig and bolt or clamp down through the test mountings to both rig beams. Check all suspensions, ensuring that they wouldn't yield in the test. Install two transducers on the front contact beam and two transducers on the rear contact suspensions, ensuring that they are perpendicular to the test bench and symmetrically positioned relative to the longitudinal axis. Dispose the transducers on body as shown in figure 1.

Imposed constraints at the left and right rear suspension supports of the car frame made it can only rotate around the x-axis (in the vehicle coordinate system, x to the right direction, y to the forward direction, z to the vertical upward), at the left and right front suspension supports of car frame constraints the translational degrees of freedom of z direction to simulate simply supported conditions. Apply 0 to 4080 Nm torque in the clockwise direction to settle body shell. Release weights. Reapply the load in increments (each increment is 1020 Nm) of four times to a maximum torque of 4080 Nm. Release in increments to zero torque.

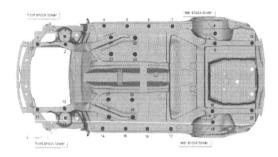

Figure 1. The positions of transducers.

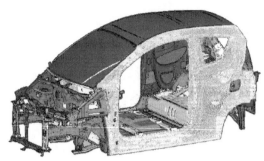

Figure 2. The finite element model of body.

Table 1. Parameters of the body-in-white.

Parameters	Value
Body length(mm)	3496
Body width(mm)	1504
Body height(mm)	1517
Wheel base(mm)	2340
Front track distance(mm)	1315
Rear track distance(mm)	1280
Body-in-white mass(Kg)	308

By measuring the z-displacement of the longitudinal bottom which in the same plane with the couple moments could get the twist angle around the y-axis, resulting attained the torsion stiffness of BIW. The z-displacement could be red from transducers.

3 THE FINITE ELEMENT ANALYSIS TO TORSION STIFFNESS OF BIW

There are two body stiffnesses: static stiffness and dynamic stiffness. Body static stiffness generally includes bending stiffness and torsion stiffness. Bending stiffness was measured by the deformation of the body. The torsion stiffness was measured by the diagonal deformation of front and rear windows and side-frame, body lock position, body torsion angle and other indicators.

The CAD model of the body-in-white was established in the software. The geometric model was simplified. The smaller radius of the holes, rounded corners and the smaller champers (rounds) between panels were ignored. The parameters of body-in-white were given in table 1.

Imported the CAD model into ANSYS software. The structure of BIW is a complex curved surface structure, using three-dimensional shell element shell93 meshing the body, divided into 565,406 units. The finite element model was shown as figure 2.

Referenced the test case, the boundary conditions was constrained all freedom degrees of front level beam and rear suspension bracket anchor. When analyzing the torsion stiffness, the load conditions were equal and opposite vertical forces which acting on the left and right front suspension spring. The couple moment was 4080 Nm as shown in figure 3.

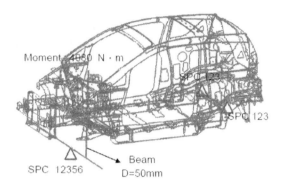

Figure 3. Boundary condition of torsion stiffness.

Table 2. The result comparison between FEA and test of torsion stiffness.

Aperture Distortion position	Nominal Length (mm)	Defamation of test (mm)	Defamation of FEA (mm)	Error %
A	1327.99	−0.90	−0.87	3.33
B	1319.67	0.84	0.80	4.76
G	903.43	0.00	0.00	0.00
C	846.75	−2.74	−2.68	2.19
D	828.76	2.44	2.39	2.05
K	645.40	0.00	0.00	0.00
T (LH)	1196.73	1.65	1.59	3.64
T (RH)	1188.73	−1.69	−1.61	4.73
E (LH)	1435.96	−1.47	−1.41	3.40
E (RH)	1436.41	1.48	1.43	3.40
H (LH)	1097.03	1.37	1.31	4.38
H (RH)	1077.55	−1.38	−1.32	4.35

The diagonal of front and rear windows and side-frame were chosen as figure 4. The comparison between finite element analysis results and test results was shown as table 2.

From table 2, we can conclude that the finite element model is reliable. Use the established and reliable finite element model analyzed the torsion stiffness of BIW without any structural adhesive. The displacement of Z-direction was shown as figure 5. The

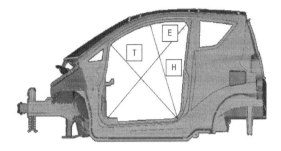

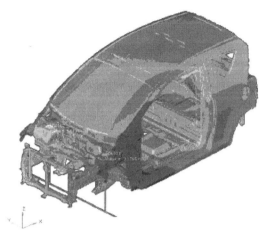

Figure 5. The displacement of Z-direction.

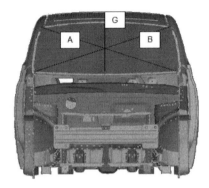

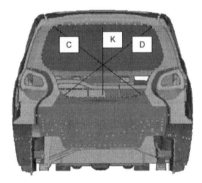

Figure 4. The door and tailgate aperture distortion position.

maximum displacement of z-direction was 3.191 mm, so the torsion stiffness was 10803.99 Nm/deg.

4 IMPROVEMENT THE TORSION STIFFNESS OF BIW BY DOW BETAMATE

Dow BETAMATE is a one component, epoxy based adhesive especially developed for the body shop. The adhesive was used in the car to increase the operation durability, the crash performance and the body stiffness. It has the following properties:

(1) Excellent process and storage stability.
(2) Excellent adhesion to automotive steels, including coated steels and pretreated aluminums with good tolerance to oils and dry lubes.
(3) Stiffness and crash stability increase of the entire car body.
(4) High durability of the adhesive and the adhesive bond.
(5) Protection of the metal and weld points against corrosion due to its sealing capability.
(6) Compatible with other mechanical and thermal joining techniques.
(7) Compatible with the electro coat process.
(8) Wash-off resistant.
(9) Procurable.
(10) Up to six weeks open time in the uncured bond.

So we intended to apply this material in automobile body instead of welding in some positions. Location as follows: floor edge, rear floor edge, firewall flange, as shown in figure 5. About 12 meters structural adhesive. E-modulus was 3,600 Mpa. Bonding line thickness in model was shell gap. Use the established and reliable finite element model analysis, the torsion displacement of z-direction shown in figure 6. Maximum displacement of Z is 3.086 mm so the torsion is 11175.27 Nm/deg.

About 30 meters structural adhesive. Location as follows: floor edge, rear floor edge, B-pillar, firewall flange, rear wheel cover, rear rail, front shock tower, rear D-ring, as shown in figure 7.

E-modulus was 2,600 Mpa. Bonding line thickness in model was 0.2 mm. Use the established and reliable finite element model analysis. Maximum displacement of Z is 2.834 mm. So the torsion is 12167.33 Nm/deg. The comparisons of the torsion stiffness of different adhesive lengths were shown in table 3.

From table 3, we can conclude that Dow BETAMATE instead weld really can improve the torsion stiffness of the car body, while can meet the requirements of light weight.

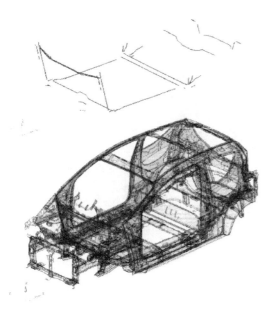

Figure 6. Locations of 12 meters adhesive.

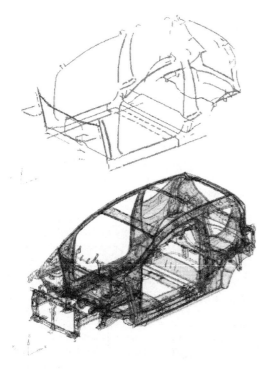

Figure 7. Locations of 35 meters adhesive.

5 CONCLUSIONS

The body stiffness is an important evaluation indicator to car performances. The torsion stiffness was studied in this paper. The CAD model of body-in-white was established in UG software, the finite element analysis of the torsion stiffness was done in the ANSYS software. The deformation of the front and rear windows and side-frame were compared to the test results, which proved that the finite element model was reliable. To improve the torsion stiffness of body-in-white, the new material DOW BETAMATE was applied instead of welding. The finite element analysis results show that the torsion stiffness was increased, while the body mass was decreased.

Table 3. The influence of Dow BETAMATE length to torsion stiffness.

Beta mate long(m)	Z-displacement (mm)	Torsion Stiffness (Nm/deg)	Stiffness Improvement
0	3.191	10803.99	0.0%
12	3.086	11175.27	3.4%
15	3.015	11442.35	5.9%
20	2.943	11722.52	8.5%
30	2.834	12167.33	12.6%

REFERENCES

[1] Gao Shenbing, Gao Weiming, Zhang Qilin, Shen Zuyan Research on the Measures for Enhancing the Torsional Rigidity of SANTANA – 2000 BIW Automotive Engineering 1996 (VOL18)No2:72–76

[2] Huang Tian-ze, Huang Jin-ling. Automotive body structure design. Beijing: Mechanical Industry Press, 2006. (in Chinese)

[3] Chen Xin, Yu Xue, Lin song Calculation of Static Stiffness of Car Body Analysis of Static Vertical Bending and Optimization Stiffness Automobile Technology 2004, 1:15–1.

[4] Wei Hong-ge, Tan Ji-jin, Ruan Ren-yu, Xu Jian-zhong Study on Stiffness of Minibus BIW Based on FEM Analysis and Test Journal Of Chongqing Jiaotong University (Natural Science) 2011. 2:147–157

[5] Huang Zong-bin, Li Jiang-liu, Cheng Dan, et al. Study on stiffness improbement of minibus body. Tractor & Farm Transporter, 2008, 35(5):34–35.

[6] The Editorial Board of Engineering Manual. Automotive engineering manual (Experimental paper). Beijing: China Communications Press, 2001. (In Chinese).

[7] Tan Ji-jin, Zhang Dai-sheng, Finite Element Analysis on Structure of Automobile. Beijing: Tsinghua University Press, 2009.

[8] Gui Liang-jin, Fan Zi-jie, Zhou Chang-lu, et al. Study on stiffness of Chang'an Star minibus white body. Chinese Journal of Mechanical Engineering, 2004, 40(9): 195–198.

[9] Beevers A, Steidler S M, Durodola J, et al. Analysis of stiffness of adhesive joints in cat bodies. Journal of Material Processing Technology, 2001, 118(1): 95–100.

[10] Huang zong-bin, li jiang-liu, cheng dan, et al. Study on stiffness improvement of minicar body. Tractor & farm transporter, 2008, 35(5):34–35. (in Chinese).

Application research on vinegar culture in vinegar packaging design

L. Zhang & S.Y. Wang
College of Mechanical Engineering, Taiyuan University of Technology, Taiyuan, China

ABSTRACT: This study analyses and thinks the application of vinegar culture in packaging through understanding and extracting the vinegar culture. Firstly, based on the vinegar culture, it researched the existing product packaging of "Meiheju, Donghu" vinegar, and then refined vinegar culture elements. Whether it is the shape, pattern, texture, color and so on, one or several aspects, had joined the vinegar culture elements in the packaging design of vinegar. Finally, by putting the vinegar culture into the packaging design of vinegar, not only to make the brand more personalized, strengthen the brand positioning, brand competitiveness, but also give consumers the visual experience and cultural infection.

Keywords: Vinegar culture; Packaging design; Culture strategy

1 INTRODUCTION

To mention taste, Chinese argument is five flavors, sour, sweet, bitter, pungent, and salty. It is also customary to always put the sour in the first place. Thousands of years ago, our country has already controlled grain vinegar technology. Now, vinegar has been used as a kind of cultural phenomenon deeply into the Chinese nation culture, forming a unique style of vinegar culture [1]. Vinegar culture is an important part of the traditional culture of China. Good packaging design is the key to successful sales, so "packaging is the best advertising of products" becomes the true saying. To promote sales and reflect the cultural values of the brand packaging design strategy is particularly important [2]. As the head of the Shanxi vinegar "Meiheju, Donghu" vinegar was hailed as "the first Chinese vinegar". In this regard, the author takes "Meiheju, Donghu" for example to analyze the vinegar culture connotation in the vinegar packaging design. Integration and analysis by shape, pattern, color, for more accurate and comprehensive understanding of the culture of vinegar, improve the vinegar packaging design, enhance the cultural added value of product have a guiding significance in reality.

2 VINEGAR AND VINEGAR CULTURAL COGNITION

2.1 *The origin of vinegar*

About the origin of vinegar, there are many different versions. Vinegar originated in the Zhou dynasty. It is the most reliable speculation. Our country earliest called vinegar as "acyl", and then "vinegar". According to the related historical records, the western Zhou dynasty appeared vinegar workshop which began to produce the vinegar. There are also have relevant legends, Du Kang invented wine, but his son called Hei Ta in charge of the wine-making after he got too old. Once, Hei Ta wasted three days because of drinking, he missed the distillation time of the wine which fermented in the wine cellar after waking, even the wine tasted sour. But after tasting, it has a distinctive flavor, the effectiveness of appetizers, have been used for cooking.

2.2 *Vinegar culture cognition*

Vinegar culture is vinegar as a carrier, vinegar through this carrier to spread all kinds of culture. It is the organic combination of vinegar and culture. It reflects some period of material and spiritual civilization. It is vinegar and folk customs, diet, etiquette, brewing process, and other related systems. It includes not only the spirit level, but also includes some related material objects.

2.3 *Vinegar and folk customs*

In the journey of Shanxi tourism, people often can see roadside erected a large sign that reads the word of "vinegar" character, the red is very eye-catching. As seen rice wine in Jiangxi, noodle in Yunnan, vinegar constitutes a beautiful landscape of Shanxi features [3]. As the development of vinegar business, vinegar culture is more and richer in folk activities. On February 13, 2007, the Shanxi province Qingxu County issued a "Chinese vinegar" logo which was approved by the state. In the same year on March 5, Qingxu County hosted the first vinegar culture festival. On August 20, 2012, "international vinegar culture festival Taiyuan (Qingxu), China" was held in Taiyuan coal trading center. During the festival, tourists could fully understand the vinegar brewing process. In the afternoon, a "Vinegar and Health" forum was held, tourists understood the health care function of vinegar, to further promote Shanxi vinegar.

2.4 *Vinegar diet culture*

Firewood, rice, oil, salt, sauce, vinegar, tea is the Chinese traditional diet. As one of them, vinegar is Chinese traditional condiment. In China, whether it is a small roadside restaurant, a large hotel ballroom, or a friend's house on the table, you'll often find such a vial filled with dark brown liquid—vinegar. According to the survey, the most popular with foreigners of Chinese food is sweet and sour pork, the second is Kung Pao chicken. But in the two dishes, vinegar has played a very important role. Especially in Shanxi, when it comes to the Shanxi people's hobby, foreigners often use a word to summarize: vinegar. Speaking of Shanxi regional famous brand, first of all, let a person think of vinegar. In Shanxi many local characteristic culture, vinegar is the most close to the people, everybody needs it, every day can not do without it [3].

2.5 *Vinegar brewing process culture*

Human edible vinegar has a long history, China mastered the grain vinegar brewing technology thousands of years ago. Vinegar with a unique brewing process to form the rich brewing culture in the history of several thousand years. After thousands of years of wind and rain, Shanxi vinegar has a long history, has formed a system, is famous for its "unique flavor, good quality". Shanxi vinegar is good, the most prominent feature is the "stuffed", it has long and superb fermentation and brewing [3]. Shanxi vinegar is made from the fermented sorghum, has more flavor than rice vinegar. In the late Ming and early Qing dynasty, "Mei he ju" was called "The first vinegar workshop". It produces vinegar with new productive technology, makes the biological factors produce new changes during the vinegar brewing process. The new changes make the health care composition which can promote the body's nutritional balance greatly increased, long favored by people.

3 INTERPRETATION AND PRESENTATION OF VINEGAR CULTURE IN PACKAGING DESIGN OF "MEIHEJU, DONGHU" VINEGAR

Packaging is the first bridge between products and consumers [4]. The connotation of vinegar culture needs these packaging design elements such as shape, pattern, material, color to interpretation and

Figure 1. "Yu guo tian qing" vinegar.

presentation. Express the connotation of vinegar culture in the form of external, to achieve internal and external unity, reasonable expression of vinegar culture.

3.1 *Cultural origins of trademark*

Trademark is a special identification of goods, is an important part of brand recognition. Trademark refers to the producers and business operators to make their own product or service differ from others, and is used on the goods and packaging. It is composed of text, graphics, letters, numbers, colors and other elements. It is a kind of visibility mark [5]. From the cultural origin, "Donghu" trademark is an iconic brand which carrying the wisdom of several generation of vinegar human. In 1957, "Donghu" trademark was named after the East Lake of Qingxu County. The trademark, approved by the State Administration for Industry and Commerce, became the first registered trademark of the Shanxi vinegar industry which enjoyed the exclusive right of autonomy [6]. So "Donghu" is "Meih ju". The basic pattern and color of trademark has no significant change until now. The trademark has a long history of cultural origin.

3.2 *Modeling design*

Type on the aesthetics, the modeling's influence is incomparable. According to the personality traits of the product, to design the unique appearance of product is very useful for helping differentiate the product from other brands of similar product. This not only can appear in a unique brand image, but also increase the sales of products [7]. "Yu guo tian qing" vinegar (fig.1), using the book-box packaging, the whole is simple and generous. The shape of the bottle is a raindrop shape of the inner packaging, which corresponds to the vinegar. "The best" vinegar packaging using the "vinegar basket". Vinegar basket is an ancient transportation of Shanxi vinegar products. It is braided with rattan outside, pig's blood and hemp paper pasted inside. It will not break by bump during transportation. It not only reflects the infinite wisdom of the ancient working people, but also reflects the historic vinegar cultural. Although the vinegar basket is made of porcelain now, it still represents the characteristic of Shanxi vinegar.

Figure 2. Health care vinegar.

Figure 4. Double bottle vinegar.

Figure 3. Century dragon vinegar.

Figure 5. Nation beauty and heavenly fragrance vinegar.

3.3 Pattern application

Pattern can be either mirrored or symbolic. Pattern to express the emotions must be based on emotion, and with the help of symbolic means can achieve the basic functions of the transmission of visual information [8]. "Donghu" Health care vinegar (fig.2) focused on an old person on the outer packaging design. The old person is generally refers to as 80–90 years of age or older who is in good health. It is consistent with the original intention of the health care vinegar. Vinegar has always been thought to have health care functions, the packaging clearly communicates this information to consumers, also fully expressed the vinegar culture. "Century dragon" vinegar (fig.3) uses the black pottery bottle of the inner packaging, black pottery is one of the Chinese four pottery. The bottle is printed with dragon patterns. The dragon is the symbol of the Chinese nation. In ancient times dragon is water. In the process of brewing vinegar, the water is particularly important. Therefore, the dragon is very important in the hearts of vinegar person. It is the symbol of good water. "Double bottle" vinegar (fig.4) uses "QingMing picture" on the outer packaging. It is used to symbolize the long history of "Donghu" vinegar. The packaging uses a hot stamping technique to make the product description. The overall design is simple and natural, the whole is one integrated mass. In the bottle design, using the shape of ancient coins. Engraving "Qing-Ming picture" on the bottle body, which correspond to the outer packaging. The product exhibits an ancient artistic conception.

3.4 Material and color selection

There are a lot of material in the packaging design can choose, different material can protect the product, but different material represents a different meaning.

Figure 6. Ancient altar vinegar.

Color is an important part of brand operation. It can establish the brand positioning, enhance the brand value, and arouse the desire to buy of the consumer [9]. We must understand the quality characteristics of the product and cultural background before determining the brand color [10]. "Nation beauty and heavenly fragrance" vinegar (fig.5) vinegar bottle is blue and white porcelain, it is Chinese core cultural elements. The bottle with red color, red is the Chinese traditional festival color. The bottle is also printed with the peony pattern, the overall is nobleness and elegant, happy and peaceful. Echoes its brand, also seems to make people feel the history of vinegar culture. "Ancient altar" vinegar (fig.6) uses the yellow color of the outer packaging. The inner packing bottle color uses black and grey. It reflects the long history of vinegar cultural. These are the consumers expect. Therefore, the color

Figure 7. The design of outer packaging.

Figure 8. The design of inner packaging.

in the packaging design for the theme, beautification products, enhance the brand value plays an important role [11].

4 VINEGAR PACKAGING DESIGN

According to the above analysis of vinegar culture and its application in vinegar packaging design, to design a vinegar packaging of "Donghu" vinegar. The design of the positioning direction is household vinegar (fig.7). Through the graphics display the complex vinegar brewing process. The packaging with yellow and brown color, brown reflects the mellow taste of vinegar, yellow reflects the sense of history and quality of vinegar. In the base map, the elements of "East" into the story of vinegar brewing. The shape of the vinegar bottle (fig.8) is extracted from the ancient vinegar jar, restores the old vinegar culture, and reflects a long history of "Donghu" vinegar. The design of the vinegar pot (fig.8) reflects the hexagon feature of "Donghu" trademark which shown in the form of three-dimensional. It makes the brand more personalized.

5 CONCLUSIONS

Apply vinegar culture to vinegar packaging design, absolutely not only use in some shape or pattern. To design a good vinegar package, designers should be carried on overall analysis and conducted brand positioning for the product after understanding vinegar culture. It will design the vinegar packaging with vinegar culture [12]. The vinegar culture strategy of vinegar packaging, only on the basis of absorption, the new situation created, combined with modern concept, and then it can be accepted by consumers. Through the use of some elements in the packaging, and the use of appropriate form of vinegar culture, or display some key attributes, and convert it to the desire to buy that consumers cannot resist.

REFERENCES

[1] Chen Li-yuan. On the Culture of Chinese Vinegar. Journal of Yuncheng University, 2003, 21(1):41–42.
[2] Huang Jing. Elements of Traditional Culture in Modern Packaging Design. Packaging Engineering, 2005, 26(1): 180–181.
[3] Song Xiao-le. Preliminary Study of Shanxi Vinegar Culture. Beijing: Minzu University of China, 2009.
[4] Gan Jin-xiu. Research on the Package Design of Woman's Cigarettes Corresponding with the Brand Culture. Design, 2012(2):74–75.
[5] Wang An-xia. Product Packaging Design. Nanjing: Southeast University Press, 2009:141–142.
[6] Ren Shan. The Best Shanxi Mature Vinegar Brewed by Cereal is Donghu. Taiyuan: Shanxi People's Press, 2006:20–21.
[7] Cheng Xiao-ting. On the Cultural Characteristics of Brand in the Package Designing of Fen Liquor. Journal of Shanxi Agricultural University, 2013, 12(2):162–164.
[8] Chen Zhan. The Application of Visual-rhetoric-based Graphic to the Packaging Design. Decoration, 2011(9): 114–115.
[9] Gavin, not Ross (Britain) Paul Harris. Zhang Fu-mei Translation. Create Brand Packaging Design. Beijing: Chinese Youth Press, 2012.
[10] Wang Zhen-hua, Wan Qing. The Application of Color Image in Package Design of Cosmetics. Decoration, 2012(5):84–85.
[11] Lin Lin, Wan Xuan. The Packaging Design of the Material Language. Art Education, 2009(3):29–30.
[12] Hu Hong-zhong, Xu Min. Discussion on the Significance of Chinese Traditional Culture Tactics in Packaging Design. Packaging Engineering, 2007(3):171–172.

Analysis of articulated chassis kinematics

Hongpeng Chi, Bing Gong, Xin Li, Huaizhuang Zang & Feng Tian
Beijing General Research Institute of Mining & Metallurgy, China

ABSTRACT: Articulated chassis motion model is analyzed. Establish motion equations of the vehicle in both cases, there is no presence of sliding and slide. Analyzes the factors affecting the articulated vehicle motion control. Established motion error equation, provides the foundation to research the self-driving articulated vehicle control technology.

Keywords: Articulated chassis; Kinematics; Motion equations; Slide

1 INTRODUCTION

Articulated chassis in the same chassis length, the turning radius can be significantly reduced when cornering, as articulated vehicles have this feature, which mostly uses articulated chassis in the trackless vehicles used in underground mining. When driving characteristics of articulated vehicles and autonomous driving technology research, we must first clarify its motion model and Kinematics.

2 VEHICLE KINEMATICS ANALYZES

Articulated vehicle motion model is different from the ordinary vehicle, movement of the articulated vehicle, there are two cases there is no sliding condition of the wheel and wheel sliding condition exists.

2.1 Motion model without considering sliding

Research articulated vehicle motion model, first, study the relationship of articulated vehicle trajectory curvature with its steering angle. In the ideal case (there is no slip), assuming the front articulated vehicle body and the after body is the same length (L1 = L2 = L). In the case, center of curvature of the path of movement of articulated vehicles, that is the intersection of the front and after axle line. When the steering angle changes, the vehicle moves around different centers of curvature with different curvatures. Fig.1, L is the length of the front and rear body of the articulated vehicle, O is the center of the articulated vehicle hinge point, α is the steering angle of the articulated vehicle, r is the curvature radius of the trajectory of the articulated vehicle, P is the center point of articulated vehicle trajectory curve.

As shown in Fig.1, the speeds of vehicle x direction and y direction are:

$$\dot{x}(t) = v \cdot \cos \varphi \quad (1)$$

$$\dot{y}(t) = v \cdot \sin \varphi \quad (2)$$

As shown in Fig.1, articulated vehicle steering angle has the following relationship:

$$\alpha = 2\eta \quad (3)$$

$$\tan \eta = \frac{L}{r}, \quad r = \frac{L}{\tan \eta} = \frac{L}{\tan \frac{\alpha}{2}} \quad (4)$$

Articulated vehicle heading angle change speed is:

$$\dot{\varphi}(t) = \frac{v}{r} = \frac{v \cdot \tan \frac{\alpha}{2}}{L} \quad (5)$$

Therefore, the motion equation of articulated vehicle can be expressed as follows:

$$\dot{x}(t) = v \cdot \cos \varphi$$
$$\dot{y}(t) = v \cdot \sin \varphi \quad (6)$$
$$\dot{\varphi}(t) = \frac{v}{r} = \frac{v \cdot \tan \frac{\alpha}{2}}{L}$$

The above motion equation of articulated vehicle is not considered the case of sliding wheels derived. Actually articulated vehicle motion exists sliding generally. Especially in the case of articulated vehicle sideslip, the above motion model will have a great error. Therefore, more accurate motion models need to build.

Figure 1. No slip case, motion relationship between the front and rear articulated vehicle body.

Figure 2. Sliding case, motion relationship between the front and rear articulated vehicle body.

2.2 Motion model considering sliding

As shown in Fig.2, consider the sliding motion of articulated vehicles, there are two new variables θ and γ will be introduced, γ and θ denote the slip angles of the front and rear vehicle body, that is the angles which between articulated vehicle speed direction under ideal motion model (perpendicular to the axle) and the real speed direction. Changes in this angle depend entirely on the sliding of the articulated vehicle. Equations of motion of point P are:

$$\dot{x}(t) = v \cdot \cos(\gamma + \varphi) \tag{7}$$

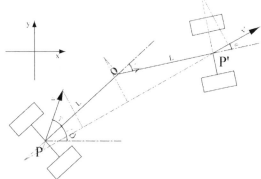

Figure 3. Sliding case, motion relationship between the front and rear articulated vehicle body.

$$\dot{y}(t) = v \cdot \sin(\gamma + \varphi) \tag{8}$$

Through the existing parameters, find the point in the vertical direction speed component of velocity, as show in Fig.3.

By the speed of point P, find a component perpendicular to v':

$$v_1 = v \cdot \sin(\gamma - \alpha - \theta) \tag{9}$$

By the rotation of point O, find a component perpendicular to v':

$$v_2 = \dot{\varphi} L \cos(\alpha + \theta) \tag{10}$$

By the rotation of point P', find a component perpendicular to v':

$$v_3 = L(\dot{\varphi} + \dot{\alpha}) \cos \theta \tag{11}$$

And by the speed in the vertical direction is equal to 0, then:

$$v_1 + v_2 + v_3 = 0 \tag{12}$$

I.e.:

$$v \cdot \sin(\gamma - \alpha - \theta) + \dot{\varphi} L \cos(\alpha + \theta) + L(\dot{\varphi} + \dot{\alpha}) \cos \theta = 0 \tag{13}$$

Calculated:

$$\dot{\varphi}(t) = \frac{v \cdot \sin(\alpha + \theta - \gamma) - \dot{\alpha} L \cdot \cos \theta}{L(\cos \theta + \cos(\alpha + \theta))} \tag{14}$$

The equation of motion of the rear vehicle body as follows:

$$\begin{aligned} \dot{x}(t) &= v \cdot \cos(\gamma + \varphi) \\ \dot{y}(t) &= v \cdot \sin(\gamma + \varphi) \\ \dot{\varphi}(t) &= \frac{v \cdot \sin(\alpha + \theta - \gamma) - \dot{\alpha} L \cdot \cos \theta}{L(\cos \theta + \cos(\alpha + \theta))} \end{aligned} \tag{15}$$

Take velocity v of point P on rear vehicle body as ωr, i.e., the wheel rotational angular velocity multiplied by the wheel rolling radius. Then we obtain the equations of motion of articulated vehicle:

$$\dot{x}(t) = \omega r \cdot \cos(\gamma + \varphi)$$
$$\dot{y}(t) = \omega r \cdot \sin(\gamma + \varphi) \qquad (16)$$
$$\dot{\varphi}(t) = \frac{\omega r \cdot \sin(\alpha + \theta - \gamma) - \dot{\alpha} L \cdot \cos\theta}{L(\cos\theta + \cos(\alpha + \theta))}$$

Thus, the change rate of the actual velocity direction of the vehicle (such as point P in the direction of $\gamma + \varphi$) depends on slip angle γ, steering angle α and angular velocity of steering angle α.

2.3 Motion error model of articulated vehicle

Comparison of the equation of motion in both cases motion model, the two models have significantly different. Especially for $\dot{\varphi}(t)$, without considering sliding, through the equations of motion, $\dot{\varphi}(t)$ only related to the steering angle. This model produces a relatively large impact on autonomic traveling control system of the vehicle. The motion model considers the sliding, if the slip angle is known, the equation of motion will be more accurate. The slip angle is the parameter can not be measured directly, but slip parameters can be estimated using the Kalman filtering technique, thus the control model of the vehicle autonomous driving and navigation system can be improved.

Since articulated vehicle is driven around, the speed of front and rear drive axle is the same, the sliding condition of the articulated vehicles can be estimated by detecting different v and v'. The relationship between v and v', as in formula

$$v = \mu \cdot v' + \eta \cdot \dot{\alpha} L \qquad (17)$$

When the vehicle without slip, i.e., u = 1, the following expression is obtained

$$v = v' + \eta \cdot \dot{\alpha} L \qquad (18)$$

In this clear special case, when the steering angle is constant or 0, the speed of front and rear vehicle body is equal, i.e.,

$$v = v' \qquad (19)$$

When α change, wheel slipping exists certainly. And when the angular velocity of the steering angle is relatively high, the articulated vehicle tires can be leaded to swipe road surface.

Through the equations of motion (1.6), (1.16), shows that the error of articulated vehicle motion model mainly come from the following parameters ($\omega, r, \alpha, \dot{\alpha}, \gamma, \varphi$) error. These parameter errors can be directly transmitted to the articulated vehicle motion model, thus affects the control accuracy of articulated vehicle autonomous. This articulated vehicle wheel rotation speed, steering angle and rotation rate of steering angle can be accurately measured. Articulated vehicles rolling radius errors at whether the vehicle load and tire wear can reach 15–20 cm. Articulated vehicle slip angles (γ, φ) cannot be directly measured by the sensor. So the error of these parameters can be estimated by the Federal Kalman filtering.

If don't consider the impact of these parameter error, it will cause the system to produce large errors, seriously affect the accuracy of autonomous driving system.

Wherein the additive noise of ω, α and $\dot{\alpha}$ can be written as $\delta\omega$, $\delta\alpha$ and $\delta\dot{\alpha}$, then there

$$\omega(t) = \overline{\omega}(t) + \delta\omega(t) \qquad (20)$$
$$\alpha(t) = \overline{\alpha}(t) + \delta\alpha(t) \qquad (21)$$
$$\dot{\alpha}(t) = \overline{\dot{\alpha}}(t) + \delta\dot{\alpha}(t) \qquad (22)$$

The errors of r, γ and φ are difficult to determine, they change dynamically. Because they involve a combination of other parameters. E.g., the slip angle will be affect by the speed, load, tire shape of the vehicle; it's a highly non-linear variable. Relatively feasible method, the integration of error white noise of these random parameter variables can be expressed by the following formulas:

$$\dot{r}(t) = \delta r(t) \qquad (23)$$
$$\dot{\gamma}(t) = \delta\gamma(t) \qquad (24)$$
$$\dot{\varphi}(t) = \delta\varphi(t) \qquad (25)$$

Assume that the noise sources $\delta\omega(t)$, $\delta\alpha(t)$, $\delta\dot{\alpha}(t)$, $\delta r(t)$, $\delta\gamma(t)$ and $\delta\varphi(t)$ are zero mean, the corresponding uncorrelated Gaussian sequence variance are σ_ω^2, σ_α^2, $\sigma_{\dot{\alpha}}^2$, σ_r^2, σ_γ^2 and σ_φ^2. The motion equation of articulated vehicles in continuous motion at time t and additional state estimation can be written as:

$$\dot{x}(t) = \omega(t) r(t) \cdot \cos(\gamma(t) + \varphi(t))$$
$$\dot{y}(t) = \omega(t) r(t) \cdot \sin(\gamma(t) + \varphi(t)) \qquad (26)$$
$$\dot{\varphi}(t) = \frac{\omega(t) r(t) \cdot \sin(\alpha(t) + \theta(t) - \gamma(t)) - \dot{\alpha}(t) L \cdot \cos\theta(t)}{L(\cos\theta(t) + \cos(\alpha(t) + \theta(t)))}$$

Wherein

$$\dot{r}(t) = \delta r(t)$$
$$\dot{\gamma}(t) = \delta\gamma(t) \qquad (27)$$
$$\dot{\varphi}(t) = \delta\varphi(t)$$

3 CONCLUSIONS

The motion characteristics of articulated vehicles chassis can be affected obviously by sliding. In the actual process of moving the vehicle wheel slip frequently exists. The actual change rate of the vehicle speed direction, depending on the slip angle, the steering angle and the angular velocity of the steering angle. Through analyze its error model, more precise control model of autonomous vehicles traveling can be established. Autonomous driving technology for intelligent articulated vehicle can be provided a favorable reference.

ACKNOWLEDGEMENTS

The authors are grateful to the financial supported by the National High Technology Research Program of China ("863" Project) (Grant No. 2011AA060405).

REFERENCES

[1] Yen, J.R. 1999. Rover Analysis Modeling and Simulation. FifthInternational Symposium on Artificial Intelligence, Robotics and Automation in Space, 249–254.
[2] Johan, L. & Mathias, B. 2006. A navigation system for automated loaders in underground mines. Springer Tracts in Advanced Robotics, 25: 129–140.
[3] Peter, R. & Peter, C. 2001. Autonomous Control of an Underground Mining Vehicle. Proc. Australian Conference on Robotics and Automation. Sydney, 14–15 November 2001. 26–31
[4] Hongpeng, C. & Kai, Z. 2012. Automatic guidance of underground mining vehicles using laser sensors. Tunnelling and Underground Space Technology. 27 (2012) 142–148
[5] Wong, J.Y. & Wei, H. 2006. Wheels vs. tracks"–A fundamental evaluation from the traction perspective. Journal of Terramechanics, 43: 27–42.
[6] Peter, R. & Peter, C. 2003. Load Haul Dump Vehicle Kinematics and Control. Transactions of the ASME. MARCH 2003, Vol. 125. 54–59
[7] Hemami, A. & Polotski, V. 1998. Path tracking control problem formulation of an LHD loader. The International Journal of Robotics Research, 17(2): 193–199.

Simulation and analysis of chatter in six-roll tandem cold rolling mills

Y.L. Gong & S.J. Wang
Cheng Steel Co. of He Bei Iron & Steel Group, ChengDe, HeBei, China

H.R. Liu
Institure of Information Technology and Engineering, Yanshan University, Qinghuangdao, HeBei, China

Y. Zhang
Institute of Electrical Engineering, Yanshan University, Qinhuangdao, HeBei China

ABSTRACT: A model of seven degree of freedom for the vertical vibration of stand rolls system in a cold tandem mill is developed. Considering the vibration factors of the upper and lower work rolls, we build a model of rolling force. Numerical simulation analysis was carried out on the stands coupling vibration model. It verifies that tension fluctuation contributes to negative damping for vertical vibration system and the vibration amplitude shows divergent patterns. It will cause phenomenon of chatter. The results of the calculation show that the vibration is greatly influenced by rolling speed and friction coefficient. With an increase in rolling speed, the mill vibration tends to be self-excited. The results discussed above provide a reference for the cold rolling mill chatter study.

Keywords: Six-roll tandem cold rolling mills; Rolling force; Tension divergence; Rolling velocity

1 INTRODUCTION

With the rapid development of the modern iron and steel industry, rolling mill technology level and rolling speed are constantly improved. With the improvement of rolling speed, rolling mill vibration problems occur frequently. Vertical chatter of stand rolls can lead to strip surface streaks and impressions. And, when this situation is occurred seriously, it can produce a lot of defective goods. So, it seriously affects the production efficiency and product quality [1].

Scholars at home and abroad have done some research work for the problem of vertical vibration of stand rolls and have achieved some results. Yun *et al.* have presented a model of modal coupling chatter from the view of the modal coupling [2]. Zhi Hao Zhai *et al.* have established nonlinear differential equations of torsional vibration and the perturbation solution of it. And, it proved that the increase of rolling speed will cause the occurrence of chatter [3]. Dian Ping Yin *et al.* have taken advantage of the integrated signal test and finite element simulation method to analysis the vibration phenomenon of double stand cold rolling. It found that the vertical vibration is generated by the simple harmonic vibration force caused by the backup roll surface defects [4]. Xu Yang *et al.* have established two degrees of freedom nonlinear self-excited vibration model after considering the influence of the nonlinear damping in rolling interface and the nonlinear stiffness between rolls and stand [5]. Dong Xiao Hou *et al.* have established vertical-horizontal nonlinear vibration dynamic model of the strip rolling mill after considering the influence of the friction factors in rolling interface. The bifurcation behavior of the coupled system was analyzed by using the theory of singularity. It provides theoretical guidance for suppressing vibration of the mill roll system [6].

On the basis of the six-roll tandem cold rolling mills stand system, nine degrees of freedom vertical vibration dynamics equation is established at this paper. Considering the mutual influence between the stand, the coupling vibration model of stand is analyzed on the basis of establishing a tension wave equation between the two stands. It verifies that the tension fluctuation will cause the negative damping effect of the vertical vibration system. And, the self-excited vibration of the vertical vibration system will be occurred with the increase of velocity. It provides a theoretical basis for chatter suppression of the six-roll cold tandem mill.

2 VERTICAL VIBRATION MODEL OF SIX-ROLL COLD TANDEM MILL

In this paper, the six-roll vibration system will be simplified a spring-mass system with nine degrees of

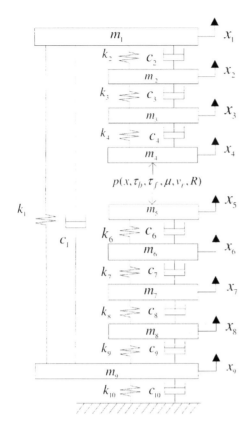

Figure 1. Model of vertical vibration of six-roll tandem cold rolling mills.

freedom to calculate the vertical vibration, as shown in Fig.1. It consists of nine mass elements, corresponding to the top and bottom housing, the top and bottom work rolls, the top and bottom middle rolls, the top and bottom backup rolls and hydraulic cylinder. The spring constant between mass element is $k_1 \sim k_{10}$. The damping constant between mass element is $c_1 \sim c_{10}$. The mass of mass element is $m_1 \sim m_9$. The rolling force is represented by P.

According to the vertical vibration dynamic model shown in Fig.1, the mechanical structure vibration equation can be written as the following matrix form.

$$[M]\{\ddot{x}\}+[C]\{\dot{x}\}+[K]\{x\}=[F] \quad (1)$$

Where M is the mass matrix of mass elements, K is the spring constant matrix, C is the damping matrix, $\ddot{x}$ is the acceleration column vector which represented speed of each mass element, $\dot{x}$ is the speed column vector which represented speed of each mass element, x is the displacement column vector which represented speed of each mass element, F is the loading matrix, $F=[0\ 0\ 0\ P\ -P\ 0\ 0\ 0\ 0]^T$,

where

$$M = \begin{bmatrix} m_1 & & & & \\ & m_2 & & 0 & \\ & & \ddots & & \\ & 0 & & m_8 & \\ & & & & m_9 \end{bmatrix}$$

$$K = \begin{bmatrix} k_1+k_2 & -k_2 & & & & -k_1 \\ -k_2 & k_2+k_3 & -k_3 & & 0 & \\ & & \ddots & & & \\ & 0 & & \ddots & 0 & \\ & & & & \ddots & \\ -k_1 & & 0 & & -k_9 & k_1+k_9+k_{10} \end{bmatrix}$$

$$C = \begin{bmatrix} c_1+c_2 & -c_2 & & & & -c_1 \\ -c_2 & c_2+c_3 & -c_3 & & 0 & \\ & & \ddots & & & \\ & 0 & & \ddots & 0 & \\ & & & & \ddots & \\ -c_1 & & 0 & & -c_9 & c_1+c_9+c_{10} \end{bmatrix}$$

3 MODEL OF THE ROLLING FORCE UNDER THE INFLUENCE OF THE VIBRATION

As the existence of vibration, the vibration displacement and velocity of the upper and lower work roll will change over time in actual rolling process. So researching the rolling force model under the influence of work roll vibration is needed. In order to reflect the reality of the rolling force, it combines the mill mechanical structure vibration model with a rolling force model. The equivalent elastic force $k_5(x_4 - x_5)$ and the equivalent damping force $c_5(\dot{x}_4 - \dot{x}_5)$ equivalent between the top and bottom work rolls are replaced by using rolling force model in the process of rolling.

The vibration equation of the top and bottom work roll can be written as the following form.

$$\begin{cases} m_4\ddot{x}_4 - c_4(\dot{x}_3 - \dot{x}_4) - k_4(x_3 - x_4) \\ -P(x,\tau_f,\tau_b,\mu,v_r,R) = 0 \\ m_5\ddot{x}_5 + c_6(\dot{x}_5 - \dot{x}_6) + k_6(x_5 - x_6) \\ +P(x,\tau_f,\tau_b,\mu,v_r,R) = 0 \end{cases} \quad (2)$$

Basing on the Hill model, the rolling force model can be written as the following.

$$P = Bl'Q_p KK_T \quad (3)$$

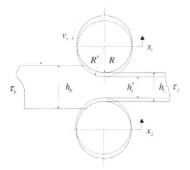

Figure 2. Model of Roll Gap.

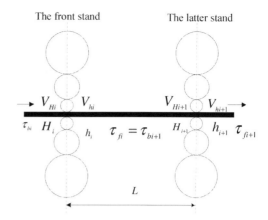

Figure 3. Model of two adjacent stands.

According to the lubrication effect coefficient K_u and rolling velocity v, the coefficient of friction μ can be written as the following.

$$\mu = K_u \left[0.07 - \frac{0.1v^2}{2(1+v)+3v^2} \right]$$

Where, K_u is lubrication influence coefficient.

In a steady rolling conditions, P is the rolling force. h_0 is the thickness of the entrance. h_1 is the thickness of the export, Δh is rolling reduction. The top roll displacement is x_1. The bottom roll displacement is x_2, h_1' is the thickness of rolled piece exports under the vibration of work roll. $\Delta h'$ is rolling reduction under the vibration of work roll. As shown in Fig.2.

Then,

$$\varepsilon = \frac{\Delta h'}{h_0} = \frac{h_0 - h_1 - x}{h_0} \qquad (4)$$

The rolling force is,

$$p(x) = 1.15\sigma B \sqrt{R'(h_0 - h_1 - x)} \left(1 - \frac{a\tau_b + (1-a)\tau_f}{K}\right) \times \left(1.08 + \frac{1.79\mu(h_0 - h_1 - x)}{h_0}\right. \\ \left. \times \sqrt{1 - \frac{h_0 - h_1 - x}{h_0}} \sqrt{\frac{R'}{h_1 + x}} - 1.02 \times \frac{h_0 - h_1 - x}{h_0}\right) \qquad (5)$$

4 TENSION INCREMENT MODEL OF TWO ADJACENT STANDS

In the actual rolling process of the cold tandem mill, the exit velocity of the front stand and the entrance velocity of the latter stand will cause small changes. That will lead to tension fluctuate between the two stands. In the process of vertical vibration, the change of the latter tension is one of the important factors causing negative damping. In order to analyze the cold tandem mill chatter phenomenon more accurately, the coupling effect between the stand cannot be ignored. So, the coupling relationship between the stand is established according to the tension wave equation between two adjacent stands. Then, a multiple stands coupling vibration model is established under considering the influence of vibration. Two adjacent stands model is shown in Fig.3.

The front Tension increment equation can be written as the following.

$$\Delta \tau_{fi} = \frac{E}{L} \int_{t_1}^{t_2} (V_{Hi+1} - V_{hi}) dt \qquad (6)$$

5 CHATTER SIMULATION OF SIX-MILL

5.1 *The influence of tension fluctuation for chatter*

The top work roll of the fourth stand which causes the chatter phenomenon easily is treated as research object. When the velocity is 35 m/s, it compares the influence of tension fluctuation and no tension fluctuation influence. The influence of tension fluctuation for chatter can be verified by simulation. Simulation diagram of time-domain is as shown in Fig.4 and Fig.5.

It can be seen from Fig.4 and Fig.5 that the vertical vibration of the top work roll under no tension fluctuation is damped. And, the vertical vibration of the top work roll under tension fluctuation is self-excited. The tension fluctuation of adjacent stand is caused by the change of the exit velocity of the front stand and the entrance velocity of the latter stand.

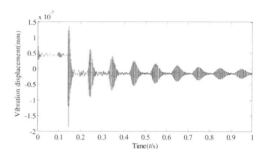

Figure 4. Simulation wave form of vertical vibration of up working roll without influence of tension fluctuation.

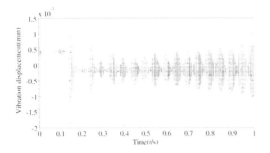

Figure 5. Simulation wave form of vertical vibration of up working roll with influence of tension fluctuation.

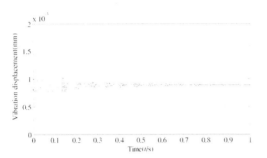

Figure 6. Simulation wave form of vertical vibration of up working roll(V = 25 m/s).

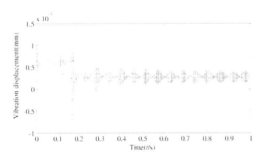

Figure 7. Simulation wave form of vertical vibration of up working roll(V = 30 m/s).

5.2 The influence of rolling velocity for chatter

Rolling velocity is the important process parameters in the process of cold tandem mill production. The influence of rolling velocity for vertical vibration of cold tandem mill is researched when V is 25 m/s, V is 30 m/s and V is 35 m/s. Simulation diagram of time-domain is as shown in Fig.6, Fig.7 and Fig.8.

According to the principle of constant metal flow, volume volatility quantity of mill entrance side increase under the same amount of roll gap fluctuation when rolling velocity is increased, that lead to increase the entrance tension fluctuation. So, as the rolling velocity is higher, the vibration of the cold tandem mill roller becomes more intense. It will reduce the stability of the rolling mill. And, that is so easy to have chatter.

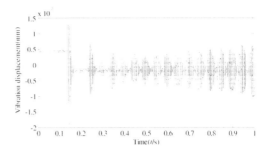

Figure 8. Simulation wave form of vertical vibration of up working roll(V = 35 m/s).

6 CONCLUSIONS

The nine degrees of freedom vertical system dynamic model of the six-cold tandem mill, the rolling force model of under the influence of the vibration and tension increment equation two adjacent stands are established in this paper. With the top work roll of the fourth stand as the research object, multiple stands coupling model is analyzed based on the rolling force model under the influence of the vibration. Some results can be obtained through the numerical simulation.

1) The negative damping effect caused by change of entrance tension can make the system unstable. And, the vibration is self-excited.

2) The change of rolling velocity will affect the change of the tension between two adjacent stands. With the improvement of rolling velocity, tension fluctuation will be increased. This vertical vibration system will gradually lose stability.

Therefore in the process of cold tandem mill production, chatter can be avoided through the appropriate process parameters are selected.

REFERENCES

[1] Qing-Long Ma, Pei-ming Shi 2013. Study of chatter strategy in the cold rolling mill of strip. *Cfhi Technology* 2: 7–9.

[2] Yun I.S., Ehmann K.F., Wilson W.R.D., et al. 1998. Chatter in the Strip Rolling Process, Part 3: Chatter Model. *Journal of Manufactoring Science and Engineering* 120(5).

[3] Zhi-Hao Qu, Shao-Kuan Chai, Qian-Yuan Ye 2006. Analysis of dynamic characteristics and chatter of a 1420 cold tandem rolling mill. *Journal of Vibration and Shock* 25(4): 25–29.

[4] Dian-Ping Yin, Zhi-Hui Sun 2010. Roller system vibration research of two-stand cold continuous rolling mill. *Metallurgical Equipment* 179(1): 21–24.

[5] Xu Yang, Jiang-Yun Li, Chao-Nan Tong 2013. Nonlinear vibration modeling and stability analysis of vertical roller system in cold rolling mill. *Journal of Vibration, Measurement & Diagnosis* 33(2): 303–307.

[6] Dong-Xiao Hou, Rong-Rong Peng, Hao-Ran Liu 2013. Vertical-horizontal coupling vibration characteristics of strip mill rolls under the variable friction. *Journal of Northeastern University (Natural Science)* 34(11): 1615–1619.

Design and verification of the current transducer based on the TMR effect

L.Y. Xue & K.J. Yi
College of Life Information Science & Instrument Engineering, Hangzhou Dianzi University, Hangzhou, Zhejiang, China

W. Huang
College of Automation, Hangzhou Dianzi University, Hangzhou, Zhejiang, China

D.Y. Song
Research and Development Department, Hangzhou Yuhang Electric Apparatus Ltd., Hangzhou, Zhejiang, China

ABSTRACT: Traditional current transformer and Hall current sensor possess several drawbacks such the remanence, magnetic saturation and core losses. A novel non-contact type of current transducer based on the Tunneling Magneto-Resistance (TMR) effect is presented to solve these problems. Due to the high sensitivity and low hysteresis of the TMR, the magnetic field concentrator can be canceled, therefore, no remanence occurs. Using PCB-IC mixed technology, a novel non-contact type of current transducer based on the TMR sensor is designed, in which the TMR sensors were configured as a Wheatstone bridge. The bridge excitation circuit with temperature compensation, and the conditioning circuit with a closed loop controller is described. In a setting −4.0A–4.0A range, the full scale output of the designed prototype is 1.30 V, the static accuracy is 0.0923% F.S., linearity is 0.0550% F.S., Hysteresis Error is 0.0712% F.S. a frequency response up to 1 MHz has been obtained.

Keywords: Current transducer; Magnetic sensor; Giant magneto-resistance effect; Tunneling Magneto-Resistance

1 INTRODUCTION

The current transducer is a significant electronic device used to offer current sensing in modern industrial applications with diverse standard. In recent years, electric systems have become more miniaturized and integrated. The current transducer is faced with new challenges regarding not only the performance requirement but also the size, weight and cost, and traditional current sensing methods are being revised and substituted by emerging technologies.

With the deepening of the research on magnetic electronics and improving of manufacturing technology on amorphous, nano-crystalline materials, especially after M. N. Baibich et al. reported for the Giant Magneto-Resistance (GMR) effect measured on Fe/Cr thin multilayer in 1988 [Baibich et al. 1988], the potentiality for development and application in position, direction and current sensing evoked great attention from academic circles, businesses, and government. New types of solid state magnetic sensor based on GMR have emerged, and successfully applied in a lot different environments [Kurlyandskaya et al. 2003, Sebastia et al. 2009, LAIMER et al. 2005, Munoz et al. 2009]. From both the physics and engineering point of view, the 2007 Nobel Prize for physics can be understood as global recognition to the rapid development and wide application prospect of GMR.

Spin-valves (SV) and Tunneling Magneto-Resistance (TMR) are two mechanisms with the practical realization of GMR device. Low cost, compatibility with standard CMOS technologies and high sensitivity are common advantages of these sensors. The research of current sensing application has been reported recently. As examples, a commercial current sensor with closed loop controller was presented in [LAIMER et al. 2005], a 0–1 MHz bandwidth, 1% temperature drift was obtained, and a specific ASIC chip has been implemented to reduce the size of sensor. A smart miniaturized power meter solution was shown in [Munoz et al. 2009]. A TMR-based Wheatstone bridge built-in IC for current self-testing was discussed and analyzed in [Le Phan et al. 2006]. A homoplastic design, a 10 μA–100 mA measuring range, 100 kHz bandwidth, more than 1 mA/V(mA) sensitivity, was demonstrated in [Cubells-Beltran et al. 2009].

For a long time, in the basis of the capability of measuring DC, AC and complex waveforms while ensuring galvanic isolation, Hall current sensors were the dominant choice for solid state magnetic sensors, and nowadays, are well established in industry [Togawa et al. 2005]. However, because of the poor sensitivity, Hall sensors need to introduce a magnetic field concentrator, usually consisted of a magnetic core and a coil around the core. The residual flux of the magnetic core induces an additional measurement offset referred to as "magnetic offset", and the bandwidth, response time and di/dt behavior are restricted to the core losses and core heating. In addition, the reliability of the Hall sensor will be reduced by overload condition, and the failure rate increases significantly as the temperature increases. In [Pelegri et al. 2003], the performance of a GMR sensor was compared with a commercial Hall current transducer within a high-frequency bi-directional three-phase rectifier. The GMR sensor displayed excellent figures regarding noisy and heat environment. The GMR-based current sensor is regarded as the potential update product of the Hall sensor.

As the magnetic core is indispensable, the traditional current transformer and Hall current sensor possess several drawbacks such as the remanence, magnetic saturation and core losses. In this work, based on the excellent magnetic detecting performance of the TMR sensor, a novel non-contact type of current transducer without magnetic core is designed.

2 DESIGN OF TMR CURRENT TRANSDUCER

2.1 Tunneling magneto-resistance (TMR) sensor

The basic layered structures of TMR consist of two or more magnetic layers of a Fe-Co-Ni alloy, as can be permalloy, separated by a very thin isolating layer. One layer of the TMR is a "pinned layer" that is not affected by the magnetic field, and the other is a "free layer", which has a magnetization that aligns parallel to the applied magnetic field. Electrons can surpass this thin film by means of the quantum tunnel effect, and the crossing probability is higher when both magnetic moments are aligned in parallel and lower when both magnetic moments are not aligned in parallel. So, a TMR sensor usually makes use of the spin-valve principle in order to fix the easy axis by means of a pinning antiferromagnetic layer. When they were configured in a crossed axis configuration, linear ranges suitable for sensor applications can be achieved, and the resistance change is proportional over a wide range of the measured field.

2.2 TMR-based current transducer

Even though a unique TMR can be used as sensing element, a Wheatstone bridge setup is always a good recommendation in the design of resistive sensors. Figure 1 shows the structure of the TMR-based current transducer.

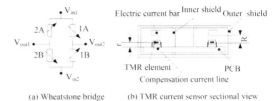

(a) Wheatstone bridge   (b) TMR current sensor sectional view

Figure 1. TMR-based transducer structure and Equivalent bridge diagram.

The multilayered structure of the TMR sensor resembles that of MRAM cells, and the current to be measured flows through a U-shaped current bar, which is arranged above two TMR chips, denoted by A and B respectively, in which two TMR elements are packaged together, and the anisotropy direction of the free layer is along the current bar. As the pinned direction is fixed for all sensor elements on a wafer, by exchanging bias direction, four elements are formed to a crossed anisotropy configuration. Equivalent bridge is shown in Figure 1(a). The compensation current line on the print circuit board (PCB) is patterned under two TMR chips, shielded from the inner and outer shield with insulation between conductors and PCB, as shown in Figure 1(b). The shields, surrounding each of the TMR sensors, play a critical role in preventing unwanted magnetic fields generated by currents in the other leg of the U-shaped conductor. In addition, the shielding prevents any external fields entering into the system.

The current sensitivity of TMR sensor, as above described, mainly comprises two components: 1) the efficiency of the magnetic field generation from the current (both the measured and compensation current), and is associated with the distance (R or r in Figure 1(b)) between conductor and sensor, the shape and location of conductors, etc.; 2) the magnetic field sensitivity of the sensor itself, is dependant on the TMR stack design and the sensor dimensions.

3 CONDITIONING AND CONTROL CIRCUITS

3.1 Biasing and conditional circuit

An appropriate circuit can set the correct bias point of the sensor. A constant voltage source can be used to feed the TMR sensor, through two opposite vertex of the bridge. The differential output voltage is taken from the remaining pins. Nevertheless, it has been demonstrated that the characteristics (temperature drifts) of spin valve based sensors are notably improved by using a constant current source for the sensor feeding. Figure 2 shows a constant current excitation circuit with temperature compensation.

In Figure 2, if R_2 is eliminated, the amplifier A_1 plays the role of a buffer only, and the amplifier A_2 acts as a voltage-current transfer. Due to the reference

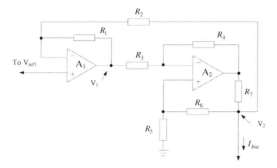

Figure 2. TMR sensoe bridge excitation circuit diagram.

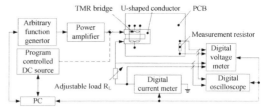

Figure 4. TMR-based current transducer characteristic test system.

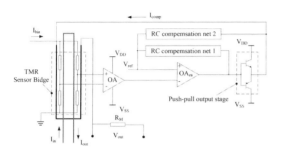

Figure 3. Closed-loop contrel circuir diagram.

voltage (V_{ref1}), coming from a band-gap voltage reference circuit, the output voltage (V_1) of A_1 is considered to be fixed, and the circuit is a constant current circuit. However, the existence of R_2 makes part of the TMR sensor bridge bias voltage (V_2) positive feed back to A_1, thus, the negative temperature coefficient of TMR bridge is compensated.

The quiescent resistance value of TMR sensor distributes in a wide range, common value is several hundreds kΩ, thus, the TMR sensor presents extremely low power consumption. On the other hand, for a resistor bridge configuration, it means that the equivalent internal resistance of the differential output signal is at large scale. Hence, the subsequent condition circuit, for amplifying and normalizing the TMR bridge output, should be designed with high performance components, and solved the application related issues such as error estimation, proper layout and so on.

3.2 Close loop controller

The closed-loop control circuit is shown in Figure 3. The operating principle of the circuit is implemented using a TMR bridge to measure a differential magnetic field generating by the U-shaped current bar and the compensation line.

When the primary current is fed through the U-shaped conductor, and creating a field gradient H_{prim} between the two sides of the conductor, the closed loop control circuit uses the TMR bridge differential output to create compensation current, fed back to the sensor chips through the compensation line. The resulting field H_{comp} exactly compensates H_{prim}. The total flux, as measured by the TMR bridge, equal to zero. In other word, the compensation current creates a flux equal in amplitude, but opposite in direction, to the flux created by the primary current.

In Figure 3, the TMR bridge and the amplifier OA act as zero flux detector, and the error amplifier OA_{ea} and RC compensation networks make up of a Proportional-Integral (PI) controller with the capability of small inertia, fast response time and good high frequency attenuation characteristic. Only one PI controller is not easy to obtain high di/dt, du/dt behavior while ensuring fast response time and broad pass-bandwidth. Therefore, another RC compensation network, linked to the output terminal of compensation current output stage (push-pull follow circuit), is introduced. By tuning RC network parameters, the match is achieved between the bridge operating parameters (the current line geometry, the ratio (R/r) of distance from conductor to sensor, etc.) and the PI controller amplitude-frequency characteristics; and the output voltage can be scaled.

4 EXPERIMENTAL VERIFICATION

4.1 Characteristic test system

The goal of the test instrument designed is to obtain an experiment verification of the TMR-based current transducer. Figure 4 shows the test system for DC, AC characterization, which includes: a personal computer, the measuring instrument required and the interface devices (power amplifier, adjustable load). The system must control and measure DC, AC currents and voltages and frequency.

In order to make the appropriate tests, the system employs a program controlled DC source and an arbitrary function generator (Tektronix AFQ310) to feed current through the U-shaped conductor and an adjustable load R_L (according to the test required, the output of function generator need power amplifier to drive R_L). The PC controls the measurement instruments (three Fluke 8846A 6-1/2 digital precision meters) and the test signal generators to achieve

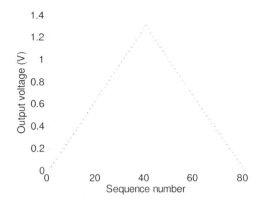

Figure 5. Output of a continuous positive and reverse stroke.

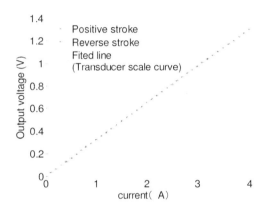

Figure 6. TMR-based current transducer scale curve.

data acquisition routine. The control interface the bus IEEE-488.2 and the RS232 interface. An oscilloscope (Tektronix TSP3052B) is used to examine the waveform of input signal and the TMR-based transducer output.

4.2 Static accuracy

As shown in Figure 4, the TMR bridge bias voltage is set to 5.0 V, the PC controls the program controlled DC source to output a stepping stroke output from 0 to 4.0 A, the step size is 100 mA, and the duration time is 1s. Figure 5 shows the output waveform of a continuous positive and reverse stroke recording by Fluke digital meter.

By several measuring, the scale curve of TMR transducer, shown in Figure 6, is obtained through fitting the measuring data based on Least Squares Method. The obtained function is $V_{out} = 0.32519I + 1.4175 \times 10^{-4}$. Table 1 shows the evaluation indexes related to the static accuracy, defined as follows:

$$K = \Delta V_{out}/\Delta I \quad (1)$$

$$\delta_L = \pm(\Delta V_{out})_{max}/Y_{FS} \times 100\% \quad (2)$$

Table 1. Evaluation indexes.

Performance Data	
Full scale output (Y_{FS})	1.3007V
Sensitivity (K)	0.3252(V/A)
Linearity (δ_L)	0.0550%
Hysteresis Error (δ_H)	0.0712%
Static Accuracy (K)	0.0923%

$$\delta_H = (1/2)\cdot \Delta H_{max}/Y_{FS} \times 100\% \quad (3)$$

$$e_s = \pm\sqrt{\delta_l^2 + \delta_u^2} \quad (4)$$

where K = Sensitivity; δ_L = Linearity; and δ_H = Hysteresis Error; e_s = Static accuracy.

Where $(\Delta V_{out})_{max}$ is the deviation of the transducer output values to the values calculated with the scale curve; ΔH_{max} is the deviation of the output values in positive stroke to the values in reverse stroke; Y_{FS} is the full sale output value.

4.3 Frequency response characteristics examination

The sine and triangle wave, produced by the function generator separately, were used as a test signal to examine the frequency response of transducer. The results (10 Hz, 1.5 kHz, 20 kHz, 500 kHz) are shown in Figure 7.

By observing the frequency response in oscilloscope and comparing the true effective value measured, a flat response up to 20 kHz with small phase shift and a zero response starting at 1 MHz have been obtained.

5 CONCLUSION

The TMR bridge based current transducer, fabricated by PCB-IC mixed technology for a galvanically isolated current measurement, is designed. The bridge excitation circuit with temperature compensation, and the conditioning circuit with a closed loop controller is described. Due to the high sensitivity of TMR sensor, there is no need for a magnetic field concentrator, therefore, no remanence occurs. The result of experimental verification demonstrates that the transducer can be characterized by high accuracy, excellent linearity, low hysteresis and wide bandwidth. In a setting 4.0A – 4.0A range, the full scale output is 1.30 V, the static accuracy is 0.0923% F.S., linearity is 0.0550% F.S., Hysteresis Error is 0.0712% F.S. The prototype of TMR current transducer possesses a wide frequency response range (0 up to 1 MHz). However, the flat response is 20 kHz. A deep study of a frequency compensation to extend its flat frequency response range is necessary.

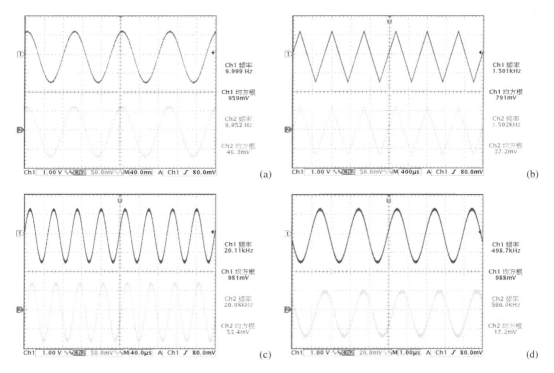

Figure 7. TMR-based transducer frequency response test (the above curve is the waveform of testing signal, the other one belongs to the transducer).

ACKNOWLEDGMENT

The work was supported by the Science Research Planning Item of Zhejiang Provincial Education Department (Grant No. 2012C21095).

REFERENCES

[1] Baibich, M.N. et al. 1988. Giant magneto-resistance of (001)Fe/(001)Cr magnetic superlattices. Phys. Rev. Lett 61: 2472–2475.
[2] Cubells-Beltran, M. et al. 2009. Full Wheatstone bridge spin-valve based sensors for IC currents monitoring. IEEE Sens. J 9(12): 1756–1762.
[3] Kurlyandskaya, G.V. et al. 2003. Advantages of nonlinear giant magnetoimpedance for sensor applications. Sens. Actuat. A: Phys 106(1–3): 234–239.
[4] LAIMER, G. et al. 2005. Design and experimental analysis of a DC to 1 MHz closed loop magnetoresistive current sensor. Applied Power Electronics Conference and Exposition 2: 1288–1292.
[5] Le Phan, K. et al. 2006. Tunnel magnetoresistive current sensors for IC testing. Sens. Actuators A–Phys 129(1–2): 69–74.
[6] Munoz, D.R. et al. 2009. Design and experimental verification of a smart sensor to measure the energy and power consumption in a one-phase AC line. Measurement 42(3): 412–419.
[7] Pelegri, J. et al. 2003. Spin-valve current sensor for industrial applications. Sens. Actuat. A: Phys 105(2): 132–136.
[8] Sebastia, J.P. et al. 2009. Vibration detector based on GMR sensors. IEEE Trans. Instrum. Meas 58(3): 707–712.
[9] Togawa, K. et al. 2005. High sensitivity InSb thin-film micro-Hall sensor arrays for simultaneous multiple detection of magnetic beads for biomedical applications. IEEE Trans. on Magnetics 41(10): 3661–3663.

Adaptive genetic algorithm for mixed-model assembly line balancing problem

Xiangjun He
School of Security and Environment Engineering of Capital University of Economic and Business, China

Jinpeng Hao
Beijing Foton Cummins Engine Co. Ltd., China

ABSTRACT: An improved genetic algorithm is put forward for the first-class assembly line balancing problem. The following aspects have been improved in genetic algorithm design: due to the strong constraint situation of job priority relationship existed in such problems, we design a unique chromosome coding and decoding rule and genetic operator, developed a genetic algorithm program, and take an experimental teaching instrument assembly lines as a research case to carried out the design and adjustment for the genetic algorithm control parameters according the characteristics of the problem, which optimized the assembly line job task assignment, and elevated the level of load balancing.

Keywords: first class assembly line balancing; mixed-model assembly line; improved genetic algorithm

1 INTRODUCTION

The essence of the assembly line balancing problem is the finite collection of processes assigned to orderly workstations under the processing sequence, and operation times of all workstations are as close as possible [1]. Nowadays, the researches at home and abroad mainly concentrate on the first-class assembly line ALB-I balancing issue, and minimize the workstation quantity by optimization under the premise of established tact and meeting job priority relationship [2]. Wherein, the tact is deduced based on historical data predictive analysis and personnel and equipment production capacity. Minimizing the workstation as its objective, a mathematical description can be [3]:

$$\begin{cases} \text{Established tact CT} \\ \text{Minimize Func1} = N \end{cases}$$

Mixed-model assembly line balancing is an ALB-I issue, and the optimizing goal of which is to seek the minimum workstations and load deviation under the condition of established tact and priority relationship constraint [4]. The following assumptions should be given before modeling:

(1) Under a given cycle time, each workstation can handle every job, and all workstations must be assigned at least one job;
(2) Job tasks is the smallest and indivisible natural operating unit with a stable operating time, and any job can be assigned to any workstation, and all jobs must be assigned to a workstation;
(3) There is only one priority relationship constraint which must be met and determined.

For involving in a number of different products, mixed-model assembly line balancing problem is more complex than single species assembly line, which can be described by the following mathematical model [5]:

Objective function:

$$\text{Min } J = w_1 \text{SST} + w_2 N \qquad (1)$$

Constraint condition:

$$\sum_{j=1}^{N} x_{ij} = 1, \quad i=1, 2, ..., Z \qquad (2)$$

$$\sum_{j=1}^{N} x_{ij} \le \sum_{i=1}^{N} l x_{kl}, \quad i, k=1, 2, ..., Z \qquad (3)$$

$$CT = PT / \sum_{m=1}^{M} D_m \qquad (4)$$

$$T_{mj} = \sum_{i=1}^{Z} t_{mi} x_{ij}, \quad m=1, 2, ..., M, j=1, 2, ..., N \qquad (5)$$

$$x_{ij} \in {0,1}, \quad i=1, 2, ..., Z, j=1, 2, ..., N \qquad (6)$$

$$CT = \sum_{m=1}^{M} q_m T_{mj} \ge 0, \quad j=1, 2, ..., N \qquad (7)$$

$$\text{SST} \sqrt{\frac{\sum_{k=1}^{N}(\sum_{m=1}^{M} q_m T_{mj} - \frac{\sum_{j=1}^{N}\sum_{m=1}^{M} q_m T_{mj}}{N})^2 N}{N}} \qquad (8)$$

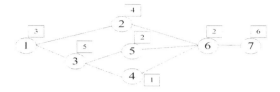

Figure 1. Assembling flowchart.

Figure 2. Job element distribution scheme and coding digital sequence.

2 IMPROVED ADAPTIVE GENETIC ALGORITHM

2.1 *Code design based on the sequence of operations*

The orderly constraint relationship among the processes on the assembly line can be described by a job priority relationship diagram. Figure 1 means the job priority relationship diagram of an assembly line, its node indicates process, the straight line connects nodes indicates the job order. The coding method of process traversal order is used for encoding on chromosome. The processes are arranged in a line based on the sequence of processes in process constraint graph, and each process corresponds to a gene position. Encoded chromosome displays a digital sequence, see Figure 2. For established digital sequence, its process position division result (including the workstation quantity and job elements in each workstation) is determined.

Coding is on the basis of priority sequence, which effectively avoids the emergence of non-feasible solution, and greatly reduces the computation amount and computing time.

2.2 *Random initialization strategy [6]*

The detailed steps are as follows:

Step 1: First select a non-preceding job element as the first gene of chromosome.
Step 2: Pick the assigned element of non-preceding job or preceding job into the alternative collection. Then, an incorporated chromosome gene position is selected randomly in the alternative collection.
Step 3: Update the alternative collection. Reciprocating like this, till all job elements are incorporated into chromosome, and result in a chromosome.

Chromosome initialization, which meets demands of randomness and priority relationship, can be generated by repeating the cycle.

Table 1. Population initialization method.

Selecting job element randomly	Alternative collection
1	2, 3
3	2, 4, 5
4	2, 5
2	5
5	6
6	7
7	

2.3 *Fitness function structure [7]*

Its mathematical expression is:

$$f(U_i) = CM - (w_1 SI_i + w_2 BL_i) \quad (9)$$

Wherein, SI is the smoothness index, BL is balancing loss coefficient, W1 and w2 are weight coefficient, CM is a constant, which is not less than (w1SIi + w2BLi), SImax, BLmax are respectively the maximum of SI, BL in each generation individuals. The balancing loss coefficient—BL indicates the empty drive time ratio due to unreasonable distribution of jobs between workstations on the assembly line, more larger of BL value, more loss time of assembly line. Its mathematical expression is:

$$BL = (N*\max(T_i) - \sum T_i) / N*\max(T_i) \quad (10)$$

SI of production line is used for indicating the load balance situation among workstations on the assembly line, and smaller SI means more balancing of load on the production line workstation, that is:

$$SI = (\sum (CT - T_i)^2 / N)^{1/2} \quad (11)$$

Wherein, CT is cycle time, Ti is the processing time of each workstation, N is the total number of workstations, max (Ti) represents the processing time of bottleneck workstations, i.e., the operating time is longest in all workstations operating.

3 CONSTRUCTION WITH FEASIBLE SOLUTION

3.1 *Select operator design [8]*

Detailed steps:

Step 1: Calculate the fitness of all individuals, and reorder according to the fitness value.
Step 2: Select the individuals with the best fitness (corresponding to the best optimization program) and directly replicate them to the next generation of community.
Step 3: Select the size of the tournament K = 2; randomly generated K individuals in the current population; compare the fitness of the selected individual values, and individuals better fitness are replicated to a new generation of individuals.

Step 4: Repeat step 3 until the number of the new generation of individuals reached population size.

3.2 Cross operator design

The method steps for crossover operation are as follows: First of all, calculate the number of chromosome pairs with the need for crossover operation by crossover probability in this population, and then conduct chromosome crossover operation randomly by pairs from the population selected; the method for each pair of chromosome crossover calculation is: randomly generate an integer that is less than the total number of genes, and take it as the location of the crossing point, then at the crossing point, divide the chromosomes into two parts. Remove the first part of one chromosome, then delete the chromosomal gene of first part in the second chromosome, and re-arrange the remaining genes in the original order. The two chromosomes obtained by this method are taken as a new individual to compose the next generation populations.

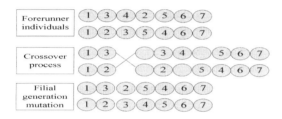

3.3 Mutation operator design

The mutation operator design also has precedence constraint problems. Genetic mutation should be developed randomly in constraint conditions. In order to avoid the destruction of the feasibility of the chromosome coding sequence, we use the shift mutation method. The detailed steps are as follows:

Step 1: According to the mutation probability, calculate the number of individual that requires mutation in the population.
Step 2: Select randomly mutated individual and mutation position k (k is an integer that is greater than the total number of genes) from the population, as shown in Operation 5 of the chromosome in the figure below.
Step 3: Search for priority matrix, and identify the feasible interval of gene mutation corresponding to position k according to priority diagram. The feasible interval of operation 2 is between operation 1 and 6.
Step 4: Insert the gene at k location into any position in the feasible interval, and complete the mutation operation.
Step 5: Repeat the above four steps according to the number of individuals in the first step until the specified number of individual mutation is achieved, see below.

3.4 Self-adaptive probability set

Crossover probability and mutative probability are self-adapted as per following formula:

$$P_m = \begin{cases} P_{m1} - \dfrac{(P_{m1} - P_{m2})(f_{avg} - f)}{f_{max} - f_{avg}} & f > f_{avg} \\ P_{m1} & f \leq f_{avg} \end{cases}$$

$$P_c = \begin{cases} P_{c1} - \dfrac{(P_{c1} - P_{c2})(f' - f_{avg})}{f_{max} - f_{avg}} & f' > f_{avg} \\ P_{c1} & f' \leq f_{avg} \end{cases}$$

Wherein: f_{avg} represents the average fitness value of each generation group; f' represents the larger fitness value of two individuals that's about to crossover. F represents the fitness value of the individual that about to crossover; gt represents the current generation number and GT represents the maximum generation number; kc, km are constant value, locating between (0, 1). Through appropriate set for kc, km parameters, operator probability can be adjusted in the computation process. When an individual's fitness value is lower than the average fitness, the individual performance is indicated as poor, and then the crossover and mutation probability will be adjusted to a higher level; if individual fitness value is higher than the average, then the individual performance is indicated as good, and the fitness value will be adjusted to a lower level; the overall adjusting is benefit for the elimination of undesirable individuals and the preservation of excellent individual. Adjustment of this linear evolution will become significant in the later phase of the group.

4 ALGORITHM DESCRIPTION AND ANALYSIS

Taking experimental teaching instrument assembly as an example, there two kinds of experimental teaching instruments are been installed on a mixing assembly line assembly, which are light chasing device A and winning game device B. These two products are produced in proportion of 1:2 on the mixing assembly line. The production plans of A and B are 3740, 1870 pieces each month respectively (calculated on 22 working days per month). The effective working time each day is 8 hours. Their mixing precedence relation can be seen from the working relation table according to operating relation Table 1 and Table 2.

The initial chromosome number of genetic algorithm parameter is 14; the iterative loop mode reaching specified algebra for ending loop iteration is used; the number of loop iteration is 40 generation, i.e. the algorithm is ended after being operated to 40 generation. The crossing-over rates are taken as: Pc1 = 0.45, Pc2 = 0.3 aberration rate, Pm1 = 0.08, Pm2 = 0.06, CM = 10.

Software Matlab-R2008a is used; the adaptive genetic algorithm solving program orientating to mixed flow assembly line balancing is designed and

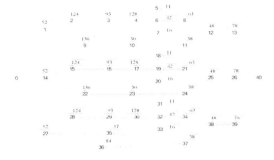

Figure 3. Mixing flow assembly drawing

Table 2. Working relation table of Prize-winning game device (A).

Basic operation	Name of working process	Working process time
A	Unpacking case	52
B	Installing rubber foot, installing 4 screws	124
C	Tightening 4 screws on rubber foot using screw driver	93
D	Installing circuit board, installing 4 nuts and tightening them	128
F	Installing rocker switch	11
G	Installing J-type aviation insert	42
H	Installing horn	16
I	Installing first row of diode socket on machine cover	63
M	Tightening nuts of diode socket using sharp nose pliers	136
N	Installing second row of diode socket on machine cover	30
O	Tightening nuts of diode socket using sharp nose pliers	38
P	Inserting diodes into two rows of diode sockets	48
Q	Closing machine cover and fixing it using screw driver	78
Total		859
Unit		Sec.

Table 3. Working relation table of prize-winning game device (B).

Basic operation	Name of working process	Working process time
A	Unpacking case	52
B	Installing rubber foot, installing 4 screws	124
C	Tightening 4 screws on rubber foot using screw driver	93
D	Installing circuit board, installing 4 nuts and tightening them	128
F	Installing rocker switch	11
G	Installing J-type aviation insert	42
H	Installing horn	16
I	Installing one row of diode socket on machine cover	63
M	Installing two button switches on machine cover	37
N	Installing one row of toggle switch on machine cover	84
O	Inserting diodes into diode sockets respectively	38
P	Tightening nuts of diode socket using sharp nose pliers	48
Q	Closing machine cover and fixing it using screw driver	78
Total		814
Unit		Sec.

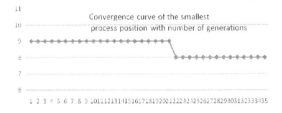

Figure 4. Convergence curve.

developed; and each process assembly time and variety assembly constraint of product A and product B is taken into simulation program; the obtained result is shown as the convergence curve in Fig. 4. We can see from the figure that algorithm rapid convergence and optimal value after a few of iterations.

In this paper, we applied adaptive genetic algorithm for optimized balance operation of mixed model assembly line, and when calculation comes to the 21st generation, the number of workstations convergence to 8, the same number with the previously calculated theoretical minimum workstation. So the operation result is correct, with faster convergence speed, and has achieved the desired state. The workstation partition on chromosome would provide the job elements and operating time for each workstation, and the workstation setup program are shown in Table 3.

The specific result:
Assembly efficiency and idle time computation after balancing optimization:

(1) CT = 339 Seconds
(2) Workstation number N = 8
(3) Total operating time

T=(52+124+93+128+11+42+16+63+136+30
+38+48+78)*2+(52+124+93+128+11+42+16
+63+37+78) =2532 seconds

(4) Total idle time $T_k = 339 * 8 - 2532 = 180$ seconds

For the mixed flow assembly line of teaching instruments after optimization, calculate balancing delay

Table 4. Distribution plan for workstation after optimization.

No. of workstation	Working process completed in workstation	Accumulated time of workstation A	No. of workstation	Working process completed in workstation	Accumulated time of workstation
1	1, 2, 3, 27	324	5	22, 23, 7	312
2	35, 28, 36, 29, 30	322	6	24, 11, 16, 37, 32, 34, 38	327
3	4, 6, 31, 5	306	7	17, 39, 18, 19, 20, 8	310
4	14, 9, 15, 33, 10	316	8	12, 21, 13, 25, 26	315

Table 5. Comparison of results before and after balancing.

	Number of Workstation	Total operating time	Total idle time	Balancing efficiency	Balancing delay rate
Before balancing	9	2532	519	83.0%	17.0%
After balancing	8	2532	180	93.3%	6.7%

rate and balancing efficiency before and after optimization; their results before and after balancing can be seen in Table 4 in detail. Obviously, the load levels of each workstation after optimization have been basically same approximately.

5 SUMMARY

Application of the production line model in the assembly of experimental teaching instrument enables diversified types and small batch of mixed flow assembly to reach a kind of optimized production organization mode for large-scale output. The parts used in experimental teaching instrument involved in this paper are all similar products, which have features of higher generalization degree and reduced amount of customized parts, which are in favor of organizing production line for mixed flow assembly; and adopted line balancing technology for optimization can greatly improve production efficiency, save costs, and enhance products competitiveness in the markets.

REFERENCES

[1] Xingzhong Pi. Assembly line balancing and research and application of simulation technology. Shanghai: Shanghai Jiaotong University, 2002
[2] Xiaofeng Chen. Applying genetic algorithm to solve balancing problem of assembly line. Computer engineering and application 2001.5

Research on the erosion of valve of positive pulse measurement while drilling and structural improvement

Changqing Zhao, Kai Liu & Xing Chai
Faculty of Mechanical and Precision Instrument Engineering, Xi'an University of Technology, Xi'an, Shaanxi, China

ABSTRACT: Valve assembly of rotary valve type Positive Pulse Mmeasurement While Drilling (PPMWD) is often used to generate pressure wave signal in logging while drilling. The sudden change of fluid direction and pressure at the valve assembly always causes erosion-corrosion, resulting in punching and breaking. To improve anti-erosion capacity and reduce the failure probability of the valve seat, the anti-erosion and anti-wear research of rotary valve was conducted in mechanism of erosion and wear, numerical simulation and structural improvement. According to the numerical analysis of the rotary valve erosion zone which mainly appears at bottom of the valve seat, the scheme for improving the rotary valve structure was formulated. No remarkable erosion occurred within the wear sleeve after the field application of the improved rotary valve. The anti-erosion effect is desirable, satisfying the requirement of field application.

Keywords: Rotary valve; valve seat; erosion and wear; numerical simulation; structural improvement

1 INTRODUCTION

Erosion, as a form of material wear, has been reported in many areas of the petroleum and natural gas industries. An example of erosion in drilling tools is the rotary valve for measurement while drilling system. Currently, positive pulse transmission (Zhu GQ et al., 2010) is the most common and stable and reliable method during the measurement while drilling and the logging while drilling system. The valve assembly are the key components of the mud pulse signal generator (Xiao JY et al., 2010; Chin DC, 2004). The motor drive system controls the rotor rotational movement, changing the relative position of the stator and the rotor to change the flow channel area of the fluid flow. As the rotor makes periodical counter rotating, the pressure signal of the upstream mud varies periodically. Its variation pattern represents the real pulse signal (Liu XS et al., 2007). It is necessary to adjust the clearance between stator and rotor or increase the flow rate of drilling fluid to improve signal strength and transmission distance. Erosion of valve assembly, used for transmitting three-phase drilling fluid, is a major problem faced by the oil and gas industry today (Wang MB, 2012). The damage caused to the exposed surface of the valve seat, due to these phenomena, can be enormous. The failure of equipment during drilling can be very expensive both from the point of view of economic losses as well as time. This will cause extra rig time in order to pull out the drill string and run it with new tools into the downhole. Hence, the determination of valve assembly is very important to predict the life of PPMWD under different operating conditions.

It is noted that the rotary valve is more eroded if the strength of pressure wave signal is very large. If we don't optimize the existing rotary valve, the life of the instrument will reduce. The existing PPMWD has the defects of high flow rates and erosion. It is important to monitoring of downhole parameters that improving the anti-erosion capability of rotary valve.

Many scholars have done lots of research on pulse both at home and abroad. Jia Peng (Jia P et al., 2010) analyzes the main factors of erosion and suggests that double orifice can reduce erosion. However, the improvement valve structure is complex. Moriarty (Moriarty KA, 2001) put forward the scheme which can increase the signal amplitude and reduce jamming. But he has not solved the problem of erosion. In order to understand and predict erosion phenomena in the rotary, the anti-erosion and anti-wear research of rotary valve was conducted in mechanism of erosion and wear, numerical simulation and structural improvement in this paper.

2 MECHANISM OF EROSION

There are several theories that can explain erosion phenomena in terms of mechanisms (Arefi B et al., 2005) of material removal. There is an agreement that brittle and plastic materials have fundamentally different mechanisms. The three important factors that are governing erosion are found to be type of material, velocity of particles and the angle of impact. All of the failure reasons of rotary valve, the most important one were found to be the inner of valve seat erosion which is caused by drilling fluid in high speed. Erosion can be defined into three types: the cavitation erosion, erosion wear and composite abrasion cavitation and erosion (Yong XY & Lin YZ, 2002).

API RP 14E recommends (API, 2007) the following threshold flow velocity equation below which acceptable erosion occurs from three phase flow in drill string in following form (Mclaury B & Shirazi S, 2008):

$$V_e = \frac{D\sqrt{\rho_m}}{20\sqrt{W}} \quad (1)$$

where D is pipe internal diameter in mm; ρ_m is drilling fluid mixture density in kg/m^3; W sand flow rate in kg/day.

The density of the drilling fluid mixture can be determined by the following empirical formula (Wylie EB & Steeter S, 2008). Then,

$$\rho_m = (1 - \beta_g - \beta_s)\rho_l + \beta_g \rho_g + \beta_s \rho_s \quad (2)$$

where ρ_l is liquid density in kg/m^3; ρ_g is gas density in kg/m^3; ρ_s is sand density in kg/m^3; β_g is gas volume fraction, dimensionless; β_s is sand volume concentration, dimensionless.

These equations are perfectly general and the erosion rate predicted by them for valve seat depends on sand flow rare. Generally, fluid erosional velocity decrease as the sand flow rate of drilling fluid is increased. Pressure wave signal will enhance with increasing flow rate of drilling fluid and decreasing clearance between the stator and the rotor. However, the increasing flow of drilling fluid will directly cause the improvement of sand flow rate; flow area and velocity gradient changes as the clearance is decreasing. Correspondingly, it will inevitably lead to the increase of sand flow rate. When the fluid velocity is higher than the erosional threshold velocity, it is a common phenomenon that high-speed fluid acting on the surface of the parts. The erosion mechanism for the rotary valve, which is an example of a hard brittle material, is caused by the higher speed. There is a significant erosion of the rotary valve as the drilling fluid velocity increases during the logging while drilling process.

3 STRUCTURE PARAMETERS AND THE COMPUTATIONAL MODEL

This paper aims at the numerical simulation of valve assembly based on rotary valve positive pulse, the core structure see Fig. 1. The rotor rotation range is $0° \leq \alpha \leq 45°$. When the rotary valve begin to close, the drilling fluids flow area becomes small accordingly. And the drilling fluids flow from the rotor gap when the rotary is closed totally. Compared with open state, when in close state, the drilling fluids impact velocity and impact angle changes a lot. This paper mainly simulated the erosion state of rotary valve when it is closed.

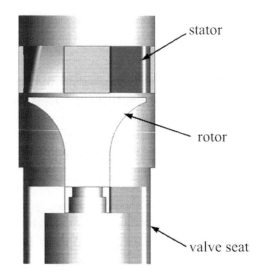

Figure 1. Structural assembly diagram of valve.

Taking drilling fluids flow value as 20 L/s, solid volume 6.25%, drilling fluids density 1200 kg/m^3. With the Euler Multiphase Flow model, control equation with standard $k - \varepsilon$, at the same time taking the critical condition of velocity entrance and pulse exit, rotor with the dynamic mesh model, time step size keep 0.001 S, rotor moves 0.18° each room size. Each time step iteration 200 times to complete the calculation.

4 CALCULATION RESULT ANALYSIS AND STRUCTURE IMPROVEMENT

The velocity vector image of drilling fluids under different states is shown as Fig. 2. And Fig. 3 is the pulse distribution chart of drilling fluids in plane under different states. From the two figures, it is known that as the angle increases, the drilling fluids velocity increases. The maximum value of drilling fluids velocity lies in the rotor basin, and the eddy current phenomenon appears when the angle is larger than 30. At this time, the impact angle made while drilling fluids impact the seat becomes bigger, and the erosion becomes obvious. Besides, it is got from the picture that the erosion mainly occurs at the rotor exit. When the rotary valve is closed, the drilling fluids angle at the rotor gap is the biggest, and the erosion is the most serious for the seat areas on the back of rotor blade.

These parts were severely eroded as shown in Fig. 4. Erosion has been observed as well as modeled in places where a change in the geometry of the valve seat that generates disturbance in the fluid flow. The calculation results agree well with the experimental data, which can forecast the vulnerable parts of the rotary valve due to erosion and provide reference for the design and regular testing of the rotary valve. At the same time, it ensures the stable and reliable transit of underground data.

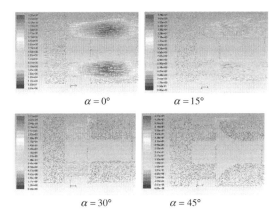

Figure 2. Drilling fluid velocity magnitude distribution with different opening of rotor.

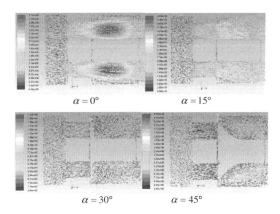

Figure 3. Drilling fluid dynamic pressure with different opening of rotor.

Based on the above numerical simulation result, this paper comes up with a scheme for improving rotary valve assembly structure. Like follows: install a wear sleeve in inner cavity of the valve seat and maximize the wear ring; increase the downstream section of seat chamber. The improved rotary valve assembly structure is like Fig. 5 shows. When the drilling fluids flow through the rotor gap, it will have direct effect on wear sleeve, so as to reduce the erosion to rotary valve and therefore can extend the life of the valve seat.

5 CONCLUSIONS

The following are the main results of this research:

1) This work shows that erosion simulation is feasible, and can utilize commercial CFD modeling components.
2) The velocity of drilling fluid is increasing during the opening process. In the valve seat, the erosion area is concentrated on the rear of the rotor and exit. The area near the rotor blade is more eroded when the rotary is fully closed.

Figure 4. The distribution of main erosion area of valve seat.

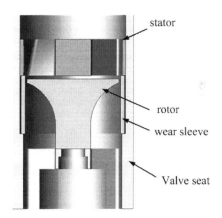

Figure 5. Structural of rotary valve after improvement.

3) The study result shows that the improvement scheme is reasonable and feasible. The improved structural is desirable, satisfying the requirement of field application.

ACKNOWLEDGMENT

The authors would like to express their appreciations to the financial support from the China National Natural Science Foundation of China under the grant No of 51275409 and Science and Technology Program of Shaanxi Grant No of 2012JQ7009. The authors also would like to thank CNPC Logging Co. Ltd for providing the continuous support during the experiment. Without their support, this work would not have been possible.

REFERENCES

API. 2007. API RP-14E Recommended practice for design and installation of offshore production platform piping system.

Arefi B., Settari A. & Angman P. 2005. Analysis and simulation of erosion in drilling tools. *Wear*, 259:263–270.

Chin D C., 2004. MWD siren pulser fluid mechanic. *Petrophysis*, 45(4):363–379.

Jia Peng, Fang Jun & Li Lin. 2010. Research on the erosion performance of the choke series continuous wave signal generator. *China Petroleum Machinery*, 38(11):11–14.

Liu Xiushan, Li Bo & Yue Yuquan. 2007. Transmission behavior of mud-pressure pulse along well bore. *Journal of Hydrodynamics*, 19(2):236–240.

Mclaury B. & Shirazi S. 2008. Generalization of API 14E for erosion service in multiphase production. *SPE annual technical conference.*

Moriarty K.A. Pressure pulse generator for measurement while drilling systems which produces high signal strength and exhibits high resistance to jamming. US 6219301B1. 2001-04-17.

Wang MIngbo. 2012. Erosion of elbows in oil and gas production system. *Advanced Materials Research*, 1129–1132.

Wylie E.B. & Steeter V.L. 1978. *Fluid Transients*. New York: McGraw-Hill international book compamy.

Xiao Junyuan, Wang Zhiming & Liu Jianling. 2010. Research status of mud pulse generator. *Oil Field Equipment*, 39(10): 8–11.

Yong Xingyue & Lin Yuzhen. 2002. Progress in study on flow-induced corrosion. *Corrosion Science and Protection Technology*, 14(1):32–34.

Zhu Guoqing & Zhang Zhaoqi. 2010. Recent advances in foreign logging while drilling technology. *Well Logging Technology*, 32(5):394–397.

/ Research on the application of storm model in fatigue life prediction for offshore platform

Xiaoshun Yan
State Key Lab of Ocean Engineering, Shanghai JiaoTong University, Shanghai, China

Xiaoping Huang
State Key Lab of Ocean Engineering, Shanghai JiaoTong University, Shanghai, China
Institute of Ocean Engineering, Jiangsu University of Science and Technology, Zhenjiang, China

Fan Liu
State Key Lab of Ocean Engineering, Shanghai JiaoTong University, Shanghai, China

ABSTRACT: Fatigue damage is one of the main failure modes for offshore structures. Thus it is essential to assess fatigue strength for an offshore platform both in design and service time. In order to predict fatigue life for a structural detail more accurately, the theory of fatigue crack propagation was applied and the simulation for wave-induced loads was investigated in the paper. Since offshore structures always cannot avoid storms, unlike ships, the storm model was firstly used on a platform. By combining the storm model and the fatigue crack growth law considering the effect of load sequence, the fatigue propagation life of a welded detail of an offshore platform was predicted. The results had shown that the fatigue strength of the discussed hot spot met the requirement. This example can be served as a reference for fatigue assessment of offshore structures.

Keywords: Storm model; semi-submersible platform; fatigue crack propagation; unique curve model

1 INTRODUCTION

The statistical survey shows that most of the accidents in offshore structures are caused by fatigue crack propagation [1]. Therefore, the research on fatigue life prediction has important practical significance in the exploration of oil and gas resources under deep water.

Many fatigue life prediction methods for marine structures have been proposed and they can be divided into: cumulative fatigue damage (CFD) theory and fatigue crack propagation (FCP) theory [2]. Since the CFD theory cannot take initial crack size, load sequence, residual stress, etc. into consideration, the fatigue life predicted based on CFD theory shows great dispersion [3]. In comparison, the FCP theory-based prediction can overcome all these difficulty and is becoming an intense field of research. Thanks to the efforts of many scholars, the behavior of fatigue crack propagation under variable amplitude loads can be accurately described. However, if the fatigue load cannot be measured or simulated rationally, the results of prediction must be inaccurate, even though an accurate propagation law is used. To simulate the wave-induced loads more realistic, Tomita et al. [4] proposed a storm model based on statistical observation. Prasetyo et al. [5] modified the original storm model by considering shipping lines, heading angles of ship and sea areas. Sumi et al. [6] applied the original storm model on a containership and successfully predicted the crack propagation life of thick deck plate. Actually, with the growing ability of marine meteorological forecasts, a ship can avoid the adverse sea conditions to some extent. Unfortunately, for an offshore platform working at a fixed place for a long time, it is too expensive to tow away a platform which results in it being exposed to the storms. Thus the storm model should be more suitable for offshore platforms while there is nearly no relevant report.

This paper investigated the storm model and crack propagation model considering the effects of the load sequence. A deepwater semi-submersible platform, working in South China Sea, was selected for this study. The storms of the South China Sea were configured and the wave-induced loads were simulated. Then fatigue life is predicted for a welded detail of this platform by using the unique curve model. Finally, the effects of the storm sequence were discussed.

2 FATIGUE LOAD GENERATION BASED ON STORM MODEL

2.1 *Storm configuration*

The storm model was firstly proposed by Tomita [4], who analyzed the onboard wave history acquired by the Japanese Marine Self Defense Force in the North

Table 1. Storm level and parameters [4].

Storm	A	B	C	D	E	F
Hs (m)	6	7	8	9	11	15
Number	42	25	12	7	6	1

Pacific Ocean. He categorized sea states into two conditions: calm condition and storm condition. And the two conditions randomly occurred at a certain probability. The basic assumptions for generating the storm model are referred to Tomita's study as follows:

(1) Storm model has isosceles triangle shape, Hs (significant wave height) increases and decreases monotonically in storm condition;
(2) Short term sea state is 2 hours, and storm duration is 3.5 days;
(3) When Hs > 2Hs$_{mean}$ (mean significant wave height), Hs is considered in storm state;
(4) When Hs < 2Hs$_{mean}$, Hs occurs randomly in calm sea state.

Based on the statistical results, Tomita gave a storm composition, shown as Tab. 1. Since the storms given were just suitable for certain sea area, Tomita suggested configure storms based on wave scatter diagram.

In this study, storms are configured based on the basic assumptions above and the wave scatter diagram of the South China Sea. The results are shown in section 4.1.

2.2 Wave-induced stress

The wave-induced stress can be calculated in the linear wave theory. The relationship between power spectral density (PSD) of wave $G_{\eta\eta}(\omega)$ and PSD of wave-induced load $G_{XX}(\omega)$ can be written as:

$$G_{XX}(\omega) = [H(\omega)]^2 G_{\eta\eta}(\omega) \quad (1)$$

where, ω is the frequency of the wave and $H(\omega)$ is the transfer function between wave height and wave-induced stress. In this study, the Pierson Moskowitz wave spectrum proposed by ITTC is applied.

If the PSD of wave-induced stress is obtained, the zero- and second-order moment can be derived:

$$m_n = \int_0^{+\infty} \omega^n \cdot G_{XX}(\omega) d\omega \quad (2)$$

In general, the random wave-induced stress ranges for a short sea can be considered as a narrowband random process, the distribution of which approximately obeys the Rayleigh distribution [7], namely

$$F_{\Delta\sigma}(\Delta\sigma) = 1 - \exp\left(-\frac{\Delta\sigma^2}{8m_0}\right) \quad 0 \le \Delta\sigma < +\infty \quad (3)$$

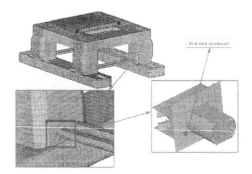

Figure 1. The location of the discussed hot spot.

3 FCP LAW CONSIDERING LOAD SEQUENCE

The amplitudes of wave-induced loads are variable, so the effects of load sequence should be taken into consideration. Although there are several FCL laws considering effects of load sequence, the unique curve model proposed by Huang et al. [9] combines simplicity and practicability. The basic expression of the unique curve model is:

$$\frac{da}{dN} = C\left[\left(\Delta K_{eq0}\right)^m - \left(\Delta K_{th0}\right)^m\right] \quad (4)$$

$$\Delta K_{eq0} = M_R M_P \Delta K \quad (5)$$

where a is the crack length; N is the number of applied cycles; C, m are the Paris parameters; ΔK_{eq0} is the equivalent SIF range corresponding to stress ratio $R = 0$; M_R, M_P are the correction factors for the effects of stress ratio and load sequence, respectively. If $M_R = M_P = 1$, the influences of stress ratio and load sequence are not considered and the unique curve model reverts to Paris law, which reflects the universality of the unique curve model. More details can be referred to [9].

4 APPLICATION ON A PLATFORM

The discussed offshore platform is working in the South China Sea. The design life is 25 years and the fatigue strength of the platform has been simply assessed in literature [8]. The hot spot discussed in this paper is the connection between a brace and the column, shown as Fig. 1.

4.1 Fatigue load spectrum

The wave scatter diagram used is the statistical data at the working location. According to the storm configuration method described in section 2.1, the storms and their numbers of occurrence in 25 years are shown in Tab. 2. The composition of wave height in each storm is shown as Fig. 2.

The results of transfer function are shown in Fig. 3. Based on the method provided in section 2.2, the wave-induced stresses for each storm and calm condition are shown in Fig. 4.

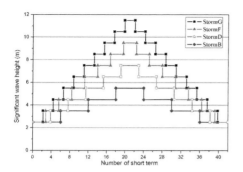

Figure 2. Wave height in each storm.

Table 2. Results of storm configuration for the South China Sea.

Storm	A	B	C	D	E	F	G
Hs (m)	4.5	5.5	6.5	7.5	8.5	9.5	11.5
Number	193	98	33	20	17	7	5

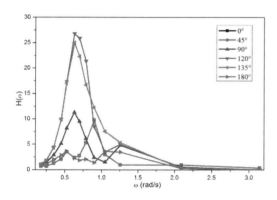

Figure 3. Transfer function of the target hot spot [8].

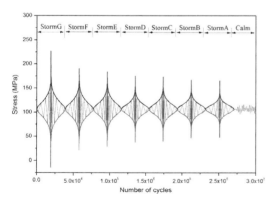

Figure 4. Wave-induced stress of each sea condition.

4.2 Calculation for crack growth curves

The material parameters in the FCP law are set as: $C = 3 \times 10^{-13}$, $m = 3$, $\Delta K_{eq0} = 63 \text{ MPa} \cdot \text{mm}^{0.5}$.

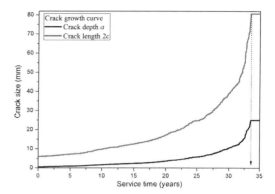

Figure 5. Crack growth curves of the target hot spot.

According to the recommendation of ABS, a semi-elliptical surface crack, the depth a of which is 0.5 mm, is supposed as initial flaw. Since the multiple cracks always initiate simultaneously, crack half length c is assumed to be 3 mm in this study. The stress intensity factors are calculated based on the formulas recommended by BS7910 [10]. The structure is failure when the crack depth propagates to the thickness of the plate.

Program to calculate the integration of FCP law cycle by cycle, and storm conditions and calm conditions randomly occur at the given probability. The crack growth curves of the target hot spot are shown in Fig. 5. It is not difficult to find that the fatigue life of the discussed hot spot is about 33 years. So the structure met the design requirement.

4.3 Discussion of load sequence

In order to investigate the effect of sequence of sea conditions on fatigue life, crack growth curve is calculated for the five cases explained below.

Case 1: Storm conditions and calm conditions randomly occur at the given probability. Try for 10 times.
Case 2: Traverse all calm conditions firstly, and then Storm A, Storm B, ..., Storm G.
Case 3: Traverse all Storm G, and then Storm F, Storm E, ..., Storm A, calm conditions.
Case 4: Storm conditions randomly occur at the given probability. No calm conditions.
Case 5: Equivalent constant stress range, which is defined by:

$$\Delta \sigma_{eq} = \sqrt[m]{\sum (\Delta \sigma_i^m \cdot n_i) / \sum n_i} \qquad (6)$$

where $\Delta \sigma_i$ is the stress range of the i-th level; n_i is the corresponding number of loading cycles; and m is the exponent of the crack propagation law.

The results are shown in Fig. 6. Comparing Case 1~Case 3, it is not difficult to figure out the sequence of the sea conditions has a great importance on fatigue life. In addition, the calm conditions have little impact on fatigue life by comparing Case 1 and Case 4. Finally,

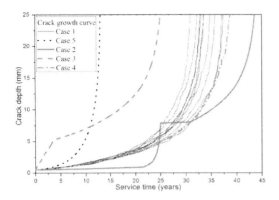

Figure 6. Results of crack growth curves for different load cases.

the equivalent constant stress range commonly used is too conservative by comparing Case 1 and Case 5.

5 CONCLUSIONS

This paper firstly applied the storm model in the fatigue strength assessment of a deepwater semi-submersible platform. The fatigue life of a welded detail was predicted by using fatigue crack propagation theory with load sequence being considered. The result showed the fatigue strength met the design requirement. In addition, the discussion on sequence of sea conditions showed the sequence was important to fatigue life prediction and it is very important to take the sequence effects into consideration.

REFERENCES

[1] Bignell V, Fortune J. 1984. Understanding Systems Failure. Manchester University Press.

[2] Cui W, Wang F, Huang X. 2011. A unified fatigue life prediction method for marine structures. Marine Structures, 24(2): 153–181.

[3] Fricke W, Cui W, Kierkegaard H, et al. 2002. Comparative fatigue strength assessment of a structural detail in a containership using various approaches of classification societies. Marine Structures, 15(1): 1–13.

[4] Tomita Y, Matobat M, Kawabel H. 1995. Fatigue crack growth behavior under random loading model simulating real encountered wave condition. Marine structures, 8(4): 407–422.

[5] Prasetyo F A, Osawa N, Kobayashi T. 2012. Study on Preciseness of Load History Generation based on Storm model for fatigue assessment of ship structures members //Proceeding of 22nd ISOPE conference IV.

[6] Sumi Y. 2014. Fatigue crack propagation in marine structures under seaway loading. International Journal of Fatigue, 58: 218–224.

[7] ABS. 2005. Guide notes on spectral-based fatigue analysis for floating offshore structures. New York: American Bureau of Shipping.

[8] Cui L. 2013. Fatigue analysis of deepwater semi-submersible platform and fatigue test on key joint. Zhejiang University. (In Chinese).

[9] Huang X, Moan T, Cui W. 2009. A unique crack growth rate curve method for fatigue life prediction of steel structures. Ships and Offshore Structures, 4(2): 165–173.

[10] BS7910. 2005. Guide to methods for assessing the acceptability of flaws in metallic structures. British Standards Institution.

Reliability analysis of machine interference problem with vacations and impatience behavior

M.J. Ma
Office of Academic Affairs, Yan Shan University, Qinhuangdao, Hebei, China

D.Q. Yue & B. Zhao
College of Science, Yan Shan University, Qinhuangdao, Hebei, China

ABSTRACT: This paper investigated the machine interference problem with failed machines which may balk and renege, where there is one repairman who takes a single vacation. In the viewpoint of reliability, the method of blocked matrix is used to derive the obvious iterative formula of the mean time to first failure of the system. The expression of Laplace transformations of the system is deduced. At last, the effect of the parameters on the performance measures about reliability of the system is carried out by numerical analysis.

Keywords: Machine interference problem; balking; reneging; vacation; reliability

1 INTRODUCTION

Machine interference problems have attracted the attention of many scholars and are widely applied in computer systems, manufacturing systems and communication systems. In the typical machine interference problems, there are m holotype machines and one repairman in the system. When the machines break down, the repairman repairs the machines in the mode of no-priority. Many scholars such as Wang & Sivazlian (1989) and Yue etc. (2006) studied the machine interference problems by combining different types of standbys and vacation policies into the typical systems in the viewpoint of reliability. On the other hand, the typical machine interference model, which is considered as the queuing model with finite capacity, is investigated based on the queuing theory. Most of the studies obtained the index about queuing theory, rather than index about reliability and availability problems of the system.

People have enriched the classic queuing model by adding the common phenomenon in the real life to the model, such as balking and reneging. In these conditions, some impatient customers may not join the queue for repair and some customers who have joined the queue depart without being repaired in the process of queuing. Most of the researchers studied in the viewpoint of queuing theory. Gupta (2004) primarily considered interrelationship between queuing models with balking, reneging and machine repair problem with warm spares. He exploited the method of queuing processing of the machine interference problems with warm standbys and obtained some index of queuing theory. Shawky (1997) discussed the single-server machine interference model with balking, reneging and an additional server for longer queues. He deduced some expression of steady-state probability and some index of queuing theory. Wang etc. (2007) studied the profit analysis of the M/M/R machine repair problem with balking, reneging, and standby switching failures. Ke & Wang (2005) discussed the cost analysis of the M/M/R machine repair problem with balking, reneging and server breakdowns. Wang & Ke (2003) studied the probability analysis of a repairable system with warm standbys plus balking and reneging. They derived steady-state probability and the mean time to first failure of the system about the reliability of system.

The phenomenon that the repairman takes secondary tasks in terms of cost for a random period when there are no failed units often happens in real life. The random period is known as "vacation" in queuing system. Gupta (1997) considered machine interference problems with warm spares, server vacations and exhaustive service. He deduced some results involving queuing theory. Ke & Wang (2007) studied vacation policies for machine repair problems with two type spares. Jain etc. (2004) discussed N-policy for a machine repair system with spares and reneging. They derived some index related to the reliability of the system.

However, there are only a few researches about queuing model involving balking, reneging and vacation, especially in the viewpoint of reliability of the system. Sun & Yue (2007) developed performance analysis of $M/M/S/N$ queuing system with balking, reneging and multiple synchronous vacations. They processed the performance analysis. Yue & Sun (2008) discussed the waiting time of $M/M/c/N$ queuing system with balking, reneging and multiple synchronous vacations of partial servers in the viewpoint of queuing theory.

In this paper, the machine interference problems combining the failed unit with balking, reneging, and the repairman taking a single vacation are studied. The method of blocked matrix is used to deduce obvious iterative formula of the mean time to first failure of the system and the expression of Laplace transformations for the system. Some numerical results are provided to show the effects of parameters on the mean time to first failure of the system.

2 THE MODEL ASSUMPTIONS

The designed model is based on the following assumptions.

(1) There are m operating units in the system and the lifetime is exponentially distributed with rate λ.
(2) There is one repairman, and the repair time is exponentially distributed with rate μ. The vacation policy is the single vacation. If there are no failed units, the repairman goes to vacation. When the repairman comes back from vacation and there is no failed unit in the system, he is vacant to waiting for failed unit. If the repairman has finished repairing the failed units, he can go to vacation again. The vacation time is exponentially distributed with rate θ. The repairman serves according to FIFO discipline.
(3) The impatient failed unit may balk with probability $1 - b$, while join the queue with b.
(4) The impatient failed unit waiting in the queue may leave without being repaired and the waiting time is exponentially distributed with rate r.
(5) The failed unit balking and reneging may be replaced by a new unit at once and are immediately send to repair. If the repairing of the failed unit is completed, the unit may become as good as new.
(6) All of the parameters are respectively independence.

3 PERFORMANCE ANALYSIS

3.1 *Steady-state probability equations*

$N(t)$ and $J(t)$ are respectively defined as the number of the failed units and the state of the repairman at time t in the system,

$$J(t) = \begin{cases} -1, & \text{(The server is vacant at service time } t) \\ 0, & \text{(The server is on vacation at the time } t) \\ 1, & \text{(The server is on the service at the time } t) \end{cases}$$

then, $\{N(t), J(t), t \geq 0\}$ is a Markov process with state space as follows:

$$\Omega = \{(0,0)\} \cup \{(0,-1)\} \cup \{(i,j)\}|i = 1,2,\cdots m; j = 0,1\}.$$

The state space of unit on working is expressed as follows:

$$W = \{(0,0),(0,-1),(1,0),(1,1),\cdots,(m-1,0),(m-1,1)\},$$

and the state space of unit failed is expressed as follows:

$$F = \{(m,0),(m,1)\}.$$

The steady-state probabilities of the system are defined as follows:

$$\pi_{00} = P_{00} = \lim_{t \to \infty}\{N(t) = 0, J(t) = 0\},$$

$$\pi_{0,-1} = P_{0,-1} = \lim_{t \to \infty}\{N(t) = 0, J(t) = -1\},$$

$$\pi_{ij} = P_{ij} = \lim_{t \to \infty}\{N(t) = i, J(t) = j\},$$

$i = 1, 2, \cdots m; j = 0, 1$

Then, according to the Markov process theory, steady-state probability equations are obtained as following:

$$(\theta + m\lambda b)\pi_{00} = r\pi_{10} + \mu\pi_{11},$$

$i = 0, j = 0$

$$[(m-i)\lambda b + \theta + ir]\pi_{i0} = (i+1)r\pi_{i+1} + (m-i+1)\lambda b\pi_{i-1,0},$$

$1 \leq i \leq m-1, j = 0$

$$(mr + \theta)\pi_{m0} = \lambda b\pi_{m-1,0},$$

$i = m, j = 0$

$$m\lambda\pi_{0,-1} = \theta\pi_{00},$$

$i = 0, j = -1$

$$[(m-1)\lambda b + \mu]\pi_{11} = \theta\pi_{10} + m\lambda\pi_{0,-1} + (\mu + r)\pi_{21},$$

$i = 1, j = 1$

$$\{[m-i]\lambda b + [\mu + (i-1)r]\}\pi_{i1}$$
$$= \theta\pi_{i0} + (m-i+1)\lambda b\pi_{i-1,1} + (\mu + ir)\pi_{i+1,1},$$

$2 \leq i \leq m-1, j = 1$

$$[\mu + (m-1)r]\pi_{m1} = \theta\pi_{m0} + \lambda b\pi_{m-1,1},$$

$i = m, j = 1$

The steady-state probability equations above can be rewritten in the matrix form as:

$$\pi A = 0,$$

where $\pi = \{\pi_{00}, \pi_{0,-1}; \pi_{10}, \pi_{11}; \cdots; \pi_{m0}, \pi_{m1}\}$.

The transition rate matrix A of the Markov process has the following blocked matrix structure:

$$A = \begin{pmatrix} A_0 & C_0 & & & & \\ B_1 & A_1 & C_1 & & & \\ & \ddots & \ddots & \ddots & & \\ & & & B_{m-1} & A_{m-1} & C_{m-1} \\ & & & & B_m & A_m \end{pmatrix}$$

where the sub-matrices of A are listed as following:

$$A_k = \begin{pmatrix} a_k & \theta \\ 0 & b_k \end{pmatrix}, \quad 0 \le k \le m$$

$a_0 = -(\theta + m\lambda b)$, $b_0 = -m\lambda$

$a_k = -[(m-k)\lambda b + \theta + kr]$,

$b_k = -[(m-k)\lambda b + \mu + (k-1)r]$, $1 \le k \le m-1$

$a_m = -[\theta + mr]$, $b_m = -[\mu + (m-1)r]$

$$B_1 = \begin{pmatrix} r & 0 \\ \mu & 0 \end{pmatrix}, \quad B_i = \begin{pmatrix} ir & 0 \\ 0 & \mu + (i-1)r \end{pmatrix}, 2 \le i \le m$$

$$C_0 = \begin{pmatrix} m\lambda b & 0 \\ 0 & m\lambda \end{pmatrix}, \quad C_j = (m-j)\lambda b I, \quad 1 \le j \le m-1$$

$$C_0^{-1} = \frac{1}{m\lambda b}\begin{pmatrix} 1 & 0 \\ 0 & b \end{pmatrix},$$

$$C_j^{-1} = \frac{1}{(m-j)\lambda b} I, \quad 1 \le j \le m-1$$

The normalized condition is:

$$\sum_{i=0}^{m} \pi_{i0} + \sum_{i=1}^{m} \pi_{i1} + \pi_{0,-1} = 1$$

3.2 Matrix solution

In the following discussion, the method of blocked matrix is used to derive the obvious iterative formula of the steady-state probability vectors.

Theorem 1 The formula of the mean time to first failure of the system is known as following:

$$MTTFF = \int_0^{+\infty} R(t)dt = \lim_{s \to 0} R^*(s) = -Q_W(0)\boldsymbol{B}^{-1}\boldsymbol{e}_W,$$

where e_W is a column vector with W components, and each component of e_W equal to one. The matrix B is obtained from matrix A, in which the last two rows and the last two columns for the absorbing states are deleted.

$$(x_0, x_1 \cdots x_{m-1})B = -Q_W(0), \quad (1)$$

where

$x_0 = (x_{00}, x_{0,-1})$, $x_i = (x_{i0}, x_{i1})$, $i = 1, 2, \cdots, m-1$,

$Q_W(0) = (1, 0, 0, \cdots, 0)$.

Eq. (1) can be expressed in the matrix form as:

$$\begin{cases} x_0 A_0 + x_1 B_1 = -(1,0) \\ x_i C_i + x_{i+1} A_{i+1} + x_{i+2} B_{i+2} = 0, & i = 0, \cdots m-3 \\ x_{m-2} C_{m-2} + x_{m-1} A_{m-1} = 0 \end{cases} \quad (2)$$

The $MTTFF$ can be expressed as following:

$$MTTFF = x_0 e_2 + x_1 e_2 + \cdots + x_{m-1} e_2$$

where

$$x_i = x_{m-1} H_i, \quad i = 0, 1, 2, \cdots m-2 \quad (3)$$

$$H_i = -(H_{i+1} A_{i+1} + H_{i+2} B_{i+2}) C_i^{-1},$$

$$i = 0, 1, 2, \cdots m-4 \quad (4)$$

$$H_{m-3} = -(H_{m-2} A_{m-2} + B_{m-1}) C_{m-3}^{-1} \quad (5)$$

$$H_{m-2} = -A_{m-1} C_{m-2}^{-1} \quad (6)$$

$$MTTFF = -(1,0)(H_0 A_0 + H_1 B_1)^{-1}$$

$$* (H_0 + \cdots + H_{m-2} + I) e_2 \quad (7)$$

Proof From the last equation of Eq. (2):

$$\boldsymbol{x}_{m-2} = -\boldsymbol{x}_{m-1} A_{m-1} C_{m-2}^{-1} = \boldsymbol{x}_{m-1} H_{m-2} \quad (8)$$

where $H_{m-2} = -A_{m-1} C_{m-2}^{-1}$.

Substituting the result into Eq. (2), when $i = m-3$, we get:

$$\boldsymbol{x}_{m-3} = -\boldsymbol{x}_{m-1}(H_{m-2} A_{m-2} + B_{m-1}) C_{m-3}^{-1} = \boldsymbol{x}_{m-1} H_{m-3} \quad (9)$$

where $H_{m-3} = -(H_{m-2} A_{m-2} + B_{m-1}) C_{m-3}^{-1}$.

Substituting the results Eq. (5) and Eq. (6) into Eq. (2), when $i = m-4$, we get:

$$\boldsymbol{x}_{m-4} = -\boldsymbol{x}_{m-1}(H_{m-3} A_{m-3} + H_{m-2} B_{m-2}) C_{m-4}^{-1}$$

$$= \boldsymbol{x}_{m-1} H_{m-4} \quad (10)$$

where $H_{m-4} = -(H_{m-3} A_{m-3} + H_{m-2} B_{m-2}) C_{m-4}^{-1}$.

Substituting the results obtained successively into the previous equation of Eq. (2), we get:

$$x_i = x_{m-1} H_i, \quad i = 0, 1, 2, \cdots m-2$$

where $H_i = -(H_{i+1} A_{i+1} + H_{i+2} B_{i+2}) C_i^{-1}$.

Substituting the results into the first equation of Eq. (2), we get:

$$x_{m-1} = -(1,0)(H_0 A_0 + H_1 B_1)^{-1}$$

Thus, x_0, x_1, x_{m-2} is expressed by x_{m-1}, and the mean time to first failure of the system is derived as following:

$$MTTFF = -(1,0)(H_0 A_0 + H_1 B_1)^{-1}$$
$$* x_{m-1}(H_0 + \cdots + H_{m-2} + I)e_2$$

Theorem 2 The transition rate matrix A of the Markov process can be expressed as:

$$A = \begin{pmatrix} B_{2m \times 2m} & C_{2m \times 2} \\ D_{2 \times 2m} & E_{2 \times 2} \end{pmatrix}$$

When the failed states of the system are absorbing states, the transition rate matrix is expressed as:

$$\tilde{A} = \begin{pmatrix} B_{2m \times 2m} & C_{2m \times 2} \\ 0 & 0 \end{pmatrix}$$

Then, we get:

$$(\dot{Q}_W, \dot{Q}_F) = (Q_W, Q_F)\tilde{A}$$

where

$Q_W(t) = \{Q_{00}(t), Q_{0-1}(t), Q_{10}(t), Q_{11}(t), \cdots, Q_{m-1,0}(t), Q_{m-1,1}(t)\}$, $Q_F(t) = \{Q_{m0}(t), Q_{m1}(t)\}$,

the following equation set is derived:

$$\begin{cases} \dot{Q}_W(t) = Q_W(t)B \\ Q_W(0) = (1,0,0,\cdots 0) \end{cases} \quad (11)$$

Eq. (11) can be expressed as:

$$(s-a_0)Q_{00}^*(s) = rQ_{10}^*(s) + \mu Q_{11}^*(s) + 1 \quad (12)$$

$$(s-a_1)Q_{10}^*(s) = m\lambda b Q_{00}^*(s) + 2rQ_{20}^*(s) \quad (13)$$

$$(s-a_i)Q_{i0}^*(s) = (m-i+1)\lambda b Q_{i-1,0}^*(s)$$
$$+ (i+1)rQ_{i+1,0}^*(s) \quad (14)$$

$$(s-a_{m-1})Q_{m-1,0}^*(s) = 2\lambda b Q_{m-2,0}^*(s) \quad (15)$$

$$(s-b_0)Q_{0,-1}^*(s) = \theta Q_{00}^*(s) \quad (16)$$

$$(s-b_1)Q_{11}^*(s) = m\lambda Q_{0,-1}^*(s) + \theta Q_{10}^*(s)$$
$$+ (\mu+r)Q_{21}^*(s) \quad (17)$$

$$(s-b_i)Q_{i1}^*(s) = (m-i+1)\lambda b Q_{i-1,1}^*(s) + \theta Q_{i0}^*(s)$$
$$+ (\mu+ir)Q_{i+1,1}^*(s) \quad (18)$$

$$(s-b_{m-1})Q_{m-1,1}^*(s) = 2\lambda b Q_{m-2,1}^*(s) + \theta Q_{m-1,0}^*(s) \quad (19)$$

From deducing Eq. (11), we can get:

$$R^*(s) = \sum_{i=0}^{m-1} Q_{i0}^*(s) + \sum_{i=1}^{m-1} Q_{i1}^*(s) + Q_{0,-1}^*(s)$$

$$= [\sum_{i=0}^{m-2} \varphi_i(s) + 1 + \sum_{i=1}^{m-1} g_i(s) + g_{-1}(s)]Q_{m-1,0}^*(s)$$

$$- \sum_{i=1}^{m-1} f_i(s) \quad (20)$$

where

$$\varphi_i(s) = \frac{1}{(m-i)\lambda b}[(s-a_{i+1})\varphi_{i+1}(s) - (i+2)r\varphi_{i+2}(s)],$$
$$i = 0, \cdots, m-4, \quad (21)$$

$$\varphi_{m-3}(s) = \frac{1}{3\lambda b}[(s-a_{m-2})\varphi_{m-2}(s) - (m-1)r] \quad (22)$$

$$\varphi_{m-2}(s) = \frac{1}{2\lambda b}(s-a_{m-1}) \quad (23)$$

$$g_{-1} = \frac{\theta}{s-b_0}\varphi_0 \quad (24)$$

$$g_1 = \frac{1}{\mu}[(s-a_0)\varphi_0 - r\varphi_1], \quad f_1 = \frac{1}{\mu} \quad (25)$$

$$g_2 = \frac{1}{\mu+r}[(s-b_1) - m\lambda g_{-1} - \theta\varphi_1],$$

$$f_2 = \frac{(s-b_1)f_1}{\mu+r} \quad (26)$$

$$g_i = \frac{1}{\mu+(i-1)r}[(s-b_{i-1})g_{i-1} - (m-i+2)\lambda b g_{i-2}]$$
$$- \frac{\theta\varphi_{i-1}}{\mu+(i-1)r}, \quad i = 3, \cdots, m \quad (27)$$

$$f_i = \frac{1}{\mu+(i-1)r}[(s-b_{i-1})f_{i-1} - (m-i+2)f_{i-2}],$$
$$i = 3, \cdots, m \quad (28)$$

$$Q_{m-1,0}^* = \frac{(s-b_{m-1})f_{m-1} - 2\lambda b f_{m-2}}{(s-b_{m-1})g_{m-1} - 2\lambda b g_{m-2} - \theta} \quad (29)$$

Proof From Eq. (15)

$$Q_{m-2,0}^*(s) = \frac{1}{2\lambda b}(s-a_{m-1})Q_{m-1,0}^*(s) = \varphi_{m-2}(s)Q_{m-1,0}^*(s)$$

Substituting the result obtained successively into the previous equation of Eq. (14), where $i = m - 2, m - 3, \ldots, 0$, to yield:

$$Q_{m-3,0}^*(s) = \frac{1}{3\lambda b}[(s - a_{m-2})\varphi_{m-2}(s) - (m-1)r]Q_{m-1,0}^*(s)$$

$$= \varphi_{m-3}(s)Q_{m-1,0}^*(s)$$

$$Q_{i,0}^*(s) = \frac{Q_{m-1,0}^*(s)}{(m-i)\lambda b}[(s - a_{i+1})\varphi_{i+1}(s) - (i+2)r\varphi_{i+2}(s)]$$

$$= \varphi_i(s)Q_{m-1,0}^*(s), \quad i = 0, \cdots m - 4$$

Thus, Eq. (21), Eq. (22) and Eq. (23) are derived.
Substituting the above results into Eq. (12) to yield:

$$Q_{11}^*(s) = \frac{1}{\mu}[(s - a_0)\varphi_0 - r\varphi_1]Q_{m-1,0}^*(s) - \frac{1}{\mu}_1$$

$$= g_1 Q_{m-1,0}^*(s) - f$$

By utilizing the same method, substituting the results into Eq. (17) and Eq. (18), we get:

$$Q_{0,-1}^*(s) = \frac{\theta}{s - b_0}\varphi_0(s)Q_{m-1,0}^*(s) = g_{-1}Q_{m-1,0}^*(s)$$

$$Q_{21}^*(s) = \frac{1}{\mu + r}[(s - b_1) - m\lambda g_{-1} - \theta\varphi_1]Q_{m-1,0}^*(s)$$

$$- \frac{1}{\mu + r}(s - b_1)f_1$$

$$= g_2 Q_{m-1,0}^*(s) - f_2$$

$$Q_{i1}^* = \frac{Q_{m-1,0}^*(s)}{\mu + (i-1)r}[(s - b_{i-1})g_{i-1} - (m - i + 2)\lambda b g_{i-2} - \theta\varphi_{i-1}]$$

$$- \frac{1}{\mu + (i-1)r}[(s - b_{i-1})f_{i-1} - (m - i + 2)f_{i-2}]$$

$$= g_i Q_{m-1,0}^*(s) - f_i, \quad i = 3, \cdots, m$$

Then, Eq. (24), Eq. (25), Eq. (26), Eq. (27) and Eq. (28) are deduced. Eq. (29) can also be derived by substituting the obtained results into Eq. (19).
Moreover, Eq. (20) can be expressed.

4 NUMERICAL EXAMPLES

In this section, some numerical results are provided to show the effects of various system parameters on the mean time to the first failure of the system.

(1) Figures 1–4 show that *MTTFF* increases drastically as repair rate μ increases, while the value of μ varies from 0 to 0.5. However *MTTFF* decreases insignificantly as μ increasing higher than 1.

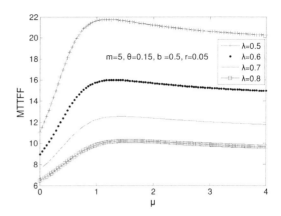

Figure 1. Effect of λ on *MTTFF*.

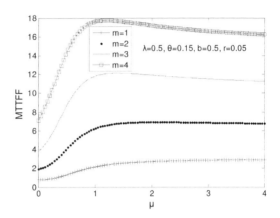

Figure 2. Effect of *m* on *MTTFF*.

(2) In Figure 1, the system parameters are assigned as $m = 5, \theta = 0.15, b = 0.5, r = 0.05$. The value of λ varies from 0.5 to 0.8. It can be seen from Figure 1 that that *MTTFF* decreases remarkable as λ increases for small values. However, the rate of decrease in *MTTFF* is not significant for large λ.
(3) In Figure 2, the system parameters are assigned as $\lambda = 0.5, \theta = 0.15, b = 0.5, r = 0.05$. The value of *m* varies from 1 to 4. It can be presented that *MTTFF* increases conspicuously as *m* increases.
(4) In Figure 3, the system parameters are assigned as $m = 5, \lambda = 0.5, \theta = 0.15, r = 0.05$. The value of *b* varies from 0.5 to 0.8. It can be observed that *MTTFF* decreases as *b* increases the same as the effect of λ on *MTTFF*.
(5) In Figure 4, the system parameters are assigned as $m = 5, \lambda = 0.5, \theta = 0.15, b = 0.5$. The value of *r* varies from 0.02 to 0.05. It can be observed that *MTTFF* increases gently as *r* increases.
(6) In Figure 5, the system parameters are assigned as $m = 5, \lambda = 0.5, b = 0.5, r = 0.05$. The value of θ varies from 0.15 to 0.3. It is presented that *MTTFF* increases slightly as θ increases for small value of μ. Intuitively, *MTTFF* increases markedly as θ increases for μ higher than 0.5.

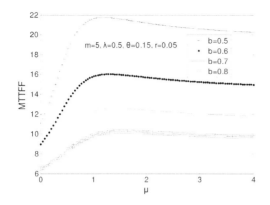

Figure 3. Effect of b on MTTFF.

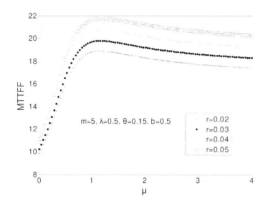

Figure 4. Effect of r on MTTFF.

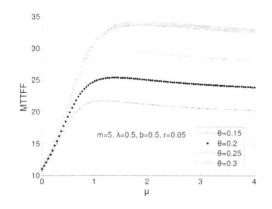

Figure 5. Effect of θ on MTTFF.

5 CONCLUSIONS

In this paper, the machine interference model where there are several operating units and one repairman was considered. The failed impatient units may balk or renege from the queuing and the repairman takes the single vacation. From the viewpoint of reliability theory, some performance analyses are presented utilizing the method of blocked matrix. Numerical analyses are carried out. It can be concluded that the effect of r on MTTFF is less insensitive than other parameters. When the value of μ is small, MTTFF changes drastically. The effect of the parameters on the system can be applied to optimize the system model. This is of interest for future research.

ACKNOWLEDGEMENTS

This work is supported in part by Natural Science Foundation of Hebei Province, China (No. G2012203136), supported in part by Key Project of Natural Research of University of Hebei Province, China (ZH2012021), and supported in part by Science and technology project of Qinhuangdao (No. 201101A067), China.

REFERENCES

Gupta SM, 1997, Machine interference problem with warm spares, server vacations and exhaustive service, Performance Evaluation 29(3): 195–211.

Gupta SM, 2004, Interrelationship between Queuing Models with Balking and Reneging and Machine Repair Problem with Warm Spares. Microelectronics Reliability 34(2): 201–209.

Jain M, Rakhee, Maheshwari S, 2004, N-policy for a machine repair system with spares and reneging, Applied Mathematical Modelling 28(6): 513–531.

Ke J, Wang K, 2005, Cost Analysis of the M/M/r Machine Repair Problem with Balking, Reneging and Server Breakdowns. Journal of the Operational Research Society 50: 275–282.

Ke JC, Wang KH, 2007, Vacation policies for machine repair problem with two type spares, Applied Mathematical Modelling, 31: 880–894.

Shawky A, 1997. The Single-server Machine Interference Model with Balking, Reneging and an Additional Server for Longer Queues. Microelectronics Reliability 37(2): 335–357.

Sun Y, Yue D, 2007, Performance Analysis of M/M/S/N Queuing System with Balking, Reneging and Multiple Synchronous Vacations, Systems Engineering-theory & Practice, 27(5): 152–158.

Wang KH, Sivazlian BD, 1989, Reliability of a system with warm standbys and repairmen. Microelectron Reliab 29:849–860.

Wang KH, Ke JC, 2003, Probability analysis of a repairable system with warm standbys plus balking and reneging, Applied Mathematical Modelling 27: 327–336.

Wang KH, Ke JB, Ke JC, 2007, Profit analysis of the M/M/R machine repair problem with balking, reneging, and standby switching failures, Computers & Operations Research 34: 835–847.

Yue D, Zhu J, Qin Y, Li C, 2006, Gaver's Parallel System Attended by a Cold Standby Unit and a Repairman with Multiple Vacations. Systems Engineering-theory & Practice 6: 59–68.

Yue D, Sun Y, 2008, Waiting Time of M/M/c/N Queuing System with Balking, Reneging, and Multiple Synchronous Vacations of Partial Servers, Systems Engineering 28(2): 89–97.

Digital simulation of sensitivity in the planar mechanism motion accuracy

Gang Hu & Shiliang Liu
Shandong Water Polytechinc, Rizhao, China

ABSTRACT: In order to acquire the reliability sensitivity of kinematical accuracy of plane mechanism, based on the research of kinematical equation and kinematical accuracy mode, combine analytical method of reliability sensitivity and computer simulation technology, propose a digital simulation method of kinematical accuracy sensitivity. Use the analysis on kinematical accuracy sensitivity of plane four-bar linkage as example, confirm the efficiency and feasibility of the method.

Keywords: planar mechanism; kinematic accuracy; sensitivity; digital simulation

1 INTRODUCTION

With the rapid development of modern science and technology, the kinematic precision of mechanism was taken seriously, especial in some precise and advanced technology [1].

Reliability sensitivity analysis on the mechanism kinematical precision indicates the important influence of design parameters change on the reliability of kinematical accuracy [2].

Consequently, in the process of design and manufacture, the parameter which has a greater influence on the kinematical accuracy reliability of mechanism should be strictly controlled and in a small change. Conversely, in order to reduce the manufacture cost, relax the parameters which has little influence on the kinematical accuracy reliability of mechanism in order to reduce the manufacture cost.

Use MFOSM to analyze the reliability sensitivity of planar mechanism, find and simulate the most sensitive design parameter. The efficiency of the method was verified with digital sensitivity based on the analysis on the kinematical accuracy of planar four-bar linkages and it provided references for mechanism scheme design.

2 MOTION PRECISION MODEL FOR THE PLANE MECHANISM

Kinematical equation for the plane mechanism with the input/output relationship [4].

$$F(U,V,L) = 0 \qquad (1)$$

Use the matrix method to analyze the accuracy of mechanism motion, and use the Taylor equation to expense the equation (1).

$$\frac{\partial F}{\partial U^T}\Delta U + \frac{\partial F}{\partial V^T}\Delta V + \frac{\partial F}{\partial L^T}\Delta L = 0 \qquad (2)$$

In the actual calculation, the input errors are often neglected that $\Delta V = 0$, so the equation can be simplified as:

$$\Delta U = -\left[\frac{\partial F}{\partial U^T}\right]^{-1}\frac{\partial F}{\partial L^T}\Delta L \qquad (3)$$

3 RELIABILITY SENSITIVITY ANALYSIS ON THE KINEMATICAL PRECISION OF MECHANISM

Performance function of the kinematic precision in mechanism is that

$$g(\delta,r) = \delta - r = \varepsilon_t - \Delta u_t \qquad (4)$$

$$\left.\begin{array}{l}\mu_g = E[g(\delta,r)] = \mu_\delta - \mu_r \\ \sigma_g^2 = Var[g(\delta,r)] = \sigma_\delta^2 - \sigma_r^2\end{array}\right\} \qquad (5)$$

The definition of reliability index show that [6]:

$$\beta = \frac{\mu_g}{\sigma_g} = \frac{E[g(\delta,r)]}{\sqrt{Var[g(\delta,r)]}} \qquad (6)$$

When the basic random parameter vector complies with normal distribution, reliability of the first-order moment estimator is [6]:

$$R = \Phi(\beta) \qquad (7)$$

In this mathematical formula, $\Phi(\bullet)$ is the cumulative distribution function of standard normal distribution.

Reliability of planar linkage mechanism motion accuracy on the organization's basic parameters sensitivity of vector mean and variance are as follows:

$$\frac{\partial R}{\partial \mu_{x_i}} = \frac{\partial R}{\partial \beta}\frac{\partial \beta}{\partial \mu_{x_i}} \qquad (8)$$

$$\frac{\partial R}{\partial \sigma_{x_i}} = \frac{\partial R}{\partial \beta} \frac{\partial \beta}{\partial \sigma_{x_i}} \qquad (9)$$

In this mathematical formula, $\frac{\partial R}{\partial \beta} = \phi(\beta)$, $\frac{\partial \beta}{\partial \mu_{x_i}} = \frac{1}{\sigma_g}$, $\frac{\partial \beta}{\partial \sigma_{x_i}} = -\frac{\mu_g}{\sigma_g^2}$, $\phi(\bullet)$ are standard normal probability density functions.

4 DIGITAL SIMULATIONS IN THE SENSITIVITY OF PLANAR MECHANISM MOTION ACCURACY

Plane four-bar linkage as shown in figure 1, $l_1 = 60\,\text{mm}$, $l_2 = 170\,\text{mm}$, $l_3 = 220\,\text{mm}$, $l_4 = 240\,\text{mm}$. And $l_1 \sim (0.06, 0.0003)$, $l_2 \sim (0.17, 0.00085)$, $l_3 \sim (0.22, 0.00011)$, $l_4 \sim (0.24, 0.0012)$. The allowable error of mechanism motion output angle (α_3) is $1.2°$. When the enter angle (α_1) is $60°$, and the out angle (α_3) fits the upper limit $\Delta\alpha_3 \leq \delta$, how much is the sensitivity of planar mechanism motion accuracy?

$U = [\alpha_2, \alpha_3]^T$, $V = \alpha_1$, $L = [l_1, l_2, l_3, l_4]^T$.

According the Taylor equation can acquire:

$$\frac{\partial F}{\partial U^T} = \begin{bmatrix} -l_2 \sin\alpha_2 & l_3 \sin\alpha_3 \\ l_2 \cos\alpha_2 & -l_3 \cos\alpha_3 \end{bmatrix}$$

$$\frac{\partial F}{\partial L^T} = \begin{bmatrix} \cos\alpha_1 & \cos\alpha_2 & -\cos\alpha_3 & -1 \\ \sin\alpha_1 & \sin\alpha_2 & -\sin\alpha_3 & 0 \end{bmatrix}$$

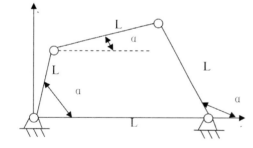

Figure 1. The sketch map of plane four-bar linkage.

According to the equation (4), the output angle error of plane four-bar linkage is:

$$\Delta\alpha_3 = -\frac{1}{\frac{\partial F_1}{\partial \alpha_2}\frac{\partial F_2}{\partial \alpha_3} - \frac{\partial F_1}{\partial \alpha_3}\frac{\partial F_2}{\partial \alpha_2}} \left\{ \left(-\frac{\partial F_2}{\partial \alpha_2}\frac{\partial F_1}{\partial l_1} + \frac{\partial F_1}{\partial \alpha_2}\frac{\partial F_2}{\partial l_1}\right)\Delta l_1 + \left(-\frac{\partial F_2}{\partial \alpha_2}\frac{\partial F_1}{\partial l_2} + \frac{\partial F_1}{\partial \alpha_2}\frac{\partial F_2}{\partial l_2}\right)\Delta l_2 \right.$$
$$\left. + \left(-\frac{\partial F_2}{\partial \alpha_2}\frac{\partial F_1}{\partial l_3} + \frac{\partial F_1}{\partial \alpha_2}\frac{\partial F_2}{\partial l_3}\right)\Delta l_3 + \left(-\frac{\partial F_2}{\partial \alpha_2}\frac{\partial F_1}{\partial l_4} + \frac{\partial F_1}{\partial \alpha_2}\frac{\partial F_2}{\partial l_4}\right)\Delta l_4 \right\}$$

Otherwise, $\Delta l = l - l^*$, l^* is the ideal value or nominal value of the linkage length of plane four-bar linkage. l is the actual value of the linkage length of plane

$$f(X) = \delta + \frac{1}{\frac{\partial F_1}{\partial \alpha_2}\frac{\partial F_2}{\partial \alpha_3} - \frac{\partial F_1}{\partial \alpha_3}\frac{\partial F_2}{\partial \alpha_2}} \left\{ \left(-\frac{\partial F_2}{\partial \alpha_2}\frac{\partial F_1}{\partial l_1} + \frac{\partial F_1}{\partial \alpha_2}\frac{\partial F_2}{\partial l_1}\right) \times (l_1 - l_1^*) + \left(-\frac{\partial F_2}{\partial \alpha_2}\frac{\partial F_1}{\partial l_2} + \frac{\partial F_1}{\partial \alpha_2}\frac{\partial F_2}{\partial l_2}\right) \right.$$
$$\left. \times (l_2 - l_2^*) + \left(-\frac{\partial F_2}{\partial \alpha_2}\frac{\partial F_1}{\partial l_3} + \frac{\partial F_1}{\partial \alpha_2}\frac{\partial F_2}{\partial l_3}\right) \times (l_3 - l_3^*) + \left(-\frac{\partial F_2}{\partial \alpha_2}\frac{\partial F_1}{\partial l_4} + \frac{\partial F_1}{\partial \alpha_2}\frac{\partial F_2}{\partial l_4}\right) \times (l_4 - l_4^*) \right\}$$

When the output angle (α_3) of plane four-bar linkage fits the upper limit $\Delta\alpha_3 \leq \delta$, the sensitivity of motion accuracy ($R = 0.9997$) can be acquired with MFOSM.

Reliability of planar linkage mechanism motion accuracy on the organization's basic parameters sensitivity of vector mean and variance are as follows:

$$\frac{\partial R}{\partial \mu_x} = \begin{bmatrix} R_{E(l_1)} \\ R_{E(l_2)} \\ R_{E(l_3)} \\ R_{E(l_4)} \end{bmatrix} = \begin{bmatrix} -0.89667 \\ -0.90110 \\ 0.36743 \\ 0.52564 \end{bmatrix}$$

$$\frac{\partial R}{\partial \sigma_x} = \begin{bmatrix} R_{\sigma(l_1)} \\ R_{\sigma(l_2)} \\ R_{\sigma(l_3)} \\ R_{\sigma(l_4)} \end{bmatrix} = \begin{bmatrix} 2.49798 \times 10^3 \\ 2.52274 \times 10^3 \\ 4.19441 \times 10^2 \\ 8.58417 \times 10^2 \end{bmatrix}$$

According to the former calculating result can get that:

sensitivity		Introductions	kinematic accuracy of plane mechanism
Mean value	$\partial R/\partial \mu_{x_i}$	The sensitivity of mean value is large when the absolute of mean value is large, The sensitivity of mean value is small when the absolute of mean value is small. If the value is positive number, the number shows that the parameter increases, the results tend to be more reliable. On the contrary, the number shows that the parameter increases, the results tend to be more unreliable (lose efficacy).	When the output angle (α_3) fits the upper limit $\Delta\alpha_3 \leq \delta$, linkage ($l_2$) has the greatest influence on the reliability sensitivity. The next is crank (l_1). When the output angle (α_3) fits the upper limit $\Delta\alpha_3 \leq \delta$, the mean value of rocker (l_3) and support (l_4) appropriate increases cause output angle (α_3) becoming more reliable. The mean value of crank (l_1) and linkage (l_2) increases cause out angle (α_3) becoming unreliable (lose efficacy).
variance	$\partial R/\partial \sigma_{x_i}$	The larger computed results are, the more sensitive the parameters are.	When the output angle (α_3) suffice the upper limit $\Delta\alpha_3 \leq \delta$, the variance of the basic random variable parameters (l_2) has the greatest influence on the reliability sensitivity of out angle (α_3).

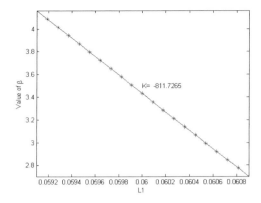

Figure 2. Visualization of β curves with L_1 changes.

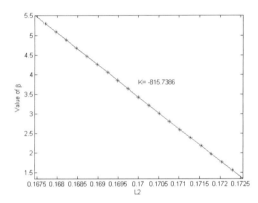

Figure 3. Visualization of β curves with L_2 changes.

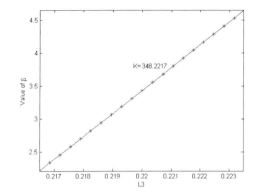

Figure 4. Visualization of β curves with L_3 changes.

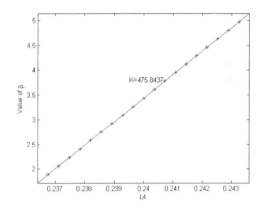

Figure 5. Visualization of β curves with L_4 changes.

Figures 2 to 5 show that simulation result in the numerical of reliability index and the change of output variable with MATLAB.

Figure 2 and figure 3 show that two negative slope of the straight line. Figure 4 and figure 5 show that two positive slope of the curve line. Otherwise, the slope of absolute value is the maximum in figure 3. It corresponds to the theoretical analysis results above.

Therefore, in the design and manufacture of four-bar linkage, the mean and variance of linkage (L2) size error should be decreased. The reliability of motion accuracy will be effectively improved.

5 CONCLUSIONS

(1) In this paper, the method of digital simulation in mechanism motion accuracy sensitivity is discussed. It demonstrates the structure parameter influences on motion precision reliability sensitivity, and it provides a theoretical foundation for the rational design and manufacture of mechanism.
(2) Take the analysis on movement precision reliability sensitivity of planar four-bar linkages as an example, the method is verified effectively. The changes of reliability index-β over output variables are simulated with MATLAB, the result is consistent with theoretical analysis, and it provides a theoretical foundation for analyzing and modifying the movement precision reliability level of planar linkage mechanism.

REFERENCES

[1] Zhang Yimin, Huang Xianzhen, He Xiangdong, et.al. Reliability Sensitivity Design of the Kinematic Accuracy of a Planar Linkage Mechanism with Arbitrary Distribution Parameters. Mechanical Science and Technology, 2008, 27(5):684–691.
[2] Zhang Yi-min, Huang Xian-zhen, HE Xiang-dong, et al. Reliable sensitivity design for kinematics accuracy of planar linkage mechanism. Journal of Engineering Design, 2008, 15(1):25–28.
[3] Zhang Jiyuan. Mechanism analysis and synthesis of solution (mechanical engineering). Beijing: China Communications Press, 2007:150–154.

[4] Zhang Yimin, Huang Xianzhen, HE Xiangdong. Reliability-based Robust Design for Kinematic Accuracy of the Shaper Mechanism under Incomplete Probability Information. Journal of Mechanical Engineering, 2009, 45(4):105–110.

[5] Xu Weiliang, Deng Jiaxian, Wu Cisheng. Matrix Method for Determining the Kinematic Errors of a Linkage. Journal of Southeast University (Natural Science Edition), 1986(3):32–34.

[6] Lu Zhenyu. The structure reliability and reliability sensitivity analysis. Beijing: Science Press, 2009:9–19.

Using the modified zeolites for treatment of industrial wastewater and synthesis of chemicals on the stage of recovery from aqueous solutions

A.S. Konovalov
Baikal Museum ISC SB RAS, Listvyanka, Russia
FSBI "Centre of an Agrochemical Service "Irkutskiy", Irkutsk, Russia

D.I. Stom
FSBEE HPE "Irkutsk State University", Irkutsk, Russia

M.V. Butyrin
FSBI "Centre of an Agrochemical Service "Irkutskiy", Irkutsk, Russia

M.N. Saksonov
FSBEE HPE "Irkutsk State University", Irkutsk, Russia

V.V. Tyutyunin
FSBEE HPE SR "Irkutsk State Technical University", Irkutsk, Russia

ABSTRACT: Using the methods of chemical analysis and bioassay, investigated the toxicity and detoxication of model solutions containing salts of arsenic. The ability of zeolites to absorb arsenic from model solutions was examined. Shown to reduce the toxic effect of pollutants under the influence of commodity humic substances ("Powhumus", "Humate-80"), as well as zeolites, activated with their application. The most effective sorbent was zeolite modified by degassing, decationization, thermal and chemical treatment.

Keywords: zeolite, wastewater, arsenic, purification, detoxication, humic substances, bioassay

1 INTRODUCTION

Arsenic and its derivatives are among the most hazardous ecotoxic materials. They could be thrown out in environment by the operations of black and nonferrous metallurgy, and also chemical industry. Toxic compounds on the basis of arsenic, engaging in cycle of matter, has a ruinous effect on microflora of soils, vegetation cover and the person (Berg, M. et al, 2007).

There are reports on capacity of humic substances (HS) to decontaminate various toxicants. Weak the moment in utilization of HS are good water solubility (Ozdoba, D.M. et al, 2001; Tan, K.H. 2003). One of the most perspective methods of excision of pollutants, in particular arsenic from solid and fluid mediums, is utilization of various sorbents (Baydina, N.L. 2011; Kireycheva, L.V. et al, 2009). As absorbents of arsenic contamination the particular interest are introduced by natural zeolites. They possess rather high capacity of absorption and at the same time differs sufficient cheapness and availability (Kragović, M. et al, 2013).

The aim of this work is to combine the above techniques to develop methods of detoxification of arsenic pollutants.

2 EXPERIMENTAL

Model arsenic contaminant was selected salt solution of 100 mg/dm^3 Na$_3$AsO$_4$ brand reagent grade. As our previous experiments (Konovalov A.S. et al, 2013), this concentration was acutely toxic.

As adsorbent we used natural zeolite of the Sokirnitsky deposit, Krasnoyarsk region (LLC "Etnakom"). Sorption capacity of zeolites was evaluated by iodometric method (*Act. charc. crush. Spec. State Standard 6217-74*).

Sources of HS were commercial preparations of "Humate-80" and "Powhumus". The first is a mixed humate K/Na, manufactured by LLC "Agricultural Technologies" Irkutsk. It is obtained by mechanochemical treatment of lignine and the mixture of K$_2$CO$_3$/Na$_2$CO$_3$. "Powhumus" is potassium humate (Humintech Ltd., Germany), produced by the standard wet alkaline extraction of oxidized coal (leonardite). (Ozdoba, D.M. et al, 2001).

Zeolite was modified thermally, degassing, decationization and impregnation by chemicals (Kan, V.M., 2013). For this purpose, 100 g of zeolite was washed with hot water (60–70°C) to remove the clay inclusions. The washed zeolite was calcined at 400–450°C and placed in a solution of nitric acid (60 g/dm^3) per day for decationization. Zeolite after decationization washed with distilled water and calcined. Zeolite thus prepared was impregnated with the respective solution (5 g of $(NH_2)_2CO$, 5 g of NH_4NO_3, 40 ml of distilled water, 2.5 ml of $MnSO_4$, 7,5 ml HS "Powhumus"). Alternatively, take a zeolite pre-prepared and impregnated with a solution "Powhumus" at 1 g/dm^3 (in some experiments, instead of "Powhumus" used "Humate-80" at the same concentrations). 100 g of zeolite was impregnated by 50 ml HS.

As control we evaluated the effectiveness of binding of arsenic by calcined zeolite without further addenda HS or other reagents.

To determine the degree of absorption of the arsenic from the solution by investigated sorbents it is placed in the column with a diameter of 20 mm and a height of 450 mm, which the test solutions was passed through for 60 minutes.

Amount of arsenic in the source and the outlet solution was measured colorimetrically with diethyldithiocarbamate of silver in chloroform (*Raw mat. and food prod. Meth. for the det. of arsenic. State Standard 26930-86.*).

Toxicity of analyzed solutions was evaluated by the change of the luminescence intensity of chlorophyll cells of algae Scenedesmus quadricauda (Turp.) Breb. Fluorescence measurement was carried out on "Fluorat-02-3M" in continuous measurement mode after three days of incubation under the luminescent lighter (*Meth. for det. the tox. of water, aq. ext. from soil, sewage sludge and waste, changes in the level of chlor. fluor. and the number of algal cells, 2007*).

As part of experiments independent experiments were carried out not less than 5 times in triplicate. Statistical analysis of the data obtained using the software package MS Excel. The significance of differences was determined using Student's t test. The tables are represented by average values for the sample and their standard deviations.

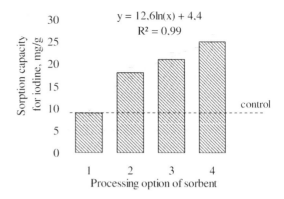

Figure 1. Sorption capacity of studied zeolites for iodine:
1. Zeolite calcinated (control);
2. Zeolite prepared and impregnated HS "Humate-80";
3. Zeolite prepared and impregnated HS "Powhumus";
4. Zeolite modified by degassing decationization and heat treatment with the addition respective chemicals.

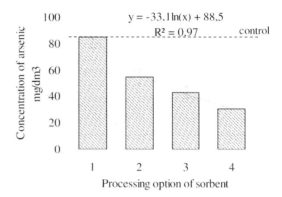

Figure 2. Concentration of arsenic in solutions after passage through sorbents (in mg/dm^3):
1. Zeolite calcinated (control);
2. Zeolite prepared and impregnated HS "Humate-80";
3. Zeolite prepared and impregnated HS "Powhumus";
4. Zeolite modified by degassing decationization and heat treatment with the addition respective chemicals.

3 RESULTS AND DISCUSSION

During the first stage of work we have estimated sorption capacity of various samples of zeolites. Sorption capacity of a preparation of zeolite preliminary modified by impregnation a solution 1 g/dm^3 "Powhumus" was above the control on 58.5 ± 6.3%, and at application of the same concentration "Humate-80" exceeded it on 39.8 ± 4.1%. But the greatest sorption capacity was at the zeolite subjected to degassing, decationization, thermal processing and modified by corresponding chemicals (($NH_2)_2CO$, NH_4NO_3, $MnSO_4$, HS "Powhumus"). It exceeded control (the calcinated zeolite) on 83.3 ± 8.1% (fig. 1).

Figure 2 shows that the zeolite impregnated "Powhumus" represents an increase of 27.2 ± 3.1%, and with the addition of zeolite "Humate-80" – at 15.7 ± 2.2% higher than the control sample. Zeolite, activated degassing, decationization and thermal treatment with the addition of respective chemicals, absorbed arsenic from a model solution at 41.2 ± 3.9% more intense than in the control embodiment.

As the solution of 100 mg/dm^3 Na_3AsO_4 which has passed through a column with calcinated zeolite without the subsequent addition to it HS or any other reagents follows from figure 3, suppressed level of fluorescence of chlorophyll more than on 80%. At the same time, zeolite, after degassing, decationization, thermal processing and addition of corresponding chemicals or impregnated "Powhumus", reduced toxic action of arsenic more, than on 60% concerning the calcinated zeolite. Thus intensity of fluorescence of cells of algae S. quadricauda has made 90.3 ± 9.1%, 79.5 ± 8.4% and 15.5 ± 2.1% accordingly. At biotesting of the solution passed through zeolite, impregnated

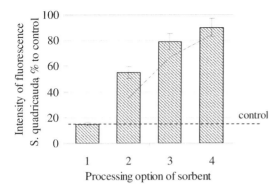

Figure 3. Influence of solutions of 100 mg/dm^3 Na$_3$AsO$_4$ on fluorescence of cages of seaweed S. quadricauda after passage through sorbents (in % to the control – dechlorinated water):
1. Zeolite calcinated (control);
2. Zeolite prepared and impregnated HS "Humate-80";
3. Zeolite prepared and impregnated HS "Powhumus";
4. Zeolite modified by degassing decationization and heat treatment with the addition respective chemicals.

HS "Humate-80", intensity of a luminescence of a chlorophyll of cells of microalgae made 55.7 ± 5.9% concerning the control (fig. 3).

4 CONCLUSION

Thus, the highest efficiency for binding model arsenic contamination (Na$_3$AsO$_4$ aqueous salt solution at a concentration of 100 mg/dm^3) showed zeolite, activated by washing, decationization, calcining and impregnated respective solution (5 g of (NH$_2$)$_2$CO, 5 g of NH$_4$NO$_3$, 40 ml distilled water, 2.5 ml of MnSO$_4$, 7.5 ml HS "Powhumus" to 100 g of zeolite). The same sample of zeolite had, accordingly, and maximum sorption capacity on iodine (Kan, V.M. 2013).

After passage through a sorbent layer, the least size of suppression of fluorescence of chlorophyll of cells of seaweed S. quadricauda observed at use of the zeolite passed processing on a method offered by us (Kan, V.M. 2013).

The authors are grateful to B.Stern "Humintech GmbH" for preparation granting "Powhumus".

Work was spent at partial support of the Russian fund of basic researches (No. I-140122090159), Strategic Development Program of Irkutsk State University and Project R211-PF-016.

REFERENCES

Active charcoal crushed. Specifications. State Standard 6217-74.

Baydina, N. L. 2011. Inactivation of heavy metals by humus and zeolites in technologically contaminated soil. *Pedology* 3(03): 121–125.

Berg, M., Stengel, C., Trang, P., Hungviet, P., Sampson, M., Leng, M., Samreth, S., Fredericks, D. 2007. Magnitude of arsenic pollution in the Mekong and Red River Deltas. *Cambodia and Vietnam Science of The Total Environment* 372: 413–425.

Camacho, L.M., Parra, R.R., Deng, S. 2011. Arsenic removal from groundwater by MnO$_2$-modified natural clinoptilolite zeolite: Effects of pH and initial feed concentration. *Journal of Hazardous Materials* 189: 286–293.

Haw-Tarn Lin, Wang, M.C., Li, Gwo-Chen 2004. Complexation of arsenate with humic substance in water extract of compost. *Chemosphere* 56: 1105–1112.

Kan, V.M. 2013. Way of detoxifying wastewater contaminated with arsenic salts (Variants) / Kan V.M. and others// Application for a patent No. 2013152455 by 26.11.2013.

Kireycheva, L. V., Ilynsky, A. V., Yashin, V. M., Nguen Suan Khay 2009. Detoxication of contaminated with heavy metals leached black soil and ancient alluvial soils using sorption materials / *Proc. Rus. Acad. Agric. Sci.* 3: 41–43.

Konovalov A. S., Stom D. I., Evsyunina E. V. 2013. Evaluation of detoxication arsenic salt solutions by humate by bioassay methods / *Proc. intern. sci.-pract. conf. "Modern scientific achievements"*: 89–91.

Kragović, M., Daković, A., Marković, M., Krstić, J., Gatta, G.D., Rotiroti, N. 2013. Characterization of lead sorption by the natural and Fe(III)-modified zeolite. *Applied Surface Science* 283: 764–774.

Li, Zh., Jean, J.-Sh., Jiang, W.-T., Chang, P.-H., Chen, Ch.-J., Liao, L. 2011. Removal of arsenic from water using Fe-exchanged natural zeolite. *Journal of Hazardous Materials* 187: 318–323.

Methodology for determining the toxicity of water, aqueous extracts from soil, sewage sludge and waste, changes in the level of chlorophyll fluorescence and the number of algal cells. – M.: Aquaros. – 2007. – 48 P.

Raw materials and food products. Method for the determination of arsenic. State Standard 26930-86.

Tan, K. H. 2003. Humic Matter in Soil and the Environment. Principles and Controversies. *Marcel Dekker Inc.*

Tarasova, A.S. 2011. Effect of humic substances on toxicity of inorganic oxidizer bioluminescent monitoring / A.S. Tarasova, D.I. Stom, N.S. Kudryasheva // *Environmental Toxicology and Chemistry* 30(5): 1013–1017.

Ozdoba, D.M., Blyth, J.C., Engler, R.F., Dinel, H., Schnitzer, M. 2001 Leonardite and humified organic matter. *In Proc Humic Substances Seminar V, Boston, MA, March 2001.*

Catalysts for purification of sulfur-containing industrial wastewater

Kochetkov Alexey Yurevich & Kochetkova Raisa Prohorovna
LLC SPA "Catalysis", Angarsk, Russia

Stom Devard Iosifovich
FSBEE HPE "Irkutsk State University", Irkutsk, Russia;
Irkutsk Technical University, Irkutsk, Russia; BM ISC SB RAS, Listvyanka, Russia

ABSTRACT: It was constructed and studied a large series of heterogeneous catalysts based on complexes of phthalocyanine with PcCo, NiO, CoO, CuO, MnO_2, PbO, V_2O_5, ZnO, Fe_2O_3. Shown to be promising the obtained catalysts for wastewater treatment in pulp and paper industry production of sulfate containing hydrogen sulfide and mercaptans.

Keywords: Complexes of phthalocyanine, heterogeneous catalysts, hydrogen sulfide and mercaptans, wastewater

A significant problem is the waste water from pulp and paper industry sulfate production (PPI) containing hydrogen sulfide (H_2S) and mercaptans (RSH). Especially perspective for treatment of PPM wastewater is the use of catalysts based on complexes of phthalocyanine with metal oxides (Loas et al, 2011; Mai et al, 2007).

For the same purposes in the LLC SPA "Catalysis" developed, produced and studied a number of new catalysts (Kochetkova R.P., authors. cert., 1993). This report shows some results of these studies.

1 EXPERIMENTAL

For the synthesis of metal complexes (MC) in the machine-homogenizer at a temperature of 130–135° C for 15 minutes, mixed cobalt phthalocyanine (PcCo), a metal oxide (or mixture thereof) with a complexing agent (CA) in a ratio of 0.1:9.0 – 0.5:9.5 respectively. To the resulting MC added high-density polyethylene (HDPE) or polypropylene (PP) in the ratio of MC: HDPE equal 1.4:2.5 – 1.2:2.5. Then, the resulting mass is stirred at a temperature of 130–135°C for 30 minutes. The resulting product is fed into a screw extruder KE-200, where the temperature is maintained at 160–175°C. Molten mass of the catalyst was fed from the extruder to granulation apparatus, wherein the granules formed GMC size from 5 to 15 mm in diameter. Study GMC activity in the oxidation of sulfide and sodium mercaptide by molecular oxygen was carried out in a column of air apparatus (diameter – 250 mm, height – 1500 mm). Oxidation of H_2S and RSH conducted by GMC in model solutions during the pilot testing in the condensates obtained from the process of evaporation of black liquor of "Baikal Pulp and Paper Mill" (pH – 7.5–12.0, concentration of H_2S and RSH 230–260 dm^3, temperature 20–800°C, air flow rate of 0.5–6.5 min^{-1}, the content of the catalyst – 50–250 g/dm^3, pressure – 0.3 MPa, during 10 min). Determination of the mercaptide, sulfide, thiosulfate, and sulfate was carried out according to known methods. The paper shows the average value of 3 parallel measurements in 5 different experiments.

2 RESULTS AND DISCUSSION

Studies have shown that the catalytic activity of the synthesized GMC increased with increasing content of PcCo. However, increasing the proportion of MC-PcCo with respect to the polymer over 25% in the case of HDPE, and up to 15% PP is impractical. In this case, as revealed exploitation, fragility of catalyst pellets is increased.

Electron microscopy of the GMC revealed, which unlike the original polyethylene structurally homogeneous microparticles MC-PcCo arranged in the GMC in the form of individual amorphous zones.

To study the oxidation of H2S and RSH was made a large series of GMC using: PcCo, NiO, CoO, CuO, MnO_2, PbO, V_2O_5, ZnO, Fe_2O_3. GMC were in the form of pellets consisted of 10% CA, 20% of the corresponding metal oxide, and only 70% of polyethylene.

The obtained data suggest that the GMC on the activity MC form the following series: Cr > Co >

Table 1. Characteristics of the various samples GMC-Fe (for pellets d = 4 mm).

The catalyst composition, %	$S_{sp.}$, m^2/g	Bulk density, g/cm^3	Degree of conversion Na$_2$S, % relatively	Degree of conversion RSH, % relatively
1. PC – 10.0; CA – 7.0; polyethylene – 83.0	1.23	0.35	65.0	70.0
2. PC – 20.0; CA – 7.0; polyethylene – 73.0	1.36	0.41	83.0	87.0
3. PC – 30.0; CA – 7.0; polyethylene – 63.0	1.25	0.48	88.0	91.0
4. PC – 40.0; CA – 7.0; polyethylene – 53.0	1.27	0.58	92.0	95.0
5. PC – 45.0; CA – 7.0; polyethylene – 48.0	1.25	0.62	91.0	95.0
6. PC – 35.0; CA – 0; polyethylene – 65.0 GMC – fragile, mechanically groggy	0.65	0.68	87.0	91.0
7. PC – 25.0; CA – 0; polyethylene – 75.0	0.92	0.65	76.0	81.0
8. PC – 35.0; CA – 3.0; polyethylene – 62.0	1.12	0.62	87.0	90.0
9. PC – 35.0; CA – 5.0; polyethylene – 60.0	1.27	0.58	91.0	94.0
10. PC – 35.0; CA – 10.0; π polyethylene – 55.0	1.25	0.45	92.0	94.0
11. PC – 35.0; CA – 15.0; polyethylene – 50.0	2.48	0.28	90.0	92.0

Note: GMC-Fe No. 6.7 – fragile, mechanically groggy.

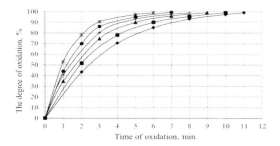

Figure 1. Dependence degree of oxidation of sulfide ions on the oxidation time of the catalyst samples of GMC-Fe, modified PcCo:
1. GMC-Fe+0.125% PcCo;
2. GMC-Fe+0.1% PcCo;
3. GMC-Fe;
4. GMC+10% PcCo;
5. GMC+5% PcCo;
6. 10% PcCo.

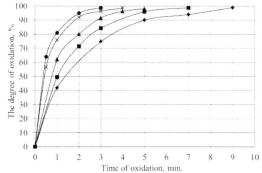

Figure 2. Dependence degree of oxidation of sulfide ions on the oxidation time of the catalyst samples of GMC-Fe, modified by cuprous oxide:
1. GMC-Fe+CuO 10%;
2. GMC-Fe+CuO 5%;
3. GMC-Fe+CuO 3%;
4. GMC-Fe+CuO 1%;
5. GMC-Fe.

Ni > Cu > Fe > Mn > Pb > V > Zn > PcCo. H$_2$S oxidation products for all GMC, except containing PcCo, were: thiosulfate (78–81%), sulfite (1.5–2.0%) and sulfate (18.0–20.0%). In the oxidation of RSH all GMC formed disulfides (RSSR).

GMC with iron oxides are less active than GMC with some other oxides, but their cost 10–15 times lower. Moreover, iron oxide can be replaced by waste of sulfuric acid production – pyrite cinder (PC) containing 68–72% iron oxide. Therefore, for the synthesis of industrial catalyst (GMC-PC), it is advisable to use the PC modifying their various additives. The catalytic activity of GMC-PC increased with increasing content of PC in the MC up to 40% (Table 1).

The optimum content of the CA in the MC-PC is 7.0–10.0 wt%, MC content in GMC – 35, 0–40, 0% by weight. Elevated levels of CA in GMC-PC significantly reduces bulk weight – less than 0.3 g/cm^3, increases the surface area – 2.5 m^2/g, and all this reduces the performance of GMC-PC (Fig. 1, 2).

Copper oxide has a high catalytic activity of oxidation H$_2$S and RSH. Therefore, as a modifying agent GMC-PC studied and copper oxide. Experiments showed that the catalyst activity increases with symbatically increasing content of CuO in the GMC-PC of up to 5% by weight. GMC composition: PC – 35%, copper oxide – 5%, CA – 10%, 50% – HDPE, at "Baikal Pulp and Paper Mill" (installation capacity of 400 m^3/h) worked in the cleaning process condensates from H$_2$S and RSH for 7 years.

This work was financially supported by the Ministry of Education of the Russian Federation within the framework of the project of the state task in the field of scientific activity (Application 1263).

REFERENCES

Authors Certificate Number 1839330 (USSR), 1993. Catalyst for the oxidation of sulfur compounds. / Kochetkova R.P., Shiverskaya I.P., Shpilevskaya L.I., Glazyrin V.V., Tikhonov G.P.

Loas, A. Broadening the reactivity spectrum of a phthalocyanine catalyst while suppresing its nucleophilic, electrophilic and radical degradation pathways // A. Loas, R. Gerdes, Y. Zhang, S.M. Gorun // Dalton Translation, The Royal Society of Chemistry. – 2011. – V. 40. – P. 5162–5165.

Mei, H. Preparation and characterization of NiO/MgO/Al_2O_3 supported CoPcS catalyst and its application to mercaptan oxidation / H. Mei, M. Hu, H. Ma, J. Shen // Fuel Processing Technology. – 2007. – V. 88. – P. 343–348.

Using microbial fuel cells for utilization of industrial wastewater

E.Yu. Konovalova, D.I. Stom, A.E. Balayan & E.S. Protasov
FSBEE HPE "Irkutsk State University", Irkutsk, Russia; Institute of Biology, Irkutsk, Russia; BM ISC SB RAS, Russia

M.Yu. Tolstoy & V.V. Tyutyunin
FSBEE HPE "National Research Irkutsk State Technical University", Irkutsk, Russia; Irkutsk Technical University, Irkutsk, Russia; BM ISC SB RAS, Russia

ABSTRACT: Investigated the effect of inoculation of microorganisms and the presence of Sodium Dodecyl Sulfate (SDS) to a Microbial Fuel Cell (MFC). The perspective of using EM preparation "Vostok EM – 1" and it was isolated strain *Bacillus cereus* for producing electrical energy MFC. SDS at concentration of 0.1%, 0.05%, 0.01% inhibited the production of electricity and the growth of microorganisms when used as a preparation "Vostok EM – 1" and isolated strain.

This work was financially supported by the Ministry of Education of the Russian Federation within the framework of the project of the state task in the field of scientific activity (Application 1263) and out with the partial financial support of RFBR (No. I-140 122 090 159), the strategic development and project ISU P211-PF-016.

Keywords: microbial fuel cell; sodium dodecyl sulfate; Bacillus cereus

1 INTRODUCTION

The microbiological purification is the main way of neutralization of sewage including industrial. But in many cases it is already cannot cope with increasing volumes of wastewater. On the other hand the depletion of natural resources leads to the fact that increasingly encouraged to apply waste as secondary raw materials. MFC allows you to directly generate electricity by utilizing components of wastewater (Bullen et al., 2006; Rabaye et al., 2005). The most important element in determining work of MFC – the microorganisms are introduced into the anode compartment. One of the major reasons that block the use of MFC in practice is that today little is known strains of microorganisms suitable for MFC (Bond et al., 2003; Gorby et al., 2005, 2006; Rabaye et al., 2004, 2005). The disadvantage of microbial cultures have used as biological agents is a relatively narrow range of substrates used (C_{TOM} et al., 2013). H. Teruo developed a unique preparation "effective microorganisms" (EM). It consists of more than 80 "balanced" species of microorganisms. Strains include both aerobic and anaerobic bacterial species. In addition to photosynthetic bacteria EM composition includes lactic acid bacteria, yeasts, actinomycetes, fungi *Aspergillus* type and *Penicillium* and other organisms (Parr, 1994; Higa, 1996). A consortium of strains EM provides the broadest metabolic capabilities. EM capable of high speed utilize a wide range of pollutants. It is no accident that the use of EM in many countries yielded some surprising practical results. There is hard foreseeable amount of data on high performance and perspectivity of EM in the struggle against environmental pollution: local systems wastewater treatment a variety of industries; for the disposal of household and agricultural waste; for bioremediation of soils; productivity growth, protect plants from pests and diseases; for accelerated composting; improve the health of domestic animals and birds; improve crop protection; health and other spheres of human activity (Jusoh et al., 2013; Szymanski et al., 2003; Ting et al., 2013). Reproduction of EM preparation undertaken by many firms in different countries.

Surface-active substances (surfactants) are critical components of industrial and domestic wastewater. In connection with the above, the aim of this work – to study the possibility of using a microbiological preparation "Vostok EM – 1" and an isolated strain *B. cereus* for producing electrical energy MFC and the effect of surfactants on the MFC.

2 MATERIALS AND EQUIPMENT

As used in this paper MFC consisted of cells separated using the proton-exchange membrane into two parts. The volumes of the anode and cathode chambers are equal and 230 ml. Geometric dimensions of chambers:

length – 55 mm, width – 55 mm, height – 110 mm. Chamber walls are made of organic glass, 4 mm thick. Carbon paper used as the electrodes (Toray, Japan). Proton-exchange membrane thickness of 210 microns (MF-4SK brand) was purchased from LLC "Plastpolymer", St. Petersburg, Russia. Rectangular window size 40 × 90 mm between the cameras. The cathode chamber equipped with devices for supply and exhaust air. The aeration of the cathode compartment was carried out by the compressor (Dezzie, D-044, China) with a rate of 1.5 l/h.

In this paper, biological agents for the production of electricity were commercial biopreparation EM "Vostok EM – 1", produced by LLC "Seaside EM Centre" (Vladivostok, Russia) and isolated from EM preparation strain *B. cereus*. Basic preparation before using activated. For this initial concentrate "Vostok EM – 1" was diluted 1:50 model waste water containing 1% glucose. The suspension was then cultured for 24 hours at 30°C and introduced into the anode compartment of the cell. Anode and cathode chambers were filled autoclaved model waste water (C_{TOM} et al., 2013). Control MFCs without adding to the anode and cathode chamber of microorganisms. All figures MFC (titer, voltage) was measured immediately and after 8, 10, 24 hours. Removing discharging curves performed at circuit electrodes on a constant load. To determine the total microbial count (TMC) sterile syringe samples were taken periodically from the cells, did dilution in sterile water and sown on the surface of the medium RPA in Petri dishes. Within experiments independent experiments were carried out not less than 5 times in triplicate. Statistical analysis of the data was obtained using MS Excel software package for Windows. The tables show the average values for the sample and their standard deviations.

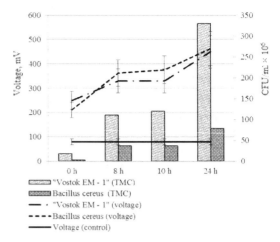

Figure 1. Voltage and TMC in MFC with preparation "Vostok EM – 1" and *B. cereus*.

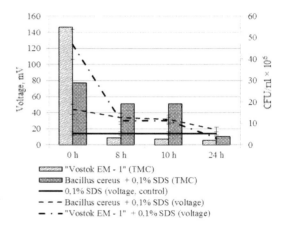

Figure 2. Voltage and TMC in the MFC in the presence of the preparation "Vostok EM – 1" and B. cereus, with the addition of 0.1% SDS.

3 RESULTS AND DISCUSSION

In the control MFC with model waste water, but without the addition of microorganisms, the substantial growth of bacteria in the day experiment was not recorded. The highest voltage in the control cell was achieved by the end of the experience – 80 ± 0.9 mV (Fig. 1). During the first 8 hours of the experiment was observed the growth a TMC in cells with "Vostok EM – 1" (from $1.8 \pm 0.2 \times 10^6$ to $1.10 \pm 0.2 \times 10^7$ CFU/ml) and *B. cereus* ($3.0 \pm 0.2 \times 10^5$ to $3.64 \pm 0.2 \times 10^6$ CFU/ml), and the rate increase of the electromotive force (EMF) "Vostok EM – 1" with 250 ± 26.9 to 330 ± 49.5 mV; *B. cereus* from 210 ± 31.5 to 362 ± 54.3 mV. After 10 hours, TMC in an embodiment with the preparation "Vostok EM – 1" was $1.2 \pm 0.2 \times 10^7$, and a day later $3.3 \pm 0.5 \times 10^7$ CFU/ml. In the case of *B. cereus* for these same time intervals varied with the number of microorganisms $3.72 \pm 0.2 \times 10^6$ to $7.8 \pm 0.2 \times 10^6$ CFU/ml. Indicators EMF in these MFC increased in "Vostok EM – 1" with 330 ± 49.5 to 450 ± 67.5 mV, *B. cereus* from 376 ± 56.4 to 463 ± 69.5 mV (Fig. 1).

In the presence of 0.1% SDS for the first 8 hours of the experiment in MFC titer reduction with all test organisms was observed. In cells with the preparation "Vostok EM – 1" TMC has dropped to $5.50 \pm 0.2 \times 10^6$ to $3.24 \pm 0.2 \times 10^5$ CFU/ml, and with *B. cereus* changed from $2.89 \pm 0.2 \times 10^6$ to $1.93 \pm 0.2 \times 10^6$ CFU/ml (Fig. 2).

During the same period we recorded decline in EMF cells with the preparation "Vostok EM – 1" under the influence of 0.1% SDS with 125 ± 18.8 and 30 ± 4.5 mV. In MFC with *B. cereus* potential difference varied from 44 ± 6.6 to 34 ± 5.1 mV. In all tested cells from 8 to 10 hours of experience indicators titer of microorganisms and EMF were consistently low. Control cells (without microorganisms, but with the addition of 0.1% SDS) we indicated EMF at ~14 mV during the entire experimental period. In the period from 10 to 24 hours in the MFCs with the preparation "Vostok EM – 1" value which reflects the number of microorganisms and generation

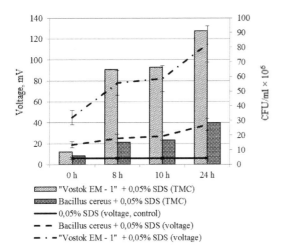
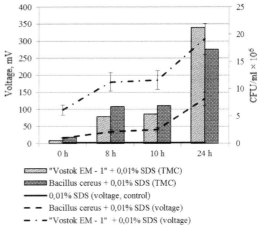

Figure 3. Voltage and TMC in the MFC in the presence of the preparation "Vostok EM – 1" and *B. cereus* adding 0.05% SDS.

Figure 4. Voltage and TMC in the MFC in the presence of the preparation "Vostok EM – 1" and *B. cereus* adding 0.01% SDS.

of electricity in the presence of 0.1% SDS, remained practically unchanged or slightly changed: TMC from $2.80 \pm 0.2 \times 10^5$ to $2.10 \pm 0.2 \times 10^5$ CFU/ml, EMF indicators from 30 ± 4.5 to 9 ± 1.4 mV. In MFCs with *B. cereus* strain TMC during the same period decreased from $1.92 \pm 0.2 \times 10^6$ to $3.92 \pm 0.2 \times 10^5$ CFU/ml, EMF 0.1% SDS changed from 32 ± 4.8 to 19 ± 2.8 mV (Fig. 2).

When added to the anode cell 0.05% and 0.01% SDS noted growing TMC and EMF in the cells with the preparation "Vostok EM – 1" and with *B. cereus*. In control MFE with 0.05% and 0.01% SDS, but no microorganisms remained EMF respectively at 6 and 3 mV. During the period from 0 to 8 hours in the anode chamber with 0.05% SDS and preparation "Vostok EM – 1" and *B. cereus* respectively TMC raised with $8.60 \pm 0.2 \times 10^5$ to $6.47 \pm 0.2 \times 10^6$ CFU/ml and $6.08 \pm 0.2 \times 10^5$ to $1.52 \pm 0.2 \times 10^6$ CFU/ml (Fig. 3). EMF index in MFC during the same period ranged from 45 ± 6.8 to 78 ± 11.7 mV with EM preparation, and cells with *B. cereus* from 19 ± 2.8 to 25 ± 3.7 mV. 10 hours from the beginning of the experiment the number of microorganisms was $6.64 \pm 0.2 \times 10^6$, and the voltage is 82 ± 12.3 for MFC with "Vostok EM – 1", through the day – TMC $9.11 \pm 0.2 \times 10^6$ and EMF 115 ± 17.3 mV. In experiments with *B. cereus* after 10 hours with the same concentration of SDS microorganisms reached $1.67 \pm 0.2 \times 10^6$ and a day later – $2.84 \pm 0.2 \times 10^6$ CFU/ml and the voltage through 10 hours was respectively 27 ± 4.0 and by $24-38 \pm 5.7$ mV.

At a concentration of 0.01% SDS TMC in cells both with EM preparation, and with *B. cereus* in comparison with 0.05% SDS remained at virtually the same level. The number of microorganisms from 0 to 24 hours increased from $5.50 \pm 0.2 \times 10^4$ to $2.12 \pm 0.2 \times 10^6$ CFU/ml for the preparation of EM and $1.18 \pm 0.2 \times 10^5$ to $1.72 \pm 0,2 \times 10^6$ CFU/ml for *B. cereus* (Fig. 4). In cells with the preparation "Vostok EM – 1" in the presence of 0.01% SDS voltage increased from 98 ± 14.7 to 306 ± 45.9 mV, and with *B. cereus* – 16 ± 2.4 to 130 ± 19.5 mV.

4 CONCLUSION

Thus the results shown perspective using the preparation "Vostok EM – 1" and isolated from it strain *B. cereus* as bioagents for MFC. SDS at concentrations of 0.1%, 0.05% and 0.01% inhibited electricity production and the growth of microorganisms when used as a preparation "Vostok EM – 1", and *B. cereus*.

REFERENCES

Bond, D. R.; Lovley, D. R. 2003. Electricity production by Geobacter sulfurreducens attached to electrodes. Appl. Environ. Microbiol., 69, 1548–1555.

Bullen, R.A., Arnot, T.C., Lakeman, J.B., Walsh, F.C. 2006. Biofuel cells and their development. Biosens. Bioelectron. 21, 2015–2045.

Gorby, Y. A.; Beveridge, T. J. 2005. Composition, reactivity, and regulation of extracellular metal-reducing structures (nanowires) produced by dissimilatory metal reducing bacteria. Presented at DOE/NABIR meeting, April 20, Warrenton, VA.

Gorby, Y. A.; Yanina, S.; McLean, J. S.; Rosso, K. M.; Moyles, D.; Dohnalkova, A.; Beveridge, T. J.; Chang, I. S.; Kim, B. H.; Kim, K. S.; Culley, D. E.; Reed, S. B.; Romine, M. F.; Saffarini, D. A.; Hill, E. A.; Shi, L.; Elias, D. A.; Kennedy, D. W.; Pinchuk, G.; Watanabe, K.; Ishii, S.; Logan, B. E.; Nealson, K. H., Fredrickson, J. K. 2006. Electrically conductive bacterial nanowires produced by Shewanella oneidensis strain MR-1 and other microorganisms. PNAS, July 25, vol. 103 no. 30 11358–11363.

Higa, T. 1996. An Earth Saving Revolution (Vol. 1: 335 p, Vol. 2 – 367 p) Sunmark Publishing; 2nd edition.

Jusoh, M.L., Manaf, L.A., & Latiff, P.A. 2013. "Composting of rice straw with effective microorganisms (EM and its influence on compost quality." Iranian Journal of Environmental Health Science Engineering, 10(1): 17.

Parr, J.F. 1994. For a sustainable agriculture and environment, International Nature Farming Research Center Atami, Japan.

Rabaey, K.; Boon, N.; Siciliano, S. D.; Verhaege, M.; Verstraete, W. 2004. Biofuel cells select for microbial consortia that self-mediate electron transfer. Appl. Environ. Microbiol., 70, 5373–5382.

Rabaey, K.; Boon, N.; Hofte, M.; Verstraete, W. 2005. Microbial phenazine production enhances electron transfer in biofuel cells. Environ. Sci. Technol., 39, 3401–3408.

Rabaye, K. and Verstraete, W. 2005. Microbial fuel cells: novel biotechnology for energy generation. Trends in Biotechnology. Vol. 23, No. 6, pp. 291–298.

Szymanski, N. and Patterson, R.A. 2003. "Effective microorganisms (EM) and wastewater systems." Future Directions for On-site System: Best Management Practice Proceedings of On-site '03 Conference. 348–355.

Ting, A. S. Y., Rahman N. H. A., Isa, M. I. H. M., Tan, W. S. 2013. "Investigating metal removal potential by Effective Microorganisms (EM) in alginate-immobilized and free-cell forms." Bioresource Technology, 147: 636–639.

Стом Д.И., Коновалова Е.Ю., Пономарева А.Л., Протасов Е.С., Толстой М.Ю. 2013. Использование в микробных топливных элементах штаммов, изолированных из препарата «Восток». Известия Самарского научного центра Российской академии наук, том 15, №3(3), С. 1153–1156.

A new type of tilting mechanism for tilting tray sorter

Mingqian Du, Xiuqing Yang, Yilong Qian & Lin Huang
Department of Technology Research and Development, Civil Aviation Logistics Technology Company Limited, Chengdu, China

ABSTRACT: The tilting tray sorter is the terminal core equipment of the baggage handling system to implement the baggage high speed automation sorting, and the tilting mechanism is the core component of the tilting tray sorter. The traditional tilting mechanism is to realize the baggage sorting by mechanical mode with low baggage handling efficiency and wide noise. A new kind of electric tilting device has been designed to meet the requirements of baggage sorting with the high handling efficiency of 5000–6000 bph (baggage per hour) in this paper. This mechanism can implement the self-hold in the initial and terminal position by the guiding roller and guiding groove framework, thus it not only increases the mechanism reliability and avoids the impact force passed to the tilting motor, but only realizes the no-load starting and braking of motor and decreases the peak current of motor to save energy.

Keywords: Tilting mechanism; Guiding roller and guiding groove; Baggage sorting system; Civil aviation airport

1 INTRODUCTION

With the quick increase of airport passengers, it puts forward higher requests to the airport baggage handling system, especially to the baggage sorting equipment in order to meet the requirements of baggage sorting, and the tilting tray sorter is a kind of advanced equipment of baggage handling system to implement the baggage high speed automation sorting at present, its sorting efficiency is about 5000–6000 bph [1]. This equipment shown in Figure 1 is composed of sorter frame, carriage, tilting mechanism, tray components, drive and control system and so on; the tilting mechanism is the core component of all to decide the sorting efficiency. The tilting mechanism [2] is to realize the baggage sorting by mechanical mode, that is to set the trigger mechanism at sorting chute, the trigger mechanism is stroked by the electric device to trigger the tilting mechanism to implement the baggage sorting when the baggage passes the scheduled sorting chute, but this kind of mechanism is to realize baggage sorting by mechanical contact mode, so the noise is big, what's more the triggering process is long, which results in low baggage sorting efficiency. The tilting mechanism [3] is to realize the baggage sorting by electric mode, but it doesn't implement the mechanism self-locking at the start and the end. The tilting mechanism [4] is to realize the baggage sorting by electric mode, but it is driven by turbine worm motor, the power consumption of mechanism is high because of the low transmission efficiency of turbine worm.

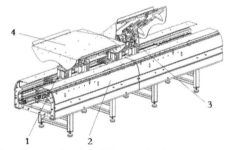

1 carriage 2 sorter frame 3 tilting mechanism 4 tray components

Figure 1. The structure of tilting tray sorter.

A new type of electric tilting device has been designed to meet the requirements of baggage sorting with the high handling efficiency of 5000–6000 bph in this paper. This mechanism can implement the self-hold in the initial and terminal position by the guiding roller and guiding groove framework, thus it not only increases the mechanism reliability and avoids the impact force passed to tilting motor, but only realizes the no-load starting and braking of motor and decreases the peak current of motor to save energy. Besides the noise of this designed mechanism and damage to baggage are small, and the sorting efficiency is higher than that of mechanical mode, so this innovational device can be better to

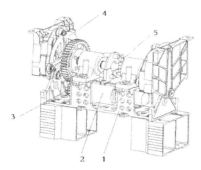

1 the base 2 tilting drive mechanism 3 gear transmission mechanism 4 guiding roller and guiding groove mechanism 5 movement detection device

Figure 2. The structure of tilting mechanism.

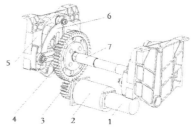

1 motor 2 motor reducer 3 small gear 4 big gear 5 guiding plate 6 transmission fork 7 principal axis

Figure 3. Transmission principle diagram of tilting mechanism.

meet the baggage sorting requirements of large and medium-sized airports.

2 GETTING STARTED GENERAL SCHEME OF TILTING MECHANISM

The speed of carriage is required to be about 2 m/s by calculation in order to meet the baggage sorting efficiency, the width of the sorting chute is usually about 2 m, so the tilting time of tilting mechanism must be less than 1 s to ensure exact sorting of baggage, namely the baggage is sorted to the scheduled chute accurately.

According to the function requirements of tilting mechanism, the designed device mainly includes tilting drive mechanism, gear transmission mechanism, guiding roller and guiding groove mechanism, movement detection device and the base, shown in Fig. 2. The drive mechanism includes motor and motor reducer, the gear transmission mechanism includes small gear, big gear, principal axis, bearing and bearing base, the guiding roller and guiding groove mechanism includes guiding plate and transmission fork, movement detection device includes proximity switch, induction block and so on.

3 THE DESIGN OF TRANSMISSION SYSTEM

The design of transmission system is the core. It covers the transmission mode, transmission structure, transmission ratio and so on. The transmission system of tilting mechanism is composed of drive mechanism, gear transmission mechanism, guiding roller and guiding groove mechanism, shown in Fig. 2.

The transmission principle diagram of tilting mechanism is shown in Fig. 3. First the motor drives the motor reducer, the motor reducer drives the small gear, the big gear is driven by meshing with the small gear, and then the fork mounted in the principal axis is turned, the guiding rollers installed in the fork move along the guiding groove mounted in the guiding plate, thus the guiding plate start to move.

From what has been discussed above, we can conclude the rotation angle of guiding plate θ:

$$\theta = N*360/60*T/I_1/I_2/I_3 \qquad (1)$$

In the formula (1), θ—rotation angle of guiding plate; N—rotation speed of motor, unite is rpm; I_1—the radio of motor reducer; I_2—the ratio of gear transmission mechanism; I_3—the ratio of guiding roller and guiding groove mechanism; T—the tilting time of tilting mechanism, unite is s.

From the formula (1), we can get I_3:

$$I_3 = N*360/60*T/I_1/I_2/\theta \qquad (2)$$

According to the function requirement of tilting tray sorter, set $\theta = 40°$. According to the speed requirement of tilting mechanism and the selected motor model, get N = 2800 rpm. According to the design requirement and the space layout of mechanism, take $I_1 = 35$, $I_2 = 2.5$. Given $T \leq 1$, shown in the section of general scheme of tilting mechanism. Take the above dates into formula (2), we can conclude I_3:

$$I3 \leq 4.8 \qquad (3)$$

The determination of I_3 is shown in the section of the design of guiding roller and guiding groove mechanism.

4 THE DESIGN OF GUIDING ROLLER AND GUIDING GROOVE MECHANISM

The guiding roller and guiding groove mechanism is the key component of the tilting mechanism, it can not only realize the self-hold in the initial and terminal position to increase mechanism reliability and avoid impact force passed to tilting motor, but only implement the no-load starting and braking of motor to decrease the peak current of motor to save energy. The guiding plate includes three grooves, the fork includes three guiding rollers. When the fork turns, the guiding

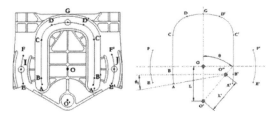

Figure 4. The structure of guiding plate and the design curve of guiding groove.

plate turns by guiding roller of fork moving along with the guiding groove of guiding plate.

The design of guiding plate is a key and difficult point, especially the design of guiding groove. The innovative design of guiding plate and guiding groove are shown in Fig. 4. When the three guiding rollers are respectively on the circular arc DD', EF and E'F', the tilting mechanism is self-locking, so the baggage impact force can't pass to the motor at this time. The circular arc DD', EF and E'F' is on the concentric circles, and the circle center is point O, so the guiding plate doesn't move when the rollers of fork move along the circular arc DD', EF and E'F' respectively. Circular arc CD and DD' is tangent, straight line segment BC is tangent to circular arc CD and AB respectively in order to realize the move smooth of guiding plate. The circle center of circular arc AB and the instantaneous movement circle center of fork are concentricity, so the guiding plate doesn't move when the roller of fork moves along with the circular arc AB. The guiding plate can implement the positive and reverse rotation by the guiding roller of fork moving along with the circular arc GA and GA' respectively.

From the design curve of guiding groove shown in Fig. 4, we can conclude the rotation angle of fork θ':

$$\theta' = 90 + \theta_1 + \theta \quad (4)$$

In the formula (4), θ_1—rotation angle of fork at the end of self-locking; θ—rotation angle of guiding plate.

According to the function requirement of tilting mechanism, set $\theta_1 = 10$. Given $\theta = 40$, shown in the section of the design of transmission system. Take the above dates into formula (4), we can get θ':

$$\theta' = 90 + 10 + 40 = 140 \quad (5)$$

So

$$I_3 = \theta'/\theta = 140/40 = 3.5 \leqslant 4.8 \quad (6)$$

From Fig. 5 we can see that (a) shows the initial position of fork and guiding plate, the mechanism is self-hold at this moment; (b) shows that the fork turns but guiding plate doesn't turn, it indicates the rollers of fork moving along with circular arc DD', EF and E'F' respectively to realize the no-load starting of the tilting motor; (c) shows that the fork continues to turn, and guiding plate starts to turn, the left and right rollers of

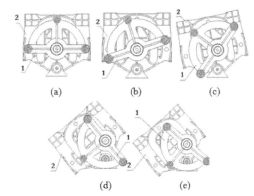

1 transmission fork 2 guiding plate

Figure 5. The motion state diagram of guiding roller and guiding groove mechanism.

Figure 6. The 3D model of tilting mechanism.

fork are out of the circular arc EF and E'F' respectively; (d) shows that the above roller of the fork moves to point B; (e) shows the state of the end of self-locking, the fork turns but guiding plate doesn't turn, it indicates the above roller of fork moving along with circular arc AB to implement the no-load braking of the tilting motor.

5 THE SIMULATION ANALYSIS OF TILTING MECHANISM

The 3D model was built by PRO/E software, shown in Fig. 6. Kinematics of tilting mechanism was simulated by PRO/E simulation module, and the displacement, velocity and acceleration of the end component guiding plate were mainly made in attention. The simulation results are shown in Fig. 7. It can be seen obviously that the design concept of guiding roller and guiding groove mechanism was proved right by the simulation results of tilting mechanism, such as self-locking, no-load starting and braking and so on.

From the displacement curve, we can see that the rotation angle of guiding plate increases from 0° to 40° at the period of 0.08 s to 0.69 s, while the angle is kept 0° from 0 s to 0.08 s and the angle is kept 40° from 0.69 s to 0.74 s; From the velocity curve, we can see that the rotation velocity of guiding plate increases

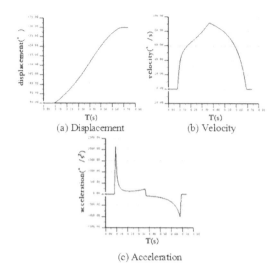

(a) Displacement (b) Velocity

(c) Acceleration

Figure 7. The simulation results of the end component guiding plate.

Figure 8. The physical prototype.

from 0°/s to 91°/s at the period of 0.08 s to 0.36 s and decreases from 91°/s to 0°/s at the period of 0.36 s to 0.69 s while the rotation velocity is kept 0°/s from 0 s to 0.08 s and from 0.69 s to 0.75 s; the acceleration of guiding plate is matching to displacement and velocity, so the detailed description isn't done at this. From what has been discussed above, we can conclude that the simulation result is consistent with the design concept of tilting mechanism.

6 THE PROTOTYPE TESTING OF TILTING MECHANISM

The physical prototype of tilting mechanism is shown in Fig. 8. The testing results indicate the tilting mechanism can meet the given requirement, for example the maximum carrying load of mechanism is about 60 Kg, and the tilting time is about 0.74 s less than 1 s. The current and voltage testing curves are shown in Fig. 9.

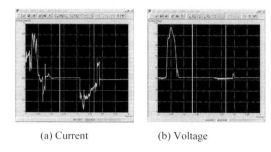

(a) Current (b) Voltage

Figure 9. The current and voltage testing curves.

7 CONCLUSION

The design of a new type of tilting mechanism was described in this paper. This kind of mechanism can implement airport baggage high-speed and automation sorting; it has many merits compared with the tilting mechanism of mechanical mode, such as small noise, high sorting efficiency and so on; it has also many merits compared with the tilting mechanism of electric mode, such as high reliability, energy saving and so on. This tilting mechanism was proved effective and can meet the given requirement by simulation analysis and prototype testing; what's more the mechanism has been batch production and applied the engineering project, besides this mechanism can apply in other areas by modifying slightly.

ACKNOWLEDGEMENT

The researches are supported by the project of "Special project of the national ministry of science and technology (2010EG125229)", "Science and technology support plan of Sichuan province (12ZC0142)" and "Science and technology major projects of the civil aviation administration of China (MHRD201058)".

REFERENCES

[1] Mao Gang, Huang Rongshun, Luo Xiao, et al. Tilting-tray baggage high-speed and automation sorter. C.N. Patent 201320176829 (2013).
[2] Yang Bo. Drop-down controllable tilting car. C.N. Patent 2513958Y (2002).
[3] Anne-Mette Hjortshoj Abildgaard, Jan Gullov Christensen. Sorting conveyer with a tilting mechanism. U.S. Patent 2004/0079618 A1 (2004).
[4] Heino Heitplatz. Method and apparatus for activating and a tipping shell sorter, as well as a tipping conveyor element. J. U.S. Patent 2004/0069593 A1 (2004).

Industrial, Mechanical and Manufacturing Science – Zheng (Ed.)
© 2015 Taylor & Francis Group, London, ISBN: 978-1-138-02656-8

Research on baggage modeling of airport baggage handling system

Yi Chen, Xiuqing Yang, Yilong Qian & Chengsheng Tang
Department of Technology Research and Development, Civil Aviation Logistics Technology Company Limited, Chengdu, China

ABSTRACT: With the fierce increase of airport baggage throughput it becomes one of the most concerning problems for the airport operator to realize the modeling and simulation of baggage handling system and to find the system question in advance, while the current simulation platform of baggage system is poor universality. Mathematics modeling of baggage is the basis to implement the general simulation, so the baggage action characters are studied and analyzed, the baggage characteristics in common at different airports are discovered, and the baggage general mathematics model is built by analyzing all kinds of distributing function and with the help of MATLAB software in this paper; what's more the built model is proved correct and reasonable by model validation method and theory. This research has laid a solid foundation to realize the general simulation of baggage handling system.

Keywords: General simulation; Mathematics modeling; Baggage handling system; Model validation

1 INTRODUCTION

The airports of more than 10 million passenger throughput reached 13 in 2009 in China, and they are expected to be 30 by 2020 [1]. The fierce increase of passenger throughput certainly will bring great tension and put forward higher requests to the baggage handling system; if the baggage handling system is out of the question resulting in baggage sorting error, plane delay and so on, it will reduce passenger satisfaction seriously and increase the operating expenses of the airport greatly. So it becomes one of the most concerning problems for the airport operator how to realize the modeling and simulation of baggage handling system and to find the system question in advance.

The evaluation and plan of system input was realized by the modeling and simulation of baggage handling system, and the simulation validation was done in the paper [2]. The simulation of baggage handling system of Shanghai Pudong international airport was implemented by the ServiceModel platform, and the simulation object is baggage, the modeling tool is petri net and the simulation strategy of event transfer was used in the simulation platform in the paper [3]. The modeling and simulation of level 5 security system was done by the Flexsim software, and the simulation results were analyzed in the paper [4]. The improved baggage flow was put forward based on the original flow, the simulation model was built by Arena software and the research methods and achievements supply a certain reference value to flow design and optimization of baggage handling system in the paper [5].

The above simulation platforms are all developed for the specific airport, the generality is poor, and a second development must be done if other airports use it. The general mathematics modeling of baggage is the first to realize the general simulation of a baggage handling system, so the baggage action characters are studied and analyzed, the baggage characteristics in common at different airports are discovered, and the baggage general mathematical model is built by analyzing all kinds of distributing function and with the help of MATLAB software in this paper, what's more the validation method and theory of model are explored and built model is proved correct and reasonable by model validation method and theory.

2 RESEARCH AND ANALYSIS OF BAGGAGE CHARACTERS

All kinds of baggage characters are analyzed, the characters include change trend of baggage check-in number, density distribution of baggage check-in, time interval of adjacent baggage check-in, time porosities of baggage check-in and so on. These characters are the foundation to build the common mathematics model of baggage.

2.1 *Character of change trend of baggage check-in number*

This character shows that the change trend of baggage check-in number within a day, the increase trend and

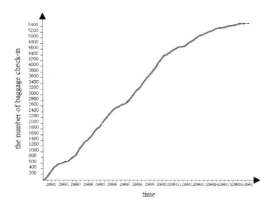

Figure 1. The curve of change trend of baggage check-in number.

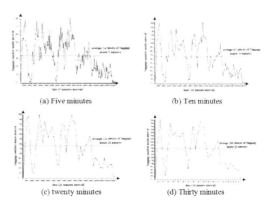

Figure 2. The curve of density distribution of different time interval.

degree of baggage check-in total number can be got and busy degree of baggage check-in work can be estimated by analyzing the slope change and change time moment of character curve.

The curve of change trend of baggage check-in number within a day of a certain airport is shown in Fig. 1. From Fig. 1 we can conclude that the number increase of baggage check-in becomes slow after two hours of morning starting check-in, and the slope change becomes small obviously after ten hours, so we can think that the work busy degree of baggage check-in becomes lower clearly.

2.2 Character of density distribution of baggage check-in

This character indicates that change trend and distribution regularity of baggage, and the baggage distribution mathematics model can be built by studying and analyzing the baggage density distribution. The distribution curve of discrete value is different according to the set different time interval, so we study the character of density distribution by setting four kinds of time interval which is five minutes, ten minutes, twenty minutes and thirty minutes, the baggage density distribution of a variety of situations are shown in Fig. 2.

From Fig. 2 we can find that the statistical points increase, the character curve shows more detail, and the overall trend performance is poorer if the time interval is shorter, on the contrary the statistical points decrease, the character curve shows less detail, and the overall trend performance is stronger if the time interval is longer, but the overall trend of curve doesn't change, that is the change trend of density distribution isn't affected by time interval.

2.3 Character of time interval of adjacent baggage check-in

The character curve of time interval of adjacent baggage check-in is shown in Fig. 3. Abscissa and ordinate

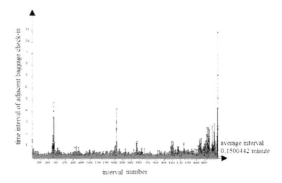

Figure 3. The character curve of time interval of adjacent baggage check-in.

indicates interval number and specific time interval respectively. The curve was formed by counting the interval time of adjacent baggage check-in and tracing point in the graph. We also side know the busy and free state of baggage check-in as well as the free time and free degree.

2.4 Character of time porosities of baggage check-in

We can get the busy and free degree of baggage check-in within a day by studying this character. The character curve of time porosities of baggage check-in is shown in Fig. 4. From Fig. 4 we can see that there is an obvious white bar between first hour and second hour after baggage check-in, it indicates free time period, in the same way there are also white bars between fourteenth hour and seventeenth hour. The other black area shows the busy time period of baggage check-in, and there is most of the busy time period in Fig. 4. In addition there are two gray bars in the middle of the figure which indicates baggage check-in frequency is between busy and free.

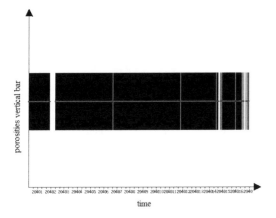

Figure 4. The character curve of time porosities of baggage check-in.

3 THE BUILDING OF BAGGAGE COMMON MATHEMATICS MODEL

The baggage general mathematics model is built with the help of MATLAB software after analyzing and studying the baggage action character, discovering the common character of baggage of different airports, comparing and analyzing all kinds of distributing function. The mathematical model shown in formula (1) includes six function expressions and three kinds of distribution function which are logarithmic normal distribution function, normal distribution function and logarithmic normal distribution function of absolute value, and that the number of normal distribution function is four.

$$\begin{aligned}
M = & A_{le} \times \frac{1}{\sqrt{2\pi}\sigma_{le} \times p_{le} \times (x - z_{le})} e^{-\frac{(\ln[p_{le} \times (x - z_{le})] - \mu_{le})^2}{2\sigma_{le}^2}} \\
& + A_{nm1} \times \frac{1}{\sqrt{2\pi}\sigma_{nm1}} e^{-\frac{[p_{nm1} \times (x - \mu_{nm1})]^2}{2\sigma_{nm1}^2}} \\
& + A_{nm2} \times \frac{1}{\sqrt{2\pi}\sigma_{nm2}} e^{-\frac{[p_{nm2} \times (x - \mu_{nm2})]^2}{2\sigma_{nm2}^2}} \\
& + A_{nm3} \times \frac{1}{\sqrt{2\pi}\sigma_{nm3}} e^{-\frac{[p_{nm3} \times (x - \mu_{nm3})]^2}{2\sigma_{nm3}^2}} \\
& + A_{nm4} \times \frac{1}{\sqrt{2\pi}\sigma_{nm4}} e^{-\frac{[p_{nm4} \times (x - \mu_{nm4})]^2}{2\sigma_{nm4}^2}} \\
& + A_{AbsLe} \times \frac{1 \times u(x - z_{AbsLe})}{\sqrt{2\pi}\sigma_{AbsLe} \times p_{AbsLe} \times |x - z_{AbsLe}|} e^{-\frac{[\ln[p_{AbsLe} \times |x - z_{AbsLe}|] - \mu_{AbsLe}]^2}{2\sigma_{AbsLe}^2}}
\end{aligned} \quad (1)$$

In the formula (1), M—baggage number distribution; σ_{le}—shape parameter 1 of logarithmic normal distribution; μ_{le}—shape parameter 2 of logarithmic normal distribution; z_{le}—abscissa offset of logarithmic normal distribution; p_{le}—scale coefficient of logarithmic normal distribution; A_{le}—amplitude efficient of logarithmic normal distribution; σ_{nm1}—shape parameter of normal distribution function 1; μ_{nm1}—abscissa offset of normal distribution function 1; p_{nm1}—scale coefficient of normal distribution function 1; A_{nm1}—amplitude efficient of normal distribution function 1; σ_{nm2}—shape parameter of normal distribution function 2; μ_{nm2}—abscissa offset of normal distribution function 2; p_{nm2}—scale coefficient of normal distribution function 2; A_{nm2}—amplitude efficient of normal distribution function 2; σ_{nm3}—shape parameter of normal distribution function 3; μ_{nm3}—abscissa offset of normal distribution function 3; p_{nm3}—scale coefficient of normal distribution function 3; A_{nm3}—amplitude efficient of normal distribution function 3; σ_{nm4}—shape parameter of normal distribution function 4; μ_{nm4}—abscissa offset of normal distribution function 4; p_{nm4}—scale coefficient of normal distribution function 4; A_{nm4}—amplitude efficient of normal distribution function 4; σ_{AbsLe}—shape parameter 1 of logarithmic normal distribution of absolute value; μ_{AbsLe}—shape parameter 2 of logarithmic normal distribution of absolute value; z_{AbsLe}—abscissa offset of logarithmic normal distribution of absolute value; p_{AbsLe}—scale coefficient of logarithmic normal distribution of absolute value; A_{AbsLe}—amplitude efficient of logarithmic normal distribution of absolute value.

The above mathematic model is a complex and plus model which is composed of several simple distribution functions, so the computing excess complexity of model is avoided, what's more the function of model is independent of each other, the baggage distribution situation of different airport can be expressed by modifying related parameters or related functions, therefore this mathematics model has strong commonality.

4 SIMULATION VALIDATION OF BAGGAGE COMMON MATHEMATICS MODEL

We select several airports randomly to validate the accuracy and feasibility of baggage common mathematics model. During the process of validation, the programming must be made at first in order to realize the synchronization between function parameter modification and the change of fitting curve.

The test method, expression, the input and output and so on are studied in order to estimate the rationality of the fitting curve after parameter modification, here we adopt the input, the output and test method as follows.

The input X = The dependent variable value of actual curve. The dependent variable value of fitting curve.

The output and its expression:
Mean value:

$$E(X) = \frac{\sum_{i=0}^{n} X_i}{n} \quad (2)$$

Mean square error:

$$D(X) = \sqrt{\sum_{i=0}^{n} (X_i - E(X))^2} \quad (3)$$

Return value:

$$reg(X) = \frac{\sqrt{\sum_{i=0}^{n} X_i^2}}{n} \quad (4)$$

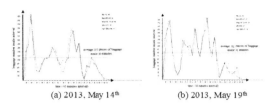

Figure 5. The sample fitting result of Tianjin airport of two days.

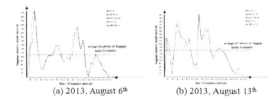

Figure 6. The sample fitting result of Qingdao airport of two days.

In the formula (2), (3) and (4), n—the number of discrete point.

We make a program to realize the computing of verification expression which is to get the output value by inputting the input value, result value display and difference curve display, what's more it is synchronous between result display and parameter modification, and the synchronization of fitting effect and verification result is implemented. The baggage fitting curves of some specific airports are shown in Fig. 5 and Fig. 6.

From the above figures we can see that the fitting curve of mathematics model and the actual curve of discrete points are almost the same, so we can conclude that the baggage general mathematics model is correct and reasonable.

5 CONCLUSION

The current simulation platforms of baggage handling system are poor generality, they are developed for the specific airport, and a second development must be done if other airports use it. The general mathematics modeling of baggage is the first to realize the general simulation of baggage handling system, and it is also the most difficult to achieve. In this paper the baggage action characters are studied and analyzed, the baggage characteristics in common at different airports are discovered, and the baggage general mathematical model is built by analyzing all kinds of distributing function and with the help of MATLAB software; what's more the validation method and theory of model are explored and the built model is proved correct and reasonable by model validation method and theory. This research has laid a solid foundation to realize the general simulation of baggage handling system.

ACKNOWLEDGEMENT

The researches are supported by the project of "Special project of the national ministry of science and technology (2010EG125229)", "Science and technology support plan of Sichuan province (12ZC0142)" and "Science and technology major projects of the civil aviation administration of China (MHRD201058 and MHRD0719)".

REFERENCES

[1] Information on http://www.caac.gov.cn/.
[2] Vu T. Le, Dr Doug Creighton, et al. Simulation-based Input Loading Condition Optimisation of Airport Baggage Handling Systems, Proceedings of the IEEE Intelligent Transportation Systems Conference, USA, 2007, pp. 574–579.
[3] Lu Xun, Zhu Jinfu, Tang Xiaowei. Airport Luggage Process Modeling and Simulation. Journal of system simulation, 2008, 20(14): 3876–3880.
[4] Xu Xiaohao, Wu Lili, Yuan Xuegong. Simulation Modeling and Analysis of Airport Baggage Five-Level Security Check System. Journal of Civil Aviation University of China, 2007, 25(5): 15–17.
[5] Yang Peng, Sun Junqing, Zhao Hui, Quan Xiongwen. Simulation Based Airport Luggage Handling Pattern Evaluation. Computer Engineering and Applications, 2011, 47(2): 217–219.

Design of automotive networking gateway based on telematics

Z.H. Dong & Z.L. Zong
Research Institute of Electronic Science and Technology, University of Electronic Science and Technology of China, Chengdu, China

ABSTRACT: This article describes the use of the Ethernet as the backbone to contract CAN networks as automotive control, Wi-Fi as vehicle inside communication and GPRS joining the controlling of center outside. After analysis, design, programming and debugging, the author realized the conversion module mainly, with better transmission performance to meet the needs.

Keywords: DSP; Ethernet; Gateway

1 INTRODUCTION

With the increasing popularity of the car, pollution and traffic congestion has become a serious problem. So researching how to control the number of cars on the road and regional distribution is very important. The use of computer and communication technology greatly enhances the safety and efficiency of future transport systems [1]. Therefore, the car could exchange data with the information platform. Moreover, entertainment system, security reversing radar and car voice calls are needed. In addition, the control system of the vehicle includes LIN-BUS, CAN-BUS, MOST-BUS and other standards. Various buses live on their own. That not only increases the cost, but also causes a safety hazard. The Ethernet has large amount of users, high communication speed and a lot of soft, hardware resources, etc. Through technical improvements, the industrial Ethernet network is expected to integrate a variety of vehicles so as to achieve complete interoperability.

Based on these reasons, the author chose Ethernet as a core agreement with CAN/GPRS/Wi-Fi as protocol conversion objects. Such agreements include the main vehicle of communication and control systems to better adapt to the basic requirements for future vehicle of things.

2 HARDWARE DESIGN

2.1 *Core chip selection*

It is known that the highest rate of CAN-BUS is 1 Mbps, GPRS is 171.2 Kbps, and Wi-Fi as a vehicle of communication can be set to a minimum of 1 Mbps. Then the total bandwidth demand is less than 2.5 Mbps. According to previous experience, the bandwidth of Ethernet load limits is 25%, so using 10 Mbps Ethernet chip can meet the basic needs, and bi-directional bandwidth of 10 Mbps can fully do.

F2812 has an eCAN module to connect CAN-BUS [2]. The dual serial ports helped the author select UART-GPRS and UART-Wi-Fi module with "AT+" commands on the market: GPRS module—GTM900B and Wi-Fi module—HLK-WIFI-M03. The hardware design also includes a memory expansion, power, reset, JTAG emulation, clock and so on.

2.2 *Ethernet chip selection*

Ethernet backbone builds a network connecting the various nodes, so it is the key factor related to whether successfully and timely communication or not. Therefore, it is the second important part only after the core chip.

RTL8019AS is designed by company of REALTEK in Taiwan and is a highly integrated full-duplex Ethernet controller chip, especially suitable for low-power DSP systems. Because of its excellent performance and low price, it takes over a considerable proportion of the market in 10 Mbps ISA bus card.

2.3 *Circuit design*

A main architecture is shown in Figure 1. ISSI61 LV6416 is an external memory, and SN65HVD260 is a CAN-BUS transceiver. Three DSP peripheral interfaces are connected to the three peripherals. Because two wireless modules work at RS232 level, so TTL level needs to be converted.

3 SYSTEM SOFTWARE DESIGN

3.1 *The main program flow chart*

In order to ensure sending and receiving data of certainty and accuracy, it is necessary to initialize the

Figure 1. Circuit Schematic.

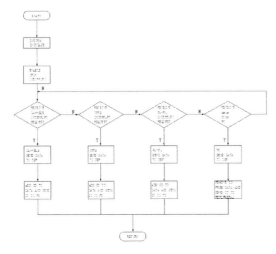

Figure 2. Circuit Schematic.

system configuration. According to TMS320F2812 memory map, the system firstly configures program space and data space; secondly defines the system stack, static variables, and external variables; thirdly creates the interrupt vector table; at last, initializes data registers, control registers, status registers and I/O ports.

Data transceiver in three protocols are all dependent on the single Ethernet cable between the PC and the DSP, so we must add the communication ID string "eCAN", "WiFi" and "GPRS" before data to show the difference.

Because Ethernet is the backbone network and has the largest amount of data, it is on the main program inside to check the data received from the PC and works in query mode. After receiving complete information, DSP removes ID and starts the corresponding peripheral transmitting interrupt and sends useful data. Conversely, when the peripheral data starts the receiving interrupt, ID is immediately added by DSP and sends frames to PC via Ethernet.

The importance of different peripherals is clearly not alike. In general, related to the mechanical control, CAN-BUS has the highest priority; GPRS communication with the external data, followed by; voice control Wi-Fi is the lowest. Therefore, the three peripheral interrupt priorities can be set accordingly. The main program flow chart is shown in Figure 2.

3.2 Protocol conversion between CAN-BUS and Ethernet

When the device located in the CAN-BUS needs sending data to a device in Ethernet, the data will be encapsulated by the CAN protocol format and then sent to the protocol gateway. After the protocol gateway receives messages, it will extract useful information from the application layer and store that into a buffer. When the data accumulates to a certain length, the gateway will add TCP or UDP header, IP header and Ethernet header in sequence and send them to PC through Ethernet together. Conversely, when the device in Ethernet sends data to that in CAN-BUS, the sequence is opposite. If the application layer data length is greater than 8, then 8 bytes of the data needed to be split for a group.

3.3 Protocol conversion between serial interface and ethernet

Because the serial modules can achieve transparent transmission mode, protocol conversion between Ethernet and GPRS or Wi-Fi will only need to be considered the conversion of serial interface. This article uses UDP transport protocol to send data into SCI_FIFO rather than TCP which is full of complex transmission mode.

Serial protocol conversion is similar to the CAN-BUS, but the lower speed of conversion is the bottleneck of communication.

4 SYSTEM TESTING AND ANALYSIS

4.1 Testing program

To facilitate the testing and data analysis, the author used a PC and a DSP gateway to achieve 2 protocols (CAN-BUS and Wi-Fi) interconnected, while GPRS is connected to the network server. The needed condition is as follows:

– One PC.
– One Ethernet card made by REALTEK and a cable.
– USB-CAN adapter made by Power Technology Co. Ltd., Wuhan Endeavour.
– Wi-Fi test needs a mini wireless transceiver made by Hi-Link company to match Wi-Fi modules working in Ad-Hoc mode.
– A web server with a fixed IP to communicate with GPRS module.
– LabVIEW programs for each device drivers, timing control, and data preservation.

During the testing phase, each individual test is only for one peripheral: author sent data through the network cable from the PC to the DSP, and then transmitted it through certain peripheral; while the corresponding device received the data, author contrasted it with that original data in PC. Similarly, the need to test in opposite direction was the reverse line of protocol conversion.

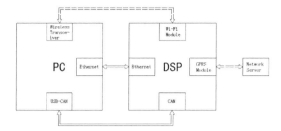

Figure 3. System Test Schematic.

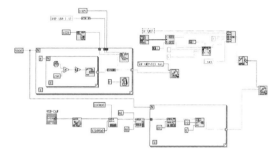

Figure 4. CAN-Ethernet Testing.

4.2 Classification test

4.2.1 Testing ethernet and CAN-BUS

According to references data in the communication is rate up to 250 Kbps, and the maximum bandwidth loss is about 500 frames/sec [3]. Therefore, the authors set the cycle time to 2 milliseconds. Below figure means sending data from the USB-CAN adapter to the DSP goes through the CAN-BUS, and forward to the PC with UDP connections. The author stored transmission data into the file "UDP-CAN", and the received one into "UDP2CAN". Then author compared the two files with WinHex tools to find out whether there was any error or packet loss. Transmitting data from Ethernet to CAN-BUS is similar.

4.2.2 Testing Ethernet and Wi-Fi

It is basically the same as the above test because of establishing a LAN connection between the PC and the DSP. The main differences are these:

- CAN-BUS is a wired connection, while Wi-Fi is a wireless connection.
- Because the PC has installed two cards, users need to set the IP addresses of two different network segments in order to avoid conflict.
- AT commands need to be used before connection, and then the module can enter the transparent mode for data transfer.

4.2.3 Testing Ethernet and GPRS

Because China Mobile has shut up the function of point to point for GPRS, therefore, GPRS couldn't be configured as a wireless ad-hoc network like Wi-Fi. Meanwhile, GPRS is assigned a dynamic IP: if no data transmission over time, users need to reconnect between two GPRS devices and find each other's IP. The usual solution is to find a computer with a fixed IP to connect. The author rented a fixed IP server, ran "PDU & UDP Test Tool" software on the specified port, and established a TCP connection on it. In the server, the author chose the option "Return the data received". Thus, GPRS module sent data to the server, but it bounced and sailed upstream to the PC through the DSP. At last, the author compared the original data and received ones.

4.3 Analysis of test results

The author used WinHex to open the sending and receiving texts, and started the comparison function in order to calculate errors and missing frames. After those tests, the correct rate can reach more than 98%, so the gateway could meet the actual needs.

5 SUMMARY

This paper describes the gateway which contains TMS320F2812 as the core chip to realize Ethernet, CAN-BUS, Wi-Fi and GPRS connectivity as well as proposes software and hardware implementation along with test methods. Through experiments it proves that this gateway can achieve the data transfer with basically meeting the design objectives.

REFERENCES

[1] Sheng, J.Z. & Ning, T.X. 2012. Architecture analysis of Telematics and its application in intelligent transportation systems. Internet of things technologies.
[2] Gu, W.G. 2011. Teaching you how to learn DSP hand by hand.
[3] Chu, D.F. 2011 Key technology research of bus driving stability.

Processing and simulation of thin-walled shell with inner ribs formed by inner spinning technology

Zhenjie Wang & Baofeng Guo
Education Ministry Key Laboratory of Advanced Forging and Stamping Science and Technology, College of Mechanical Engineering, Yanshan University, Qinhuangdao, Hebei, China

Shicheng Ma
Special Material and Technic Institution of Aerospace, Beijing, China

ABSTRACT: The formation of a thin-walled shell structure with inner ribs by inner spinning technology was simulated using the finite element simulation software ABAQUS. The spinning force and deformation condition were calculated and analyzed in this simulation. Results indicate that a thin-walled cylindrical shell with uniform deformation can be obtained using the inner spinning process with good forming effect. For all the regions, the radial load was found to be greater than the axial load, with the radial load in the range of 0–270 kN. The skin and the back frame sections experience load peaks, which is in the expansion capacity of −200 mm for the skin section. The peak load in the rear frame section is stable in a 15 mm stroke, which then gradually decreases to zero. The forming load in the front frame increases with the increase in the feed load of the spinning wheel, leading to the possibility of fracture in this region. The axial spinning load is rather small in the range of 0–47 kN. Based on the simulated values of these parameters, we demonstrated the formation of good quality 2A12 aluminum alloy products using the inner spinning technology.

Keywords: inner Spinning; simulation; thin-walled shell with inner ribs; processing

1 INTRODUCTION

Spinning is an important manufacturing technology that is widely used in various industries, including the aviation, aerospace, shipbuilding, automobile, and engineering machinery industries [1–3]. The structure of the thin-walled shell with transverse ribs is rather complex, consisting of surfaces with curved generatrix and spheroidicity. In general, the wall thickness of a workpiece is between 1.5 mm to 3 mm, with the wall thickness tolerance zone of the order of 0.2 mm. Similarly, the diameter of a workpiece is generally between Φ360 mm to Φ570 mm, with the diameter tolerance zone of 0.4 mm. Typically, processing of such workpieces is often associated with issues related to forming difficulty and low-yield. But, given the current requirements of military equipment technology, it is of critical need to develop suitable methodologies to process workpieces to form thin-walled shell structure with inner ribs.

Therefore, in this study, we have analyzed the use of spinning technology for forming thin-walled shell structure with inner ribs in the workpieces of aluminum alloy 2A12. The inner spinning technology was adopted to study the small end within the spiral spin forming technology that is used for manufacturing missile guidance tank shell parts. Finite element simulation was also used to reveal the deformation characteristics, and the major process parameters affecting the forming technology.

2 FINITE ELEMENT MODELING

2.1 *Determination of the quality scaling control parameters*

Some smaller units can control the stable time increment in quasi static analysis or dynamic analysis. The mass scaling method is typically performed using ABAQUS/Explicit to improve the computational efficiency. The method of automatic mass scaling is used for the analysis of metal formation, on the basis of grid geometry and initial conditions [4, 5].

In the finite element simulation performed in this study, we used the control mode of the automatic mass scaling factors provided by the ABAQUS software system [6, 7]. The mass scaling control parameters were determined by iterative calculations, as shown in Figure 1.

In the contact algorithm of software platform, the default tolerance was 10% of the minimum length of the element. Accordingly, we used the minimum element length of 2 mm in the current model. Comparison of the wall thickness determined by simulation

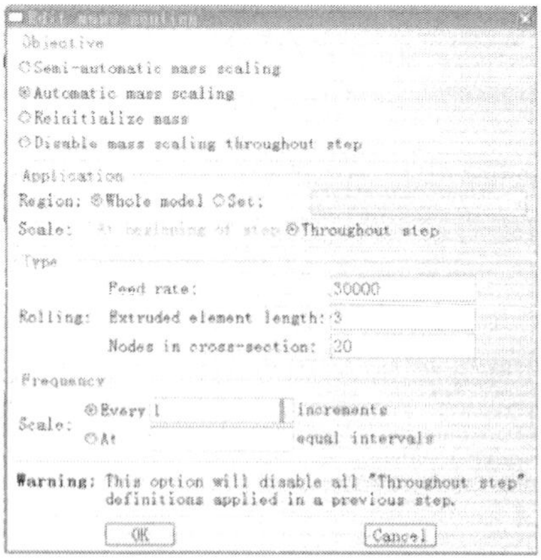

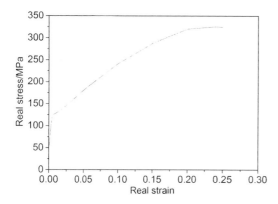

Figure 1. Mass scaling control parameters.

Table 1. Mechanical properties of 2A12.

σ_b (MPa)	E (GPa)	$\sigma_{0.2}$ (MPa)	δ (%)	μ
180	72.4	75	20	0.33

Figure 2. The tensile stress-strain curve of 2A12.

and experiment indicates a system error of 0.2 mm. This error can be eliminated in the analysis of the calculation result [8, 9].

2.2 Mechanical properties

In this study, both the experiments and simulations were performed using the aluminum alloy 2A12. The tensile stress-strain curve of the aluminum alloy 2A12 is shown in Figure 2, and the corresponding mechanical properties used in the simulation are shown in Table 1.

2.3 Technology development and parameters

The spinning product includes three parts, namely, the front frame, skin, and the back-end frame. The

Table 2. The technology parameters.

Technology parameters	Feed rate (mm/min)	Spindle speed (rpm)	Spinning roller roundness radius (mm)
First time of skin spinning	20	30	6
Second time of skin spinning	20	30	6
First time of back-end frame spinning	15	50	6
First time of front frame spinning	15	50	6

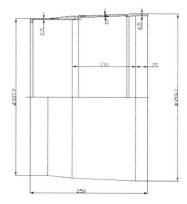

Figure 3. The drawing of blank.

simulation of the inner spinning process, beginning at the small end, is performed via two working steps, namely, pre-forming and final forming. The pre-forming technique is depicted as follow: Firstly, the skin part is obtained by spinning two times. Subsequently, the back-end frame part is obtained by spinning one time. Finally, the front frame part is obtained by one time. Similarly, in the final forming step, the skin part, back-end frame part and front frame part are obtained by sequentially performing the spinning process for one time.

The parameters of each working step are shown in Table 2. The geometry and dimension of the blank, mold are shown in Figures 3 and 4 respectively.

2.4 Establishing the model

The 3-D computational model established using the ABAQUS software is shown in Figure 5. In the simulation, the workpiece was defined as an elastic-plastic deformation body, and the mold and spinning roller were defined as rigid bodies. The contact and separation between the two bodies were determined using the contact theory. The friction coefficient of 0.1 between the mold and the workpiece, as determined from the coulomb law of friction, and the friction between the spinning roller and the workpiece was ignored.

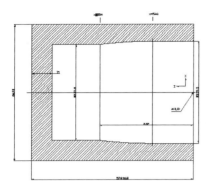

Figure 4. The drawing of mold.

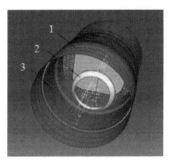

1. mold 2. spinning roller 3. workpiece

Figure 5. The model in simulation.

Figure 6. The mesh generation of pre-forming workpiece.

During the practical spinning process, the spinning roller is moved along a given path, while the workpiece makes rotational motion, as driven by the mold. However, in the simulation process, the mold is fixed, and the workpiece is fixed with the mold at the small end, owing to the limitations of the calculation technology. The spinning roller solely determines the spinning roller. The spinning roller is moved along a given path, simultaneously making a rotational movement. In this study, simulations were performed using the dynamic explicit finite element algorithm. In the typical process, we selected a workpiece with 8 nodes hexahedron. To reduce the integral entity unit, 9460 elements and 19360 nodes were divided in the model. Figure 6 shows the grid mesh of the workpiece.

(a) distribution of equivalent stress

(b) distribution of equivalent plastic strain

Figure 7. The distribution of equivalent stress and equivalent plastic strain.

Table 3. Simulation results.

	Value		
Item	Rear frame	Skin section	Front frame
Max equivalent stress	343.27	326.34	291.88
Max equivalent plastic strain	0.84	0.59	0.19
Max axial direction spinning force (kN)	36.25	29.19	46.21
Max radial direction spinning force (kN)	199.39	131.19	267.00
Wall thickness reduction (mm)	0.39	0.76	—
Average wall thickness reduction (mm)	1.86	3.24	—
Average springback (mm)	0.46	0.24	—

3 SIMULATION RESULTS

Figure 7 shows the simulation results indicating the distribution of equivalent stress and equivalent plastic strain in the workpiece. The values of maximum equivalent plastic strain and maximum equivalent stress in the front frame forming process are 0.19 and 291.88 MPa, respectively. Similarly, the values of maximum equivalent plastic strain and maximum equivalent stress in the back-end frame forming process are 0.59 and 326.34 MPa, respectively. The values of maximum equivalent plastic strain and maximum equivalent stress in the skin forming process are 0.84 and 343.27 MPa, respectively. The main simulation results are shown in Table 3.

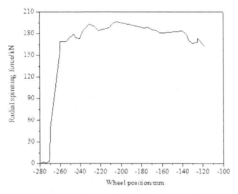

(a) radial direction spinning force of skin part

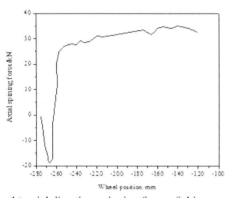

(b) axial direction spinning force of skin part

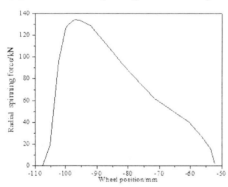

(c) radial direction spinning force of back-end frame

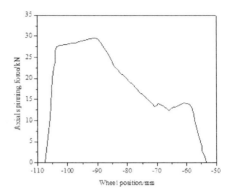

(d) axial direction spinning force of back-end frame

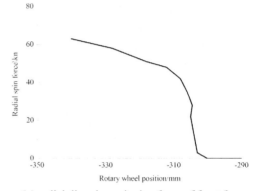

(e) radial direction spinning force of front frame

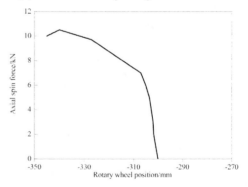

(f) axial direction spinning force of front frame

Figure 8. The spinning force curves.

Overall, the wall thickness distribution of the skin after spin formation is uniform. However, the wall thickness distribution at the ends tends to fluctuate due to the changes in the wall thickness of the blank. The wall thickness distribution in the back-end frame is rather uniform. The stick mode effect of the workpiece after spin formation of the front frame is good. The inner wall of the workpiece is relatively flat. However, the warping phenomenon is detected at the ends due to the influences of the relative rolling reduction and the end effect.

As evidenced from Figure 8, the spinning force along the radial direction is greater than the axial direction in all the areas. However, the changing trend of spinning force is different in each part.

The spinning force of the skin part tends to stabilize after rapidly increasing to a certain value. On the other hand, the spinning force along the radial direction firstly increases and then decreases in the stable stage. The peak value of spinning force is at the position of about −200 mm. This could be attributed to the maximum values of expansion-ratio and deformation

Table 4. The spinning parameters.

Item	value
spinning roller roundness radius (mm)	R10
spinning roller diameter (mm)	200
mounting angle (°)	0
Feed rate (mm/min)	25
spindle peed (rpm)	50

at this position. The observed minor difference in the spinning force along the radial direction is result of the small difference in the expansion-ratio. The spinning force along the radial direction is kept stable in the first 15 mm path of the back-end frame forming process, which then gradually decreases to zero. This could be attributed to the end effect, wherein the spinning force gradually decreases to zero as the spinning roller moves to the end. The gradually decrease of wall thickness along the axial direction leads to the gradual increase of relative rolling reduction in the front frame forming process, together with a gradual increase in the spinning force. The spinning process should be stopped when the maximum forming force reaches a certain value in the late stages of the processing.

At the beginning of the skin spinning process, there appears a negative pressure along the axial direction. This could be attributed to the absence of space between the blank and the mold at the position $X = 277$ mm, and the initiation of instantaneous axial stress at the instant when the spinning roller gets in contact with the workpiece.

Comparing the spinning force in each part, it can be realized that the difference in axial spinning force is almost negligible within the range of 0–47 N. However, the radial spinning force in the parts is different. The maximum radial spinning force appears in the front frame forming process, which obviously increases with the movement of the spinning roller. Consequently, the higher spinning force induces fracture in the front frame. In contrast, the skin and the back-end frame parts have lesser possibility for fracture, considering the closely peak value of radial spinning force existing in the spinning process of these two parts.

4 EXPERIMENT

The test piece is a curved generatrix workpiece with end frames. The techniques adopted for both the experiment and the simulation are the same. Quenching was adopted during the heat treatment process. The parameters used for the spinning process are shown in Table 4.

The products thus obtained are shown in Figure 9. As is seen, the surface integrity of the products is good, and the mechanical properties of the products meet the technical requirements, as shown in Table 5.

Figure 9. The experimental results.

Table 5. The results of mechanical properties tests.

Mechanical property	Tensile strength σ_b (MPa)	Yield strength $\sigma_{0.2}$ (MPa)	Elongation rate δ (%)
Technical indicators	420	250	10.0
Test results	480	390	12.2

5 CONCLUSIONS

(1) The formation of thin-walled shell with inner ribs by inner spinning technology was simulated using the finite element simulation software ABAQUS. The simulations were performed by considering explicit dynamic FEM and rational mass control parameters. Results indicate that cylindrical thin-walled shell structures with good forming effect and uniform deformation can be obtained using the inner spinning technology. The spinning force along the radial direction is greater than that along the axial direction in all areas, but the changing trend of spinning force in each part is different. The peak value of radial spinning force existed in the skin and back-end frame parts, but the spinning force increased obviously in the front frame part, which led to the formation of fractures in the front frame. The difference in axial spinning force was almost negligible among all the parts.

(2) Based on the simulation parameters, we demonstrated the formation of thin-walled shell structures in aluminum alloy 2A12 using the inner spinning technology. The products thus obtained had good surface quality, with mechanical properties meeting the required technical indications.

REFERENCES

[1] Wu Tongchao, Zhan Mei, Jiang, Huabing, et al. Exploring Effect of Spinning Gap on Forming Quality of Second Pass Spinning of Large-Sized Complicated Thin-Walled Shell. Journal of Northwestern Polytechnical University, 2011, 29(1):74–81.

[2] Wong C C, Danno A, Tong K K, et al. Cold rotary forming of thin-wall component from flat-disc blank. Journal of materials processing technology, 2008, 208(1): 53–62.

[3] Li Xin-he, Yang Xin-quan, Wang Yan-fen. The development of spinning fundamental research and processing technology progress of the thin-wall tubes. Forging Technology, 2011, 36(1):7–12.
[4] Zhang Chenai, Cheng Chunmei, Li Ruiqin. Numerical Simulation and Processing Parameter Optimization on Spinning of Tube Parts. New Technology And New Process, 2010 (7):98–101.
[5] Tian Hui. The finite element simulation and process analysis of Spinning forming cylindrical parts. Master degree thesis of science in material processing engineering Dalian technology university, 2010:21–24.
[6] Hu Wenjun. Numerical simulation and experimental research of Spinning forming thin-walled curve part. Master degree thesis of science in material processing engineering XiangTan technology university, 2011: 35–38.
[7] G. Sebastinani, A. Brosius, R. Ewers, M. Kleiner, C. Klimmek, Numerical investigation on dynamic effects during sheet metal spinning by explicit finite-element-analysis. Journal of Materials Processing Technology, 2006, 177(1–3):401–403.
[8] E. Quigley, J. Monaghan. Enhanced finite element models of metal spinning. Journal of Materials Processing Technology, 2002, 121(1):43–49.
[9] Zhang Tao, Liu Zhichong, Ma Shicheng. Technologic research and numerical analysis of spinning of cylinders with inner ribs. Journal of Mechanical Engineering, 2007, 43(4): 109–118.

Analysis of process parameters influence on electrical discharge grinding PDC cutting tool processing

Y.H. Jia
Beijing Institute of Electro-Machining, Beijing, China

C.Z. Guan
China Unicom Inner Mongolia branch, Inner Mongolia, China

ABSTRACT: Electrical discharge grinding is one of the most widely used methods to machine polycrystalline diamond compact cutting tool. It is a type of self-excited pulsed electric discharge machining process based on the principle of discharge erosion. Polycrystalline diamond compact samples with different crystal granularity are the research object. Electrode rotating speed, open-circuit voltage, peak current are selected as the main process parameters. The material removal volume and electrode loss set as the evaluation index of productive efficiency; workpiece surface roughness value sets as an evaluation standard of processing quality. With electrical discharge grinding experiments, combined with scanning electron microscopy observation and roughness tester, the influences of the main process parameters on electrical discharge grinding are analyzed. Polycrystalline diamond compact electrical discharge grinding process is summarized.

Keywords: Polycrystalline Diamond Compact (PDC); Electrical Discharge Grinding (EDG); Electrical discharge machining parameter; Material removal volume; Workpiece surface roughness

1 INTRODUCTION

Polycrystalline diamond crystals are arranged disorderly, isotropic, no cleavage plane, have a very high hardness, good impact toughness, resistance to abrasion uniformity and excellent heat resistance, thermal conductivity and other properties. PCD compact is the cemented carbide substrate of the composite sheet, uniform particle filter synthetic diamond crystals, with the various formulations binder, sintering at high temperature and pressure, which not only retains the diamond hardness, wear resistance, thermal conductivity, etc., but also has cemented carbide high impact toughness and good welding performance, especially suitable for processing non-ferrous metals, hard and brittle materials, oil drills. The PCD cutting tool has been extensively used in automotive, aviation, aerospace, electronics, wood processing, construction and other industries in domestic and overseas[1].

There are two chief kinds of machining methods of PCD cutting tool. One is the mechanical grinding. It is difficult to achieve complex shapes of the tapered edge and edge machining tool range of applications affected because of difficulties in shaping wheel, wheel serious loss, high cost and other issues[1,5]. The other is electrical discharge grinding (EDG). It is one kind of discharge machining process based on the pulse discharging electric corrosion principle. EDM uses thermal energy to machine electrically conductive solid material parts regardless of their geometry[2,3]. Because of the tool electrode simply forming and small loss, electrical discharge machining can be effectively used in the complex shape cutting edge of PCD cutting tool.

There were plenty of valuable research results on EDM techniques. In literature[6], the authors reviewed a variety of EDM electrodes and electrode processing methods. X200Cr15 and 50CrV4 steel were used only for the study of the influence of machining parameters on the surface integrity in electrical discharge machining. The results of the study showed that increasing energy discharges increase instability and therefore[7]. The surface quality of steel after electrical discharge machining was discussed in literature[8]. S.H. Tomadi in his article analyzed the electrical discharge machining parameters on the workpiece surface the quality[9]. Ko-Ta Chiang in his paper indicated to us that the core two significant factors on the value of the material removal rate (MRR) were the discharge current and the duty factor. The discharge current and the pulse on time also have statistical significance on both the value of the electrode wear ration (EWR) and the surface roughness (SR)[10].

Table 1. Physical properties of compax diamond tool blanks.

Property	Compax 1300	Compax 1500
Compression Strength (GPa)	7.5	7.5
Elastic Modulus (GPa)	950	1100
Transverse Rupture Strength (GPa)	1.4	0.85
Thermal Conductivity (W/mk°)	525	600
Electrical Resistivity (Ω/mm)	2.0	4.0
Density (g/(cm)^3)	4.0	3.9

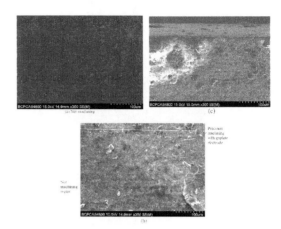

Figure 1. SEM of PCD compact samples surface.

2 EXPERIMENTAL EQUIPMENT AND METHOD

In general, rotation disk copper or graphite electrode is used as a tool electrode in electrical discharge grinding. Different from the traditional EDM, PCD electrical discharge grinding belongs to precision machining, so the machining parameters value is small. In addition, the working medium of electrical discharge grinding generally utilizes injection processing.

There are many process parameters of EDM, we can only choose to study a few of the most important discharge parameters, such as: the electrode speed v, the open circuit voltage U, the peak current i_p, the pulse width t_{on}, the pulse interval t_{off}. While, electrode loss V_E, workpiece material removal volume V_W and workpiece surface roughness W_{Ra} that are associated with the processing efficiency and processing quality are selected as evaluation index.

PDC (Compax 1300 & 1500) produced by DI Corporation are chosen as the experimental sample. The physical properties of workpiece material are shown in Table 1. The electrical discharge machine named BDM-903 is chosen as experimental equipment. The TR240 surface roughness tester and S-4800 scanning electron microscope (SEM) are selected as experimental analyzer.

3 WORKPIECE SURFACE ROUGHNESS

A SEM photograph of unprocessed Compax 1300 is shown in figure 1(a). Figure 1(b) is part of Compax 1300 surface SEM comparison charts of electrical discharge grinding. Figure 1(c) is the SEM photograph of Compax 1500 electrical discharge grinding surface. We can see from these figures, the PCD compact sample surface is flat and smooth before electrical discharge grinding, but after machining, a number of loose and hole, high temperature melting traces, local squamous lines and molten material are shown in the sample surface, which is graphite products after melting of the product and the binder is removed and the adhered matter, not flush surface.

Polycrystalline diamond is not conductive, but the binder that is surrounded by the diamond is electrically conductive. Therefore, it is widely accepted that PCD compact machining in the EDM process is hot melt adhesive, which results in the overall loss of polycrystalline diamond particles and realize the grinding removal. But in addition to this process, under the effect of instantaneous high temperature, phase transformation is happening, a large amount of graphite is generated in PCD surface, unconductive body is changed into a conductive body. PCD is also removed. So we cannot simply be considered as falling in the high-temperature melting binder. Experiments found that the workpiece surface roughness is related to electrical discharge machining parameters, diamond particle size and distribution of the diamond particles.

4 THE EFFECT OF TOOL ELECTRODE SPEED ON PROCESSING EFFICIENCY AND PROCESSING QUALITY

In order to analyze the effects of electrode speed on machining efficiency and machining quality, we set the discharge parameters as follow, open circuit voltage U = 120 V, peak current i_p = 4 A, pulse width t_{on} = 20 μs, pulse interval t_{off} = 15 μs. The test results are shown in figure 2.

From figure 2, we can see the workpiece material removal rate is the lowest and the workpiece surface roughness value is larger when the electrode does not rotate. However, as the electrode rotating, the machining speed and machining quality is significantly increased. This results from the fact that the electrode rotation improved discharge machining conditions, accelerated the workpiece material erosion rate. With the increase of the electrode rotation speed, one is that the workpiece material removal is increased, but the increase rate becomes slow, the other is that the electrode loss also increases. This is due to electrode loss mainly occurred in electrical discharge machining early stage, along with the increase of electrode

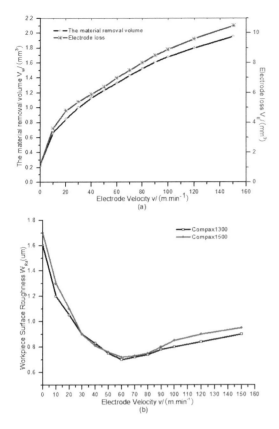

Figure 2. The effect of electrode velocity on machining efficiency and quality.

rotation speed, the discharge point move quickly, discharge frequency increased sharply, so the electrode loss increase finally. With the increase of the electrode rotation speed, the workpiece surface roughness value decreased first, and then increased, but the electrode rotation speed of 30 m/min to 150 m/min, the workpiece surface roughness value changed little. It can be seen from the figure 2 also, the electrode rotation speed impact on workpiece surface roughness has nothing to do with the granularity of the PCD compact. Thought from the two aspects influencing of the electrode rotation speed on the machining speed and machining quality, we can find that processing integrated works best when the electrode rotation speed is 40~80 m/min.

5 THE EFFECT OF OPEN CIRCUIT VOLTAGE AND PEAK CURRENT ON PROCESSING EFFICIENCY AND PROCESSING QUALITY

5.1 Open circuit voltage

In order to analyze the effects of open circuit voltage on machining efficiency and machining quality, we set the discharge parameters as follow, open circuit voltage U = 60, 90, 120, 180, 280 V, peak current $i_p = 3$ A, pulse width $t_{on} = 20\,\mu s$, pulse interval $t_{off} = 20\,\mu s$, electrode speed v = 60 m/min. The test results are presented in Figure 3.

We can be seen from figure 3, with the increase of the open circuit voltage, material removal volume; electrode loss and surface roughness value are increased. Where material removal increased by about four times, and the electrode loss is changed about three times. Surface roughness value changes smoothly in the open circuit voltage of 210 volts, but when the open circuit voltage is higher than 210 volts, the surface roughness values quickly increased. Therefore, for processing PCD material that is with high hardness, high resistance and high melting point, the machining efficiency can be improved by increasing the open-circuit voltage method, while the surface roughness of the workpiece is not affected (open circuit voltage at 210 volts below). Because to improve the open circuit voltage is also increasing breakdown voltage, thus increasing the single pulse energy, moment increased discharge explosive force, accelerate the shedding of polycrystalline diamond particles, improves the removal volume of PCD material, if the other parameters are remain in constant. The open circuit voltage for the workpiece surface roughness effects is basically consistent with machining different granularity PCD material. When the open circuit voltage is below 180 volts, the surface roughness value of different granularity PCD samples vary slightly; but the open circuit voltage greater than 180 volts, different granularity PCD samples surface roughness value has great difference, large granularity PCD sample surface roughness value is bigger than.

5.2 Peak current

In order to analyze the effects of peak current on machining efficiency and machining quality, we set the discharge parameters as follow, open circuit voltage U = 120 V, pulse width $t_{on} = 20\,\mu s$, pulse interval $t_{off} = 15\,\mu s$, electrode speed v = 60 m/min. The test results are presented in Figure 4.

Figure 4 shows that with the increase of the peak current, material removal volume increase, electrode loss decrease, the surface roughness value of the workpiece material also grows with the increase of peak current. But trends have distinction, when the peak current is less than 10 amperes, workpiece material removal and surface roughness are slowly increased, and the electrode loss is decreased dramatically; while the peak current is greater than 10 amperes, the workpiece material removal volume and workpiece surface roughness values are rapidly increasing, slowly reduce electrode loss. Meanwhile, it can be seen from figure 4, when the peak current is below 8 amperes, PCD granularity size has no effect on machining efficiency and machining quality; but when the peak current is greater than 8 amperes, PCD granularity on the machining efficiency and machining quality of workpiece has certain influence.

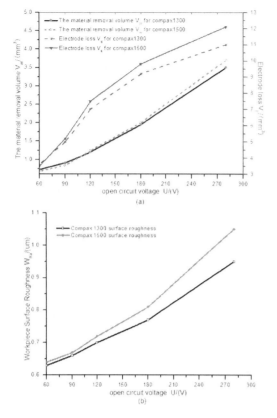

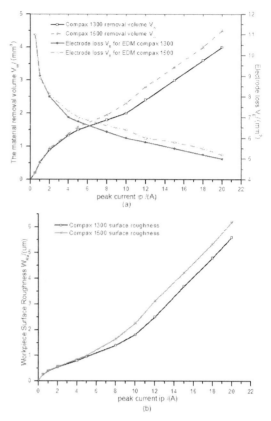

Figure 3. The effect of open circuit voltage on machining efficiency and quality.

Figure 4. The effect of peak current on machining efficiency and quality.

6 CONCLUSIONS

Through the test and analysis of the above, we can get the conclusion that:

(1) PCD crystal granularity size has some effect on the electrical discharge grinding, but the influence range is not wide;
(2) With the increase of the electrode rotation speed, the workpiece material removal volume is raised, but the increase rate becomes slow. Electrode loss is further increased with the increase of the electrode rotation speed. When the electrode rotation speed is 40~80 m/min, processing integrated works best thought from the machining speed and machining quality.
(3) Material removal volume, electrode loss and workpiece surface roughness value are increased with the increase of the open circuit voltage. Generally, we choose 120 volts or 180 volts as open circuit voltage.
(4) With the increase of the peak current, material removal volume and the surface roughness value of the workpiece material increase, electrode loss reduces. The impact of peak current on the material removal volume is more than the open circuit voltage. The arc discharge will occur and processing will be in instability if the peak current is too large, therefore, we generally set peak current at 16 amperes.

ACKNOWLEDGEMENTS

This lesson is supported by Beijing Natural Science Foundation the Grant No. 3133036. The authors would also like to thank the anonymous reviewers whose comments greatly helped in making this paper better organized and more presentable.

REFERENCES

[1] Cook, M.W. & Bossom, P.K. 2000. Trends and recent developments in the material manufacture and cutting tool application of polycrystalline diamond and polycrystalline cubic boron nitride. *International Journal of Refractory Metals & Hard Materials* 18(2):147–152.
[2] Cao, F.G. 2010. *Special Processing Handbook*. Beijing: China Machine Press.
[3] Yu, J.S. 2011. *Theory of electrical discharge machining*. Beijing: National Defence Industry Press.
[4] Kunieda, M. 2005. Fundamentals of electrical discharge machining and future prospects. *Journal of the Japan Society for Precision Engineering* 71(2):189–194.

[5] Jia, Y.H. & Li, J.G. 2011. Study on EDM Machining Technics of Polycrystalline Diamond Cutting Tool and PCD Cutting Tool's Life. *Advanced Materials Research* 268–270:309–315.
[6] Kechagias, J. & Iakovakis, V. 2008. EDM electrode manufacture using rapid tooling: a review. *J Mater Sci* 43(11):2522~2527.
[7] Boujelbene, M. & Bayraktar, E. 2009. Influence of machining parameters on the surface integrity in electrical discharge machining. *Archives of Materials Science and Engineering* 37(2):110–115.
[8] Bleys, P. & Kruth, J.P. 2006. Surface and Sub-surface Quality of Steel after EDM. *Advanced Engineering Materials* 8(1):15–25.
[9] Tomadi, S.H. & Hassan, M.A. 2009. Analysis of the Influence of EDM Parameters on Surface Quality. *Material Removal Rate and Electrode Wear of Tungsten Carbide* 18–20:289–295.
[10] Chiang, K.T. 2008. Modeling and analysis of the effects of machining parameters on the performance characteristics in the EDM process of AL_2O_3+TiC mixed ceramic. *Int J Adv Manuf Technol* 37(5):523–528.

Research on the wear and failure mechanism of tools in machining CFRP with CVD diamond film coated tools

J.D. Liu, D.M. Yu, J.W. Jin, X.Z. Ye & X.F. Yang
College of Mechanical and Energy Engineering, Jimei University, Xiamen, China

ABSTRACT: Based on the experiment of milling Carbon Fiber Reinforced Plastics (CFRP), the cutting performance and wear failure mechanism of cemented carbide tools and that of CVD diamond film coated tools were investigated contrastively. The result shows that: under the same cutting parameters, the lifetime of CVD diamond film coated tools are 8.6 times than that of cemented carbide tools. The wear failure mechanisms of CVD diamond film coated tools are abrasive wear and coating peeling, which are caused by micro-cutting of hard carbon fiber and the impact action of mechanical load and thermal stress.

Keywords: CVD diamond coated tools; CFRP; Cutting performance; Wear mechanism

1 INTRODUCTION

Carbon Fiber Reinforced Plastics (CFRP) has been widely used in aerospace, automobile manufacturing, ships, textile machinery, petrochemical industry, building, sports equipment and other areas, owing to it having excellent properties such as high strength, high heat resistant, low specific gravity, low thermal expansion, etc. However, the CFRP consists of the soft glutinous matrix, the high strength and high hardness carbon fibers, which is diphase or multiphase structure. Its mechanical property is anisotropy, some phenomena like delamination, dilaceration, burr, fiber pull-out and bursting crack are prone to happening in the cutting process, and cutting tool wears heavily, and tool life is low. Therefore, research on machining CFRP arouses broad interest at home and abroad. But the research on CFRP machining mainly focuses on drilling, grinding, turning, and cemented carbide tools, PCD cutting tools, PCBN cutting tools, etc are basically adopted. Due to it having excellent properties like good wear-resistance, thermal conductivity as well as coating on complex tool face, CVD diamond film coated tools hold great promise for use in areas where machine nonferrous difficult-to-cut materials. It is a new superhard coated tool and has a development outlook. However, there is less study on the tool cutting performance in milling CFRP with CVD diamond film coated tools at present. In the paper, based on the experiment of milling carbon fiber reinforced plastics, the cutting performance of cemented carbide tools and CVD diamond film coated tools were investigated contrastively, and the wear failure mechanism of diamond coated tools were analyzed.

Table 1. Workpiece material properties.

Reinforced material	Substrate	Fiber orientation
T700	Epoxy resins	90°
Linear density	Elongation	Tensile strength
800 mg/m	2.1%	4.9 GPa
Fiber volume fraction	Density	Tensile modulus
60 ± 3	1.80 g/cm³	230 GPa

(a) (b)

Figure 1. Experimental tool: (a) Cemented carbide tools, (b) CVD diamond film coated tools.

2 EXPERIMENTAL MATERIALS AND METHODS

The workpiece material used in the tests was CFRP composite laminate T700, and the overall dimensions of the CFRP laminate were 300 mm × 300 mm × 1.2 mm (length × width × height). The material properties are listed in Table 1. The milling tests were performed on a CNC machining center of the type TJ-700. The cutting parameters were 5000 rpm, 1 mm and 250 mm/min for spindle speed, cutting width and feed rate. Also the way of milling CFRP was taken by

Table 2. The experimental tool parameters.

	Cemented carbide tools	CVD diamond film coated tools
Diameter (mm)	8	8
Number of teeth	4	12
Rake angle (°)	7	5
Relief angle (°)	15	20
Helix angle (°)	30 (right)	40 (left), 15 (right)

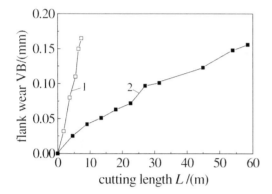

Figure 2. The wear curves of tools: 1-cemented carbide tools, 2-CVD diamond film coated tools.

reciprocating movement. The experimental tools were CVD diamond film coated tools and cemented carbide tools (YG8), as shown in Figure 1. The tool parameters are listed in Table 2. The Keyence VK-X100 laser microscope was used to observe the wear morphology of flank face and measure the wear mass loss (VB).

3 EXPERIMENTAL RESULTS AND ANALYSIS

Figure 2 shows the wear curves of tools. Figure 3 and Figure 4 show the wear morphology of flank faces of cemented carbide tools and CVD diamond film coated tools at different cutting length. As shown in Figure 2, CVD diamond film coated tools wear slowly compared with cemented carbide tools. If the wear loss of flank face (VB = 0.15 mm) was taken as tool wear criterion, when arrived at tool wear criterion, the cutting distance of CVD diamond film coated tools was 54 m, and the cutting distance of uncoated cemented carbide tools was 6.3 m. Therefore, the lifetime of CVD diamond film coated tools are 8.6 times than that of cemented carbide tools under the same cutting parameters.

The analysis suggests that: compared with uncoated cemented carbide tools, (1) CVD diamond film coated tools adopt ultra-fine grained (UFG) diamond coating, which has little diamond grit (<0.5 μm), high hardness (HV8000-10000), good thermal conductivity, low friction coefficient and coefficient of expansion, high abrasion resistance, good toughness. Even though CVD diamond film coated tools have appeared wearing at the cutting length of 4.5 m, tool wear is slowed efficiently due to "hard diaphragm" of diamond coating. (2) CVD diamond film coated tools have more teeth so that the teeth of cutting simultaneously increase in the machining process; at the same time, larger relief angle and helix angle of blade increase the actual cutting relief angle, reducing frictions and grinding of rebounding carbon fiber and powdered cuttings on the flank face. (3) Cross edge is adopted by CVD diamond film coated tools. The multiple-cutting-edge chip breaker grooves are beneficial to remove chip, reduce the friction and extrusion on machined surface, decrease cutting resistance and cutting temperature. Therefore, the useful life of CVD diamond film coated tools is longer than that of cemented carbide tools during milling CFRP.

4 THE ANALYSIS OF WEAR AND FAILURE MECHANISMS OF THE CVD DIAMOND FILM COATED TOOLS

Figure 5 shows the wear morphology of flank faces of CVD diamond film coated tools at the cutting length of 54.5 m. As shown in the figure, the wear zone of flank face appears large-scale mechanical furrow, small-scale smoothing area as well as coating peeling at the boundary. The reason is that: the hardness of carbon fiber reaches up to 53-65HRC. At the beginning of cutting, the cutting-off carbon fiber springs back, and extrudes, rubs, micro-cutting diamond coating of flank face, which leads the surface of diamond coating to wear owing to fatigue strength, and then makes coating peel in the flank face near the cutting edge. As cutting continues, rebounding carbon fiber grinds cemented carbide substrates constantly, and grooves are ploughed out in the flank face; in addition, powdered cuttings entering into flank face grind and polish the wearing surface, which makes wearing zone appear mechanical furrow and smoothing phenomena.

Meanwhile, under the action of cutting temperature, thermal stress will be produced between coating and the substrate, owing to different thermal expansion between diamond coating and cemented carbide; cobalt elements spread from the substrate to diamond coating/substrate interface, which makes carbon elements of diamond be dissolved, and then reduces bonding strength between diamond coating and the substrate. In addition, there is a different wear speed between coating and substrate, the flank face forms a wearing bevel with negative relief angle. It makes the coating surface of flank face and flow direction of carbon fiber and powdered cuttings form a certain angle. As cutting continues, the angle increases, the impact of contact pressure and thermal stress also increases between flank face and workpiece surface. Diamond coating will be peeled because of fatigue strength reducing, and the phenomena of coating peeling appear. Coating peeling further exacerbates the

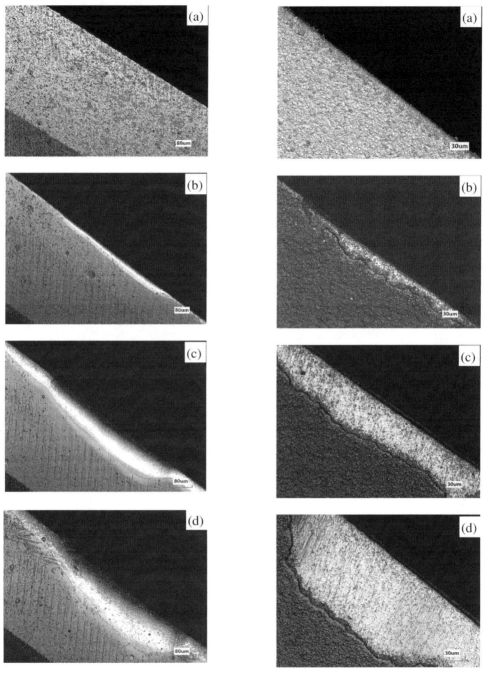

Figure 3. The wear morphology of flank faces of cemented carbide tools at the different cutting length (L): (a) $L = 0$ m, (b) $L = 1.8$ m, (c) $L = 5.4$ m, (d) $L = 6.3$ m.

Figure 4. The wear morphology of flank faces of CVD diamond film coated tools at the different cutting length (L): (a) $L = 0$ m, (b) $L = 4.5$ m, (c) $L = 18$ m, (d) $L = 40.5$ m.

abrasive wear in the flank face, reciprocating cycle, and leading to tool failure eventually.

It follows that the main wear failure modes of CVD diamond film coated tools are abrasive wear and coating peeling in this experiment.

5 CONCLUSIONS

In this experiment, the lifetime of CVD diamond film coated tools are 8.6 times than that of cemented carbide tools under the same cutting parameters.

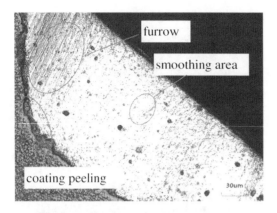

Figure 5. The morphology of flank faces of CVD diamond film coated tools ($L = 54.5$ m).

During milling CFRP with CVD diamond coated tool, the wear zone of flank face appears mechanical furrow, smoothing area as well as the phenomena of coating peeling; the main wear failure modes of coated tools are abrasive wear and coating peeling. Abrasive wear is caused by micro-cutting of hard carbon fiber, and coating peeling is mainly caused by the impact of mechanical load and thermal stress.

ACKNOWLEDGMENTS

The authors are grateful to NNSF of China (Grant No. 51175225), Science and Technology Major Special Projects of Fujian Province (Grant No. 2012HZ0001-1), Foundation for Innovative Research Team of Jimei University (Grant No. 2009A001), Science and technology planning project of Jimei District (Grant No. 20128C01) for support of this project.

REFERENCES

Chen, W.C. 1997. Some experimental investigations in the drilling of carbon fiber-reinforced plastic (CFRP) composite laminates. *International Journal of Machine Tools and Manufacture* 37(8): 1097–1108.

David-West, O.S., Nash, D.H. & Banks, W.M. 2008. An experimental study of damage accumulation in balanced CFRP laminates due to repeated impact. *Composite Structures* 83(3): 247–258.

Koplev, A., Lystrup, A. & Vorm, T. 1983. The cutting process, chips and cutting forces in machining CFRP *Composites* 14(4): 371–376.

Selzer, R. & Friedrich, K. 1995. Influence of water uptake on inter-laminar fracture properties of carbon fibre-reinforced polymer composites. *Journal of Materials Science* 30: 334–338.

Selzer, R. & Friedrich, K. 1997. Mechanical properties and failure behaviour of carbon fibre-reinforced polymer composites under the influence of moisture. *Composites* 28A: 595–604.

Tsao, C.C. 2008. Thrust force and delamination of core-saw drill during drilling of carbon fiber reinforced plastics (CFRP). *International Journal of Advanced Manufacturing Technology* 37: 23–28.

Weinert, K. & Kempmann, C. 2004. Cutting temperatures and their effects on the machining behaviour in drilling reinforced plastic composites. *Advanced Engineering Materials* 6(8): 684–689.

Xu, H.H., Xu, Q. & Liu, D. 2009. Experimental study on the high speed milling of carbon fiber reinforced polymer. *Machinery Design and Manufacture* 12: 167–169.

Zhang, H.J., Chen, W.Y. & Chen, D.C. 2004. Study on defect of elliptic delanination of hole exit zone in drilling carbon fibre reinforced plastics. *Chinese Journal of Mechanical Engineering* 40(12): 145–149.

Zhang, X.H., Meng, Y. & Zhang, W. 2004. The state of the art and trend of carbon fiber reinforced composites. *Fiber Composites* 1: 50–53.

Study on the Human-Machine interaction interface of the elderly scooter

X.Y. Fu, Y.Z. Zhu & J.F. Wu
School of Art Zhejiang University of Technology, Hangzhou, Zhejiang, China

ABSTRACT: This article is concerned with the interface of the scooter for a special population, the elderly. After a series of experiments, it puts forward some design recommendations based on the interface design principles of the existing scooter and enriches the existing design forms, which aims to design a more comfortable and natural scooter interface for the elderly.

Keywords: The Elderly; Human-Machine Interaction; Travel Tool; Design of Interface

With the rapid development of society and science as well as the rise of people's living and medical standards, the average life expectancy has increased, which makes the aging problem more serious. Old people's physical characteristics discourage their traveling quality to a certain extent, so the elderly scooter has become an important means of transport for them. However, the existing elderly scooter hasn't been totally accepted or used by the elderly because of unfamiliar or discontented with the use of the elderly scooter, so it still needs to undertake further studies of the elderly scooter interface.

1 BACKGROUND

1.1 Interface of existing scooters

According to the related market researches, the existing scooters cover American pride, INVACARE, Suzuki and Honda from foreign countries as well as Wisking, Repow and other major domestic brands. According to the collected data, the interface of the elderly scooter mainly displays power, velocity, battery, horn, light, movement and lighting information. It may also include some other functions, such as a built-in radio, fault indicating lamps, warning lights and so on.

1.2 Cognitive analyses of the elderly

With the increase of age, the unique characteristics of the elderly appear gradually, especially the physiological decline of the structure and function of various organs. Visual, auditory, tactile and other sensitivities are reducing, intelligence is also declining, so their ability to learn is also falling. The old may face some difficulties while trying new things. We need to consider an operation mode that is easy to be understood and a displaying and expressing way that is easy to be identified so as to reduce the operating accuracy and complexity of the product, reduce the amount of information, increase the acceptable level of information and minimize the burden of memory on the elderly. Some old people may suffer from slow response, poor memory and other problems, so in the design process, a relaxed operating position and an easily operating system must be considered so as to prevent greater application of forces and other operating problems[1].

Besides these unique characteristics of the elderly, old people may be influenced by physical and psychological factors, personal backgrounds, environment, etc. when they operate the product. Due to the degradation in cognitive abilities, the elderly may face a series of difficulties in interface operation, such as misunderstanding the functions of the interface, operating delay, making operational mistakes and so on.[2,3]

1.3 Study on the human-machine interaction

The study subject of ergonomics covers the design of the relationships between human and products, the overall design of the man-machine system, the design of workplaces and information transmission devices as well as the design of environmental control and safety protection. Its design includes the designs of display and manipulation. Based on the characteristics of human, ergonomics study the way, accuracy and reliability of information transmission, people's reading speed and accuracy, the shape, size, location and operation mode of the interface, the conformity of people's psychology, physiology and living habits as well as other issues.[4]

2 EXPERIMENTS

2.1 Experimental procedure

According to the actual situation of the interface of the elderly scooter, combined with the existing

Table 1. Basic information of study subjects

Basic information	Y-1	W-2	Y-3	Basic information	W-4	W-5	W-6
Age	49	53	62	Age	45	46	58
Gender	Male	Male	Male	Gender	Female	Female	Male
Education level	High-school diploma	With no high-school diploma	With no high-school diploma	Education level	High-school diploma	A level above high-school diploma	High-school diploma
Experience	Yes	No	Yes	Experience	No	No	No

interface design principles, first it makes experimental designs, including determining the experimental objective, experimental methods and experimental subjects, developing experimental tasks and experimental procedure, determining the required laboratory equipments, laboratory location and so on. It also expects the potential problems before experiments, summarizes the problems appeared in practical experiments, detects possible problems, amends experimental tasks and proposes them in formal experiments.[5]

2.2 Experimental objective and methods

Subjects in the experiment were asked to use the interface of the existing elderly scooter to complete required tasks. Observers found and put forward questions according to the exchange among subjects, then made subjective evaluations to the interface of the elderly scooter after experiments, thus obtaining the data of the availability of the interface. According to the nine usability[6] research methods described above and the actual needs of the interface of the elderly scooter, it selects three appropriate experimental methods[7], specifically, saying while operating, observation and questionnaire.

2.3 Study subjects

Study subjects include 6 people aged between 45–65 years old, who have normal hearing and vision.

2.4 Required experimental environment

Experiments need to be carried out in a low-noise place, where they won't be bothered by the external environment. In the experimental process, it needs to use cameras and video cameras to record voices and images for analyzing after experiments.

Experiment record form is a form recording the results related to the experimental tasks.

Questionnaire is a subjectively evaluating questionnaire to know the satisfaction of the availability of the product after the study subjects complete the required tasks.

Task list can be used by the study subjects to describe the specific experimental tasks. Task 1 talks about the meaning and usage of icons and buttons. Task 2 inserts the key and turns it. Task 3 describes

Figure 1. Experimental location and equipments.

Table 2. List of experimental tasks.

Task	Task content	Question
1	Describe the interface	1. What icons and buttons can you see on the interface? 2. What are they used to do? How?
2	Start	1. Do you know where is the keyhole? 2. How to use the key to start the scooter?
3	Check the remaining battery	1. How much the remaining battery is it? 2. Does it need to be charged?
4	Accelerate while moving forward	How to accelerate?
5	Honk while moving forward	How to honk?
6	Turn left while moving forward	How to turn left?
7	Turn on the headlight	How to turn on the headlight?

how much the remaining battery there is and determines whether it needs to be charged or not. Task 4 turns on or presses the acceleration control button. Task 5 presses the horn button. Task 6 presses the left directional control button. Task 7 presses the button of the headlight.

2.5 Experimental contents

Experiment 1. Interface recognition. Recognize all the icons.

Experiment 2. Task operation, including starting the scooter, moving forward and backward, accelerating, honking, turning left or right and turning on the headlight.

Table 3. Basic information list of the study subjects.

Basic information		Number	Percentage %
Gender	Male	4	66.6
	Female	2	33.3
	Total	6	100
Education level	A high-school diploma	3	50
	A high-school diploma	2	33.3
	A level above high-school diploma	1	16.7
	Total	6	100
Experience	Yes	2	33.3
	No	4	66.6
	Total	6	100

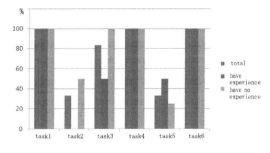

Figure 2. Data of experimental tasks.

Figure 3. Likert scale.

Table 4. Subjective evaluation results.

Content/Number	Average
Ease of learning	4.16
Satisfaction	4

Experiment 3. Questionnaire. It uses the Likert scale. The survey covers the basic information of the study subjects, ease of learning as well as the satisfaction of the appearance and use of the interface.[8]

2.6 Experimental procedure

Step 1 Let the study subjects know the experimental purposes and methods.

Step 2 Talk about the interface. After introduction, ask the older people to retell the meaning of the icons, the use of the buttons as well as the normal operating procedures. In this step, respectively rerecord the use of the icons and buttons for the two groups.

Step 3 Researchers observe the potential doubts or difficulties and talk about the potential problems with the subjects. In this step, researchers need to record the time that the objects spend in completing the tasks, the problems and difficulties appeared in the process of carrying out the tasks as well as the mistakes they made and the times. This step requires a whole shooting with cameras.

Step 4 Explain the contents and purpose of the questionnaire. Ask the subjects to fill in the subjective evaluation questionnaire to know their satisfaction with the availability of the elderly scooter.

3 ANALYSES OF RESULTS

3.1 Data analyses

In this experiment, we choose 6 old people.

Based on the tasks, analyze the cause of mistakes and compare the mistakes made by the experienced group and the inexperienced group. Task 2 is to check the remaining battery. Four subjects made mistakes because of confusing the display screen of the remaining capacity with that of the velocity. Since most of the subjects use the elderly scooter first time, they are not familiar with the functions and use of the interface and cannot accurately understand the battery display although they have been introduced to the battery icon. Task 3 is to accelerate while moving forward. Among them, only one subject in the experiment stopped the scooter before accelerating. This is because he isn't familiar with the operation of the elderly scooter. Task 5 asks the subjects to turn left while moving forward. Some of them made mistakes because they forgot to turn on the turn signal light before turning.

The comparison of the mistakes made by the experienced group and the inexperienced group shows that while checking the remaining battery, the experienced subjects would respond directly based on the previous use experience, therefore the design of the battery display on the interface misled the subjects. In Task 3, the experienced subjects stopped the scooter before accelerating. This may be because they are not accustomed to the operation of this kind of elderly scooter. Previous experience also has influence on this operation. Task 5 indicates that the previous experience is good for improving the accuracy in this task.

At the end, subjects were asked to fill in a questionnaire based on the Likert scale (1–5).

Subjects made subjective evaluations on the ease of learning of the icons and buttons on the interface as well as the satisfaction of the use of interface by giving corresponding scores as shown in Table 5.

The average scores of the ease of learning is 4.16 points, which means that it is relatively easy to learn. What's more, from the recognition of the icons on the interface and the learning of the inexperienced group, we can see they learned fast and well and only made little mistakes in the recognition of the headlights.

We can see half of the subjects gave 5 points. On the one hand, it suggests the elderly scooter is easy to be learned. On the other hand, it may be affected by self-esteem. Older people may want to prove their strong learning ability by giving a high score, which needs further research. From the above table, we know the satisfaction gets 4 points, which means that the subjective evaluation of the interface is satisfied. In Experiment 2, they gave 4 points to Task 2, 3 and 5 although there are some problems. They are worried about giving people bad impressions if they give lower scores, which is also a problem needing to be studied.

3.2 *Design recommendations*

Some design requirements on the interface design of the elderly scooter are proposed based on the above information and experimental analyses: (1) Select interface contents as required. According to the special needs of the elderly scooter as well as the analyses, this information should be displayed on the interface: velocity, horn, light and lighting information. (2) Optimize the arrangement of the interface contents. The overall interface needs to be amplified within a reasonable range, so the contents (icons, operation keys) on the interface also need to be amplified. According to the frequency of use as well as the importance, the contents of the interface should be arranged clockwise around the visual center, from left to right, from top to bottom. (3) Rationally design the specific contents. Redesign the icons that display battery, accelerating, headlights and warning lights on the basis of the experimental conclusion so as to meet the cognitive level of the elderly.

4 CONCLUSION

This article studies the interface of the scooter for the special elderly population. It puts forward some recommendations for improving the existing interface based on a series of experiments. According to the interface design principles of the existing scooter, it proposes some design suggestions and presents more design forms in order to design a more comfortable and natural scooter interface for the elderly.

ACKNOWLEDGEMENT

This work is supported by National Natural Science Foundation of China (Grant No. 61103100), Major Program of Zhejiang Provincial Natural Science Foundation of China (Grant No. 2011C11052) and Special Foundation of Cultural and Creative Industries of Hangzhou (Grant No. 201201123).

REFERENCES

[1] Botwinick J. 1982. Aging and behavior (3rd ed.). New York: Springer Publishing Co.
[2] CD Wickens, J. D. Lee, 2007. An Introduction to Human Factors Engineering. Shanghai: East China Normal University Press.
[3] Huang Zhehui. 2007. A study on the Usability of Treadmill Console Interface for Middle-Aged and Elderly Users. Taiwan: National Taipei University of Science and Technology.
[4] Liu Zhengjie. 2004. Usability Engineering. Beijing: Machinery Industry Press.
[5] Liu Wei, Yuan Xiugan. 2008. Design and Evaluation of Human-machine Interaction. Beijing: Science Press.
[6] Nielsen J. 1993. Usability Engineering. Morgan Kaufmann.
[7] Xiang Yinghua. 2008. Ergonomics. Beijing: Beijing Institute of Technology Press.
[8] Zhu Zuxiang. 2001. Industrial Psychology. Hangzhou: Zhejiang Education Press.

Experimental study on the fabrication of metal fibers to render micro-fin structures

M. Guo, Z.P. Wan, Z.G. Yan, W.C. Tang & X. Zheng
School of Mechanical & Automotive Engineering, South China University of Technology, Guangzhou, Guangdong, China

ABSTRACT: The fabrication of metal fiber that has micro-fin structures is done through a cutting process that uses a newly invented multi-tooth cutting tool on a Be-bronze bar. In this paper, three different micro-fin structures were studied. An analysis of the relationship between cutting parameters and the shape of micro-fin structures shows that micro-fin structures can be effectively controlled by changing the cutting parameters. These micro-fin structures can serve as a nucleus in composite materials to induce crystal formation. In addition, they can act as filler, reinforcement or porous materials with unique advantages.

Keywords: Micro-fin structures; metal fiber; lathe cutting process

1 INTRODUCTION

Metal fiber has been widely used since the 1970s. It not only encompasses all of the inherent advantages of a metal material but also possesses some special properties of non-metallic fiber. Metal fiber facilitates a very large, specific surface area. Its strong properties of internal structure, magnetism and heat resistance are extremely advantageous. In addition, it is flexible and has good thermal and electrical conductivity. It is widely applied in high-tech textiles, machinery, metallurgy, the chemical industry and other areas, such as aerospace, electronics and national defense. Metal fiber plays a very important role in almost every corner of industrial applications (Xi et al. 1998, Roger & Lowell 2000, Xi 2000).

At present, there are several methods available to obtain metal fibers (Liu 1994, Yuan 1987, Zhang 1987, Yao 1981), such as melt spinning, lathe cutting, single wire drawing, bundle drawing, etc. Among these, the single wire drawing method generates uniform fiber diameter. Uniform diameter is essential to generate continuous metal fiber; however, it is costly to apply the single wire drawing method. In addition, the quality of the metal fiber produced by the single wire drawing method is not favorable. Metal fiber produced by the bundle drawing method is continuous and the cost is low; however, the bundle drawing method is not convenient to use due to its technological complexities. It is difficult to generate continuous fiber with uniform diameter using the lathe cutting method; however, it has advantages that other methods do not possess. Under specific conditions, lathe cutting can produce metal fiber with micro-surface structures, which facilitates high polymer to nucleate on the surface of metal fiber. This kind of metal fiber can be used in conductive polymer composite materials (Ning & Li 1998, Tan et al. 1998). If this kind of fiber is used in friction materials, the micro-fin structure will significantly increase the high temperature oxidation and the energy absorption capacity, because of its larger, specific surface area (Farouk & Dwight 1999, Takayuki 1995).

The lathe cutting process overcomes the drawbacks of the other aforementioned methods. This study included the development of a new method that leverages the advantages of the lathe cutting process to produce metal fiber with improved strength and durability through the fabrication of micro-fin structures, which increases the specific surface area of the metal fiber. This new method can produce Be-bronze-based metal fiber that has micro-fin structures. In addition, the relationship between cutting parameters and the physical properties of micro-fin structures were comprehensively investigated in this study.

2 EXPERIMENTAL PROCEDURES

Lathe cutting was applied to fabricate Be-bronze-based metal fiber to have micro-fin structures. Before the cutting process, Be-bronze bars were washed clean (any oil on the surface of the bars should be removed). Then, a Be-bronze bar work-piece was fixed on a lathe. Before cutting the metal fiber, the oxide layer and other contaminants on the surface of the bar were removed by an external cylindrical cutter.

Figure 1 shows the experimental facilities. This experiment was carried out on a lathe (NO: C6132). The diameter of the Be-bronze bar was 50 millimeters

Figure 1. Facility used in cutting process.

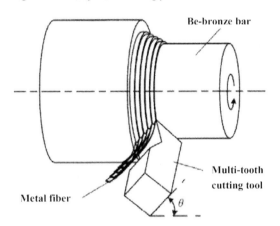

Figure 2. Schematic diagram of cutting process.

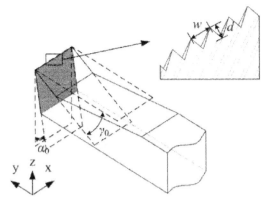

Figure 3. Schematic diagram of multi-tooth cutting tool.

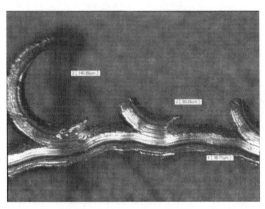

Figure 4. Metal fiber with macro- & micro-fin structure.

(mm). The cutting tool was a multi-tooth cutting tool invented by Wan et al. (2003).

Figure 2 shows the cutting process. After completing the pre-cutting process described above, a multi-tooth cutting tool replaced the external cylindrical cutter. In the determination of the cutting parameters, the largest possible value of a_p was first selected. Then a greater value of the feed rate (f) was selected. Lastly, a reasonable value of v_c was selected according to the tool life and machine power. After this proper selection of the appropriate cutting parameters, the micro-fin cutting process was carried out.

Figure 3 is a schematic diagram of the multi-tooth cutting tool. The rake angle (γ_0), clearance angle (α_0) and cutting edge inclination (θ) of the multi-tooth cutting tool were set at 30°, 8° and 45°, respectively. Both tooth spacing (w) and tooth height (d) can range from 0.2 mm to 0.4 mm. In this experiment, the tooth spacing and tooth height were 0.3 mm.

After the multi-tooth cutting tool was anchored on the lathe, the cutting process discussed above was repeated; then, multiple, continuous metal fibers were generated.

If the cutting parameters changed, the breadth (width) of the metal fiber also varied, and under the specific conditions mentioned below, metal fiber that has micro-fin structures was generated, as shown in Figure 4.

There are specific conditions to generate micro-fin metal fiber: the spindle speed (n) was set at 180 revolutions/minute (remained constant in this study); the back engagement of the cutting edge (a_p) was set at 0.10–0.45 millimeters; and there were three feed rates (f): 0.13 millimeters/revolution; 0.14 mm/r; 0.16 mm/r (different feed rates lead to different micro-fin structures).

3 EXPERIMENTAL RESULTS AND DISCUSSIONS

A single factor analysis method was applied during the experimental process (only one cutting parameter changed at a time and other cutting parameters remained constant) to obtain groups of experimental data and then the relationship between cutting parameters and the shapes of micro-fin structures was studied.

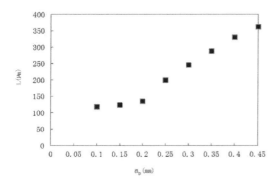

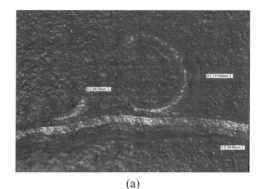

(a)

Figure 5. Effect of the back engagement of the cutting edge on fin length.

3.1 Effect of a_p on micro-fin length

According to the experimental data, as n (180 r/min) and f (0.13 mm/r) remained invariant and only the back engagement of the cutting edge (a_p) varied, it can be safely concluded that the back engagement of the cutting edge and length of the micro-fin structure have a positively correlated relationship. In addition, the data shows that as a_p increased, the fin length also increased. Figure 5 shows the average length of the micro-fin structures. It should be noted that sometimes macro-fin structures were generated. Macro-fin structures have a larger specific surface area than micro-fin structures. The longest micro-fin structure generated was 1.03 mm (when $a_p = 0.45$ mm).

As shown in Figure 6, with the increase of the back engagement of the cutting edge, the spiral degree of the micro-fin structure also increased. When a_p was 0.1 mm or 0.15 mm, the micro-fin was a kind of incomplete spiral structure. When a_p reached 0.2 mm, this incomplete structure developed into a complete spiral and if a_p increased further, then the length of the micro spiral micro-fin structure grew.

3.2 Effect of f on micro-fin shape

It can be observed from the experimental data that the feed rate has a great impact on the shapes of the micro-fin structures (Figure 7).

To investigate the two parameters, n (180 r/min) and the back engagement of the cutting edge ($a_p = 0.15$ mm) were kept constant this time (according to the investigation, only when n was 180 r/min and a_p was 0.05–0.45 mm was this kind of micro-fin structure generated). When a single factor analysis method was applied (only f varied), a spiral micro-fin structure was generated on the condition that f was 0.13 mm/r. Therefore, this parameter was selected and used in the investigation of the influence of the feed rate on the shape of the micro-fin structures. This experiment was conducted on lathe (NO: C6132), and there were seven values for f (from 0.10 mm/r to 0.19 mm/r). At first, f was set at 0.10 mm/r and the aforementioned cutting process was performed. Nothing was generated except

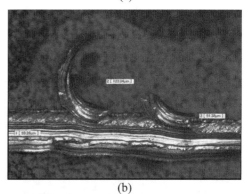

(b)

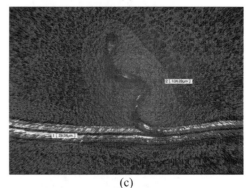

(c)

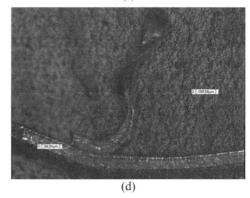

(d)

Figure 6. Photographs of fin structure with four different values of a_p: (a) 0.1 mm; (b) 0.15 mm; (c) 0.2 mm; (d) 0.25 mm.

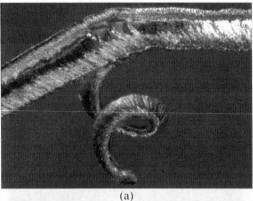

(a)

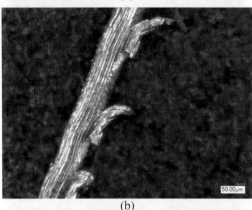

(b)

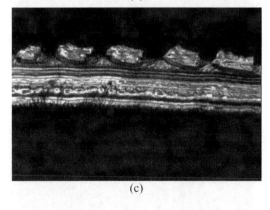

(c)

Figure 7. Photographs of fin structure with three different values of f: (a) 0.13 mm/r; (b) 0.14 mm/r; (c) 0.16 mm/r.

for a single Be-bronze-based metal fiber. There was no micro-fin structure discovered. Then, f was changed to 0.19 mm/r and 0.17 mm/r, and the same cutting process was carried out. Fin structures at the edge of the metal fibers still were not generated. Therefore, a conclusion can be drawn that the proper range of f to generate micro-fin structures is 0.13 mm/r to 0.16 mm/r. As shown in Figure 7, when f was 0.13 mm/r, the micro-fin structure generated was a complete steric spiral micro structure that was uniformly distributed at the edge of the metal fibers. Then, f was increased to 0.14 mm/r and the same cutting process was carried

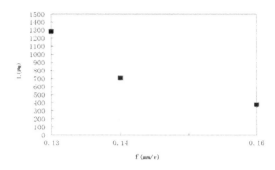

Figure 8. Effect of feed rate on spacing of fin structure.

out again. Continuous micro-fin structures with hook-like spiral shapes were found. The spiral degree of this kind of micro-fin is less than that of the complete spiral micro-fin structures. Finally, this cutting process was repeated on the condition that f was 0.16 mm/r; micro flake-like micro-fin structures were generated. These structures looked like the tentacles of an octopus. The spiral degree of the micro-fin structure decreased as the feed rate f increased.

Based on the above discussions, it can be safely concluded that the proper feed rate to generate micro-fin structures ranges from 0.13 mm/r to 0.16 mm/r. Within this range, three different kinds of micro-fin structures can be generated by selecting different f values. The relationship between f and the spiral degree of the micro-fin structure is one of negative correlation, which means that if f increased, then the spiral degree decreased.

3.3 Effect of f on micro-fin structures located at the edge of metal fibers

The relationship between the feed rate (f) and the shape of the micro-fin structures has already been discussed. It should also be noted that, according to the investigation, the feed rate also affects the distribution by changing the spacing of micro-fin structures at the edge of Be-bronze-based metal fibers.

As shown in Figure 8, the cutting process was carried out with the spindle speed (n) at 180 r/min, the back engagement of the cutting edge (a_p) at 0.15 mm and the feed rate (f) at 0.13 mm/r. Consequently, it was found that the spacing of the complete spiral micro-fin structure at the edge of metal fibers was 1283 μm. Next, only the feed rate (f) was changed to 0.14 mm/r and the same cutting procedure was carried out again. The spacing decreased to 708 μm. Finally, the feed rate (f) was increased to 0.16 mm/r and the same cutting procedure was carried out again. The spacing of the micro-fin structure at the edge of metal fiber decreased to 375 μm.

A conclusion can be drawn from the findings that as the feed rate (f) increased, the spacing of the micro-fin structure decreased.

4 CONCLUSIONS

This paper presented a new cutting process to produce Be-bronze-based metal fibers that have micro-fin structures uniformly distributed at the edge of the fibers. This cutting process utilizes a multi-tooth cutting tool to fabricate metals fibers that have micro-fin structures. The length of the micro-fin structures varied from 117.63 μm to 1030 μm. If the back engagement of the cutting edge (a_p) increased, so did the length of the micro-fin structures. It was also found that the spacing of the micro-fin structures decreased from 1283 μm to 375 μm as the feed rate (f) increased. With the change of feed rate (f), different micro-fin structures were generated. When $f = 0.13$ mm/r, complete spiral micro-fin structures were generated; when $f = 0.14$ mm/r, fin structures that had a hook-like spiral shape were generated; when $f = 0.16$ mm/r, the shape of the micro-fin structures were flake-like and looked like the tentacles of an octopus. In summary, the desired micro-fin structures can be generated by changing the cutting parameters.

ACKNOWLEDGMENTS

Financial support of this experiment was provided by the Student Research Project of South China University of Technology. Guidance from Professor Zhenping Wan is acknowledged.

REFERENCES

[1] Farouk, E.H. & Dwight, A., 1999. Effect of fiber length and coarseness on the burst strength of paper. *TAPPI Journal* 82: 202–203
[2] Liu, G.T., 1994. *Rare Metal Material and Engineering* 23:10–15
[3] Ning, F.L., & Li, Y.B., 1998. A preliminary study of low carbon fiber/high density polyethylene cross-crystal interface. *Plastics Industry* 26: 131–133
[4] Roqer, D.B. & Lowell, O., 2000. Enchanted world of metal fibers. *International Journal of Powder Metallurgy* 36
[5] Takayuki, M., 1995. Study of the durability of a paper-based friction material influenced by porosity. Journal of Tribology, *Transactions of the ASME* 117: 272–278
[6] Tan, S.T., Zhang, M.Q., & Rong M.Z., 1998. Research of Metal fiber/polymer conductive material. *Materials Engineering* 12: 15–17
[7] Wan, Z.P., Ye, B.Y., Tang, Y. & Zhang, F.Y., 2003. The Cutting Mechanism of Multi-tooth Cutting Tool and its Application in Manufacturing Metal Fibers. *Mechanical Science and Technology* 6: 95161–65
[8] Xi, Z.P., Zhou, L. & Li, J., 1998. Development Status and prospects of metal fibers. *Rare Metal Materials and Engineering* 27: 317 320
[9] Xi, Z.P., 2000. Application of metal fibers. *Metal world* 5: 3
[10] Yao, S.K., 1981. *Shanghai Jinshu* 3: 65–70
[11] Yuan, Y.H., 1987. *Chayeyong Fangzhipng* 5: 38–43
[12] Zhang, J.J., 1987. *Kuangye Gongcheng* 8: 51–55

Determination of the optimum initial operation pressure of a steam turbine unit based on a SVM and GA

Guanghui Wu
Guang Zhou Henlee Security Testing Technology Co. Ltd., Guangzhou, China

Xiaowei Peng, Zhaosheng Yu & Xiaoqian Ma
School of Electric Power, South China University of Technology, Guangzhou, China

ABSTRACT: To determine the optimum of the main steam pressure of a 300 MW sub-critical steam turbine unit, on the basis of the practical running data of DCS (Distributed Control System), a model controlling the sliding pressure characteristics of a steam turbine unit was established by using a SVM (Support Vector Machine). What's more, the GA (Genetic Algorithms) was adopted to seek the optimum of the main steam pressure by using the model controlling the sliding pressure characteristics of the unit had established. As a result, there exists a relatively difference between the practical main steam pressure and the optimum of the main steam pressure. Under the condition of various loads and relevant restrictions, the standard coal consumption of the unit after the optimization can reduce by 0.4~10 g/(kW·h).

Keywords: steam turbine; optimum initial pressure; SVM; GA; standard coal consumption

1 INTRODUCTION

As the peak-valley difference of electrical network increases day by day, a lot of thermoelectricity generator units with high running parameters and large capacity have to take part in changing the load. When the units under the non-full rated operating state, the efficiency and security will inevitably be affected. In order to ensure that the unit can operate efficiently, the power plants widely use composite sliding pressure operation mode at present [1].

At present, the main steam pressure of the coal-fired power plant is generally refer to the design value of manufacturers or calculated by heat balance [2]. In the process of actual operation of power plant, due to the problems of running environment and unit assembly, there is a big deviation between design values and the actual operation parameters of the units. The main steam pressure of the unit is not at the most optimal level. To deal with that, when using composite sliding pressure operation mode, the problem of how to find the main steam pressure of the unit to reduce the coal consumption is worth studying.

There are two main methods to determine the optimum of the main steam pressure at present: Field test [3]~[4] and mathematical modeling [5]~[6]. Field tests cost lots of time and efforts, and can only get limited typical operating conditions parameters, which do not apply to need of the operation of overhaul unit. This paper adopts the method of mathematical modeling, and takes a 300 MW sub-critical steam turbine unit as an example, using historical data of DCS in power plant. It adopts the model of sliding pressure operation characteristics of turbine of SVM. This model has many unique advantages in solving the problems such as small sample, nonlinear high-dimensional inputs [7]~[8]. It can accurately reflect the sliding pressure operation characteristics of steam turbine unit. We use the genetic algorithm (GA) to find the optimum for the model to get the optimum of the main steam pressure of the unit under any operating conditions.

2 THE SUPPORT VECTOR MACHINE (SVM) AND GENETIC ALGORITHM (GA)

2.1 *The Support Vector Machine (SVM)*

The basic idea of SVM [9]~[10] is to find a classification hyperplane to separate the two kinds of training samples, which should be as far as possible from the classification hyperplane (as shown in Fig. 1). H is the optimal classification plane and the distance between H1 and H2 m is intervals for classification.

The optimal classification function of a two-kinds problem is:

$$f(x) = \mathrm{sgn}(w \cdot \phi(x) + b)$$
$$= \mathrm{sgn}(\sum_{i=1}^{n} a_i y_i \phi(x_i) K(x_i \cdot x) + b) \quad (1)$$

Where, $w = \sum_{i=1}^{n} a_i y_i \phi(x_i) K(x_i \cdot x) = \phi(x) \cdot \phi(x_i)$

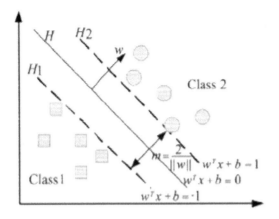

Figure 1. The optimal classification plane.

To separate the two kinds of training samples and to make its interval big enough, the classification problem can be converted to a typical quadratic programming problem:

$$Q(a) = \sum_{i=1}^{n} a_i - \frac{1}{2} \sum_{i=1}^{n} \sum_{j=1}^{n} a_i a_j y_i y_j (\phi(x_i) \cdot \phi(y_j))$$

$$s.t. \sum_{i=1}^{n} a_i y_i = 0 \quad 0 \leq a_i \leq C, i = 1, \cdots, n$$

(2)

The problem has a unique extremum, which can use standard Lagrange multiplier method to calculate. Only one part of the solution will be non-vanishing, and its corresponding sample is support vector. In the actual process of solving the optimization problem and calculating the classification plane, we just need to calculate the kernel function K(x · xi). Common kernel functions are polynomial, RBF, Sigmoid, etc., and the kernel function has been proved to be suitable for most of the nonlinear classification problem [11].

2.2 *Genetic Algorithm (GA)*

The idea of the genetic algorithm [12] is similar with the idea of natural evolution, which is a problem solving method based on the ideology of survival of the fittest. Adopting GA to solve the optimization problem doesn't directly solve the problem of the solution space, but uses coding to express the solution of the problem. Therefore, it must use a kind of data structure when solving the problem, and convert each solved parameter into coding genes. The different combination of gene can form different chromosomes, and a chromosome corresponds to an individual organism. What's more, the biological groups are expressed by a specific number of chromosomes. In GA, fitness function is used to distinguish groups of criteria for the individual, and is the only basis for natural selection. Through the internal structure of adaptive change groups, the selection mechanism of high adaptability chromosome individuals have a higher probability of survival. After confirming the adaptive metrics, the individuals have a high adaptability will be chosen as male parent.

The main steps of solving problems are coding, selection, crossover and mutation. Selection is to retain good "genes"; Crossover is to get better "genes" through exchange and combination between good "genes"; Mutation is to generate new "genes" in order to prevent the local best. For a given space, as the initial population and crossover and mutation are random, as long as there are proper initial population and enough evolution (selection, cross and mutation) times, we will find relatively satisfactory results. Genetic algorithm, as a global optimization search method, has many merits such as simple, parallel processing mathematical equations and derivative expressions which do not need a clear target and are not liable to fall into local optimal solution.

3 THE MODEL OF SLIDING PRESSURE OPERATION OF UNITS

3.1 *Model structure*

As for a certain operation condition, there exists a main steam pressure that makes the standard coal consumption minimize, which is called the optimum of main steam pressure. A lot of factors can affect the thermal efficiency of the steam turbine unit. After the weight analysis of main steam temperature, reheat steam temperature, reheat steam pressure loss, exhaust steam pressure, circulating water inlet temperature, etc., the power supply load, main steam temperature and circulating water inlet temperature are chosen as the input of SVM and the standard coal consumption chosen as output of SVM.

By adopting the function of forecasting regression of SVM, the mapping relationship of input and output of SVM was established, which was chosen as the fitness function of GA. And then use GA to find optimum of the main steam pressure. First, determine the sets of population and solution; and then make genetic operations (such as selection, crossover, mutation and so on) and individual evaluation; at last, find the reasonable optimal initial steam pressure until reaching the number of iterations or the best fitness. The specific model structure is shown in Fig. 2.

3.2 *Predict result of SVM*

The data of predictive model of SVM is got from the DCS of the power plant, which is shown in table 1. There are total 173 sets of historical operation data, and the forest of 73% is chosen as predicted data and the latter of 27% is chosen as validation data. Because the sliding pressure characteristics of unit are variable along with the operation load, the operation load must cover the range of the actual operation load.

As shown in Fig. 3, the actual operation data is contracted with forecast data of SVM. As a result,

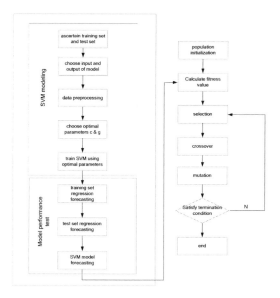

Figure 2. The model structure of SVM and GA.

Table 1. The historical operation data from DCS of power plant.

Item	P_{el}/MW	p_0/MPa	t_w/°C	$B/g \cdot (kW \cdot h)^{-1}$
1	91.6	6.1	27.0	341.7
2	112.2	7.1	27.1	331.5
...	...	...	...	...
44	122.0	8.1	25.2	333.8
45	131.3	8.7	25.4	333.0
...	...	...	...	...
99	140.3	8.2	27.4	327.1
100	155.2	9.0	26.8	331.5
...	...	...	...	...
133	192.3	12.6	26.7	312.6
134	207.0	13.2	26.4	309.8
...	...	...	...	...
172	289.6	15.4	32.47	308.7
173	302.2	14.8	31.5	304.9

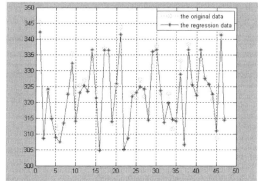

Figure 3. The comparison of original data and regression data.

Table 2. Range of the values of the input parameters.

	floor	ceiling
generated output /MW	90	300
main steam pressure /MPa	6	16.7
circulating water inlet temperature /°C	24	32

Table 3. The comparison of coal consumption before and after the optimization.

Unit load		main steam pressure/ MPa	coal consumption/ $g \cdot (kW \cdot h)^{-1}$
112.2	operation value	7.1	331.5
	optimal value	11.8	321.4
207.0	operation value	13.2	309.8
	optimal value	13.3	309.4
235.9	operation value	14.8	309.0
	optimal value	15.9	305.7

mean square error is 0.1915% and the correlation coefficient is 96.78%, which shown that the model has better approximation ability and generalization ability.

4 THE OPTIMAL MAIN STEAM PRESSURE

According to the predictive model of SVM, chosen as the fitness function of GA, the optimum of main steam pressure can be expressed as the Optimization problem as:

$$B_{\min} = f_B(N_{el}, p_0, t_{w1})$$
$$s.t. \begin{array}{l} N_{el\,\min} \leq N_{el} \leq N_{el\,\max} \\ p_{0\,\min} \leq p_0 \leq p_{0\,\max} \\ t_{w1\,\min} \leq t_{w1} \leq t_{w1\,\max} \end{array} \quad (3)$$

where, B, N_{el}, p_0 and t_{w1} is the standard coal consumption $(g \cdot (kW \cdot h)^{-1})$, generated output (MW) main steam pressure (MPa) and circulating water inlet temperature (°C), respectively. The range of the values of the input parameters is shown in Table 2.

The comparison of various parameters before and after the optimization is shown in Table 3. From the comparison of results, the standard coal consumption of unit can be reduced effectively by adjusting the main steam pressure.

5 CONCLUSION

(1) On the basis of the practical running data of DCS, a model controlling the sliding pressure characteristics of a steam turbine unit was established by using a SVM. As a result, mean square error is 0.1915% and the correlation coefficient

is 96.78%, which shown that the model has better approximation ability and generalization ability.
(2) Genetic algorithm, as a global optimization search method, has many merits such as simple, parallel processing mathematical equations and derivative expressions which do not need a clear target and are not liable to fall into local optimal solution, which can be chosen to find the optimum of the main steam pressure.
(3) The optimum of main steam pressure can be got by using GA. As a result, the optimum of main steam pressure has a big difference with the actual operation value, and the standard coal consumption of the unit after the optimization can reduce by 0.4~10 g/(kW·h).

REFERENCES

[1] Kang Song, Yang Jian-ming, Xu Jian-qun. *Steam turbine theory*. Beijing: China Power Press, 2000.
[2] Zhang Chun-fa, Wang Hui-jie, Song Zhi-png, et al. Quantitative research of the optimum operation initial pressure of a single unit in a thermal power plant. *Proceedings of China Electric Machinery Engineering*, 2006, 26(4): 36–40.
[3] Sun Yong-ping, Tong Xiao-zhong, Fan Yin-long. Experimental study of the optimization of sliding pressure operation modes of a 600 MW unit. Thermal Power Generation, 2007, (8): 66–68.
[4] Xu Shu. Cost-effectiveness analysis of the sliding pressure operation of a 300 MW steam turbine unit. Hunan Electric Power, 2007, 27(4): 19–21.
[5] Tao Jian-guo. An analysis of the operation characteristics of the fixed and sliding pressure operation of a 600 MW unit. *East China Electric Power*, 2000, 28(12): 49–51.
[6] Yong Li, Hairong Wang, Turbine operating characteristic equation based on BP neural network, *J. Eng. Therm. Energy* 17 (2002): 268–272.
[7] Zhang Chun-fa, Wang Hui-jie, Song Zhi-png, et al. Quantitative research of the optimum operation initial pressure of a single unit in a thermal power plant. *Proceedings of China Electric Machinery Engineering*, 2006, 26(4): 36–40.
[8] Hu Bing, Zhang Zhong-min, Wu Yong-yun, et al. Optimum initial pressure of a steam turbine unit for sliding pressure operation determined by using a BP neural network. *Thermal Power Generation*, 2003, 32(7): 24–26.
[9] Vapnik V. N. *The Nature of Statistical Learning Theory*. New York: Springer-Verlag, 1995.
[10] Zhang Xuegong. Introduction to statistical learning theory and support vector machines. *Acta Automatica Sinica*, 2000, 26(1): 32–42.
[11] Burges C. J. C. A tutorial on support vector machines for pattern recognition. *Knowledge Discovery and Data Mining*, 1998, 2(2): 121–167.
[12] Holland J. H. *Adaptation in natural and artificial systems*. Cambridge: MIT Press, 1975.

Research vehicle collision avoidance warning system based on CompactRIO and LabVIEW

Bo Peng, Shan Li & Fan Bai
School of Electrical Engineering, Guizhou University, Guiyang, China

ABSTRACT: As a vehicle active safety system, researchers domestic and foreign automobile industry attach importance to vehicle collision warning system. The use of radar speed detection in the system is FMCW (IVS—179), radar signal acquisition using NI company's CompactRIO analog signal acquisition, data processing by using LabVIEW. Analysis of the principle of radar speed detection, the modeling of the automobile braking process, determination of parameters. Show in the experiment platform of test results, it can measure the vehicle accurately the distance, also can measure the speed of the vehicle in front, and can prompt the driver to make the appropriate treatment, to ensure safety.

Keywords: Vehicle collision warning; radar speed detection; CompactRIO; LabVIEW

1 INTRODUCTION

With the development of Chinese economic and constant improvement of the level of people's life, cars are fastly coming into people's daily life, and accidents have increased. Among traffic accidents, the most common is the rear-end collision. The reason why there are rear-end collisions is mainly because the driver did not realize the danger timely, or they are failing to maintain safe distance when they drive the car. As an active safety system of automotive, the automotive collision avoidance alarm system is valued by researchers in auto industry from domestic and foreign. It can accurately measure the distance between the vehicle and the vehicle in front and the speed of the vehicle in front, it can prompt the driver to make the appropriate action to ensure drive safely.

2 HARDWARE DESIGN

According to the function system need, the whole system can be divided into the following parts: distance, speed detection, data processing, sound and light tips, operating tips, operation is performed to detect. The system block diagram is shown in Figure 1.

Each part of the principle of the system is as follows.

2.1 The generating of FMCW radar modulation signal

The modulation signal used in the system is the triangular wave signal of 100 Hz, using STC89C51 microcontroller and DAC0832, and LM358, in the figure, DAC0832 is a straight way, schematic shown in Figure 2.

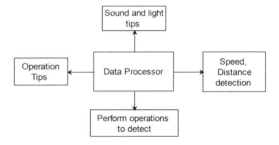

Figure 1. Block diagram of collision avoidance system.

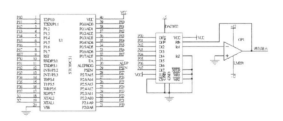

Figure 2. Schematic of radar signal modulation.

2.2 Detection of radar output signal demodulation, amplification and action execution

The received echo signal of radar is very small, we should amplify the signal. Received echo signals have been modulated, we need to demodulate the echo signals. The system uses demodulation circuit that VIS – 719 data recommend, circuit diagram shown in Figure 3.

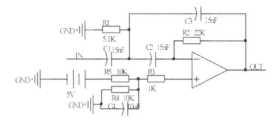

Figure 3. Schematic of radar signal demodulation.

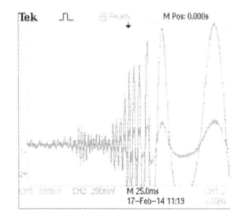

Figure 4. Radar signal in oscilloscope.

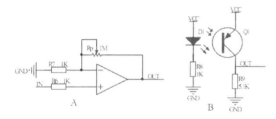

Figure 5. Demodulation signal amplification circuit diagram.

Demodulated signal is very weak, needing to be zoomed as shown in Figure 4. Small amplitude of the signal is demodulated and the large amplitude signal is a radar signal after amplification.

In the system, the operational amplifier is LM358, amplifier circuit shown in Figure 5A. In Figure 1.6, the amplitude of the signal is large radar signal after amplification.

After the system alarmed, the system starts to detect whether the driver execute the recommended action. Installing two infrared detection modules in the clutch and brake at the car, the output of module is digital, its circuit shown in Figure 1.7B. When depressing the clutch and brake at the same time, inputting by digital module NI9421 of CompactRIO. The system is very easy to be able to detect whether the driver execute the recommended action. If the hazard warning is lifted, then the system is no longer detected, if the alarm exists, and the driver does not perform the appropriate action recommended, the system alerts the driver executing the recommended action all the time.

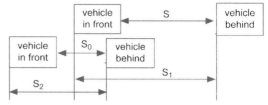

Figure 6. Schematic of two vehicles ahead.

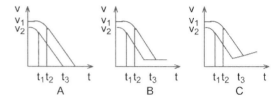

Figure 7. v-t diagram when two cars began braking.

3 SYSTEM MODELING AND MODEL PARAMETERS DETERMINED

Car crashes often occur when the front car decelerates, the car behind has no time to slow down, therefore, focusing on the system model when the front car slows down. When the car is decelerating forward, if the vehicle in front decelerates to a stop, then the car should slow down to a stop and ensure a safe distance, when the front vehicle has a uniform speed after decelerating and the vehicle has a uniform speed after decelerating, the two cars should ensure a safe distance. When the front car slows down, the schematic of two cars ahead is shown in Figure 6.

S_1 is the distance that the car behind has gone after braking, S_2 is the distance that the front car has gone after braking, the figure shows a distance alarm is $S_1 - S_2 + S_0$, two cars' v-t is shown below.

From the figure we can see, situation B and situation C is just the displacement of the two vehicles did not equal. We just need to analyze the situation A, the latter two cases can be derived from the case one. In case one, the speed of the vehicle in front decelerates to 0, then:

$$S_2 = \frac{v_2 t_2}{2} + \frac{v_2^2}{2a_2} \quad (1)$$

where: a_2 is acceleration of the vehicle in front, and by the vehicle decelerates from v_1 to a stop, then:

$$S_1 = v_1(t_1 + \frac{t_2}{2}) + \frac{v_1^2}{2a_1} \quad (2)$$

At this case, for figure 7, using integration, we can obtain their displacement, that is alarming and reminder distance are:

$$S_t = v_1 t_1 + v_{rel}\frac{t_2}{2} + \frac{v_1^2}{2a_1} - \frac{v_2^2}{2a_2} + S_0 \quad (3)$$

Table 1. Acceleration and braking road adhesion coefficient conversion table.

Road Nature	Dry road Adhesion coefficient	Braking acceleration	Wet road Adhesion coefficient	Braking acceleration
Cement	0.60~0.75	5.88~7.35	0.45~0.65	4.41~6.37
Asphalt	0.55~0.70	5.39~6.86	0.40~0.65	3.92~6.37
Snow	0.20~0.35	1.96~3.4		

$$S_v = v_1(t_h + t_1) + v_{rel}\frac{t_2}{2} + \frac{v_1^2}{2a_1} - \frac{v_2^2}{2a_2} + S_0 \quad (4)$$

Assuming under the ideal condition, the vehicles have the same braking effect, acceleration is a. At this time, the equation above can be simplified to:

$$S_t = v_1 + v_{rel}\frac{t_2}{2} + \frac{v_{rel}(2v_1 - v_{rel})}{2} + S_0 \quad (5)$$

$$S_v = v_1(t_h + t_1) + v_{rel}\frac{t_2}{2} + \frac{v_{rel}(2v_1 - v_{rel})}{2a} + S_0 \quad (6)$$

Speed of the vehicle in front can be measured though FMCW radar by the formula 1.10, the vehicle speed can be obtained from the vehicle system, therefore, the relative speed can be easily obtained, t_h is the reaction time for the driver.

driver reaction time is generally 0.3–1.5 s, which is closely related to the driver's age, gender, physical condition, speed and other factors, it is difficult to accurately determine. According to the tests, under normal circumstances, when the speed of 40 km/h, the driver's reaction time is about 0.6 s; when the speed increased to 120 km/h, the reaction time is increased to about 1.5 s. To be cautious, the model selected for the driver's reaction time t_h 1.5 s.

Braking growth acceleration time typically 0.2 s, and in accordance with the provisions of the European Economic Community, when car at the speed of 80 km/h, So brake Coordinated Universal Time = 0.2 s, security pitch is the brake to stop the car after the two should be kept a certain distance, usually taken as 2–5 m. To be cautious, the model selected safety distance S_0 is 3 m.

Table 1 is based on the average, in the dry asphalt and concrete pavement braking acceleration is 6 m/s^2, a wet road braking acceleration is 5 m/s^2, a snow and ice braking acceleration is 2.5 m/s^2.

4 SOFTWARE DESIGN BASED ON COMPACTRIO AND LABVIEW

CompactRIO is a reconfigurable embedded control system produced by the National Instruments (NI).

Table 2. The results of distance test.

Test distance (m)	Actual distance (m)	Error
19.9	20	0.5%
29.8	30	0.7%
49.6	50	0.8%

Table 3. The results of speed test.

Test distance (m)	Actual distance (m)	Error
24.6	25	1.6%
44.5	45	1.1%
65.7	65	0.5%

Users can employ graphical development tools of LabVIEW program for CompactRIO, and apply it into various types of embedded control and monitoring applications. CompactRIO is based on the LabVIEW graphical system design platform for development. LabVIEW contains thousands of high-level functions created specifically for industrial measurement and control applications. These powerful tools can easily implement advanced signal processing, frequency analysis, and digital signal processing. For example: the Fast Fourier Transform (FFT), time-frequency analysis and control design and simulation. In the automotive collision avoidance system, we only use the CompactRIO's analog input, and digital input module.

5 THE EXPERIMENTAL RESULTS

The results of the distance test of system is shown in Table 2; as it can be seen from the data in the table, when the detected distance becomes larger, the error of system is also becomes larger. Table 3 is a result of the speed test, as the speed increases, the error of test result reduces.

When the two cars' speed both 40 km/h run the interface shown in Figure 8. The figure shows the distance between the two vehicles is 35 m, according to Equation (6), a safe distance parameter calculation after substituting for 21.887 m, the actual distance is greater than the safe distance, the system displays the status of the security. System status lights is green. Driver can keep this pace to continue driving.

Figure 9, shows the vehicle speed is 40 km/h, the speed of the vehicle in front of 20 km/h, the distance between the two vehicles is 31 m, calculated according to the distance formula (6) is 30.152 m, the actual distance is very close to the safe distance, the driver needs to slow down. The system displays the status of the alarm system status lights is yellow. Warns the driver must slow down.

When prompted alarm, the car slow down immediately to 25 km/h, the speed of the vehicle in front of 20 km/h, the system state shown in Figure 10, the

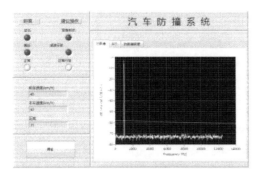

Figure 8. Experiment I test interface.

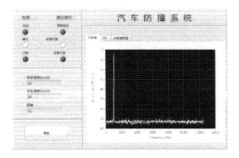

Figure 9. Experiment II test interface.

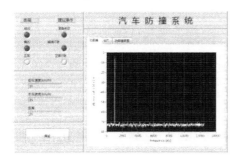

Figure 10. Experiment III test interface.

safety distance is calculated according to the formula (6). At this speed is 13.27 m, measured from the system is 20 m, more than a safe distance, so the driver can maintain this pace to continue driving.

Figure 11 shows the vehicle speed from 40 km/h, before the car is stationary, between the two vehicles is 27 m. Safety distance calculated according to the formula (6) is 33.284 m, greater than the actual distance, the status lights is red, alarm system in need of emergency braking.

As it can be seen from the interface, when the vehicle speed reaches 40 km/h, and the distance between the two vehicles is only 30 m, collision avoidance system has been seriously warned and needs to slow down.

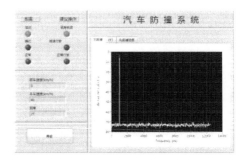

Figure 11. Experiment IV test interface.

6 CONCLUSION

At the point of view of test effect of system, the system can satisfy the requirements of practical applications. With CompactRIO and LabVIEW collecting signal and processing data, improved system's real-time. In the LabVIEW environment, FFT algorithm can be converted into standard C language, it can be easily integrated into other controllers and make collision avoidance system more practical.

ACKNOWLEDGEMENT

Project funds: Guizhou University Graduate Innovation Fund (Research Institute of 2013004) support.

REFERENCES

Yi Zeng. Research of automotive collision warning system based on ARM + FMCW radar South China University of Technology, South China University of Technology Library, 2011, 11–24 pages.

Cuiping Yang, Huifeng Guan. Research on safe driving distance of automotive collision avoidance on highway "Automation Instrumentation" 2008, Vol 29, No. 9.

Song-Qiang Liu digital signal processing system and its application Beijing: Tsinghua University Press, 2003: 261–263.

Dabiao Zhang, Hualong Yu. Design of automotive collision avoidance warning system based on LabVIEW. Based Computer Engineering and Applications, 2008, 44 (21).

Xianchao Wang, Xianchuan Wang. Research and Realization of FFT algorithm Based on LabVIEW. Instrumentation 2006.

ABOUT THE AUTHORS

Peng Bo male born in March 1986, Guizhou University Graduate School of Electrical Engineering, Research: computer control technology.

Li Shan male born in March 1987, Guizhou University Graduate School of Electrical Engineering, Research: computer control technology.

An improved algorithm based on LTE downlink channel modeling

Yanhua Jin & Qiulian Xu
Institute of Aeronautics and Astronautics, University of Electronic Science and Technology of China, Chengdu, China

ABSTRACT: This thesis builds a space-time model of LTE downlink, based on the power spectrum of angle, the Doppler spectrum, the power delay spectrum and the structure of the sending and receiving antennas. It puts forward an improved algorithm that, during the modeling process, adds random Gaussian phase to eliminate the relevance of each path and adds a low-pass filter to reduce spectral leakage, which makes the simulation of the independent fading channel much better. The simulation platform is based on MATLAB, and it simulates the channel model of the LTE 3GPP standard. The simulation results show that the channel characteristic which is produced under the method of this paper is very close to the theoretical results. And the mean error value of improved algorithm is 0.032, the mean error of traditional algorithm is 0.045. The improved algorithm can make the error value reduce 1.35 dB. It proves that the method provided by this paper is capable of simulating LTE channel's characteristics.

Keywords: LTE down link; MIMO channel; Doppler spectrum; PDP; Correlation properties

1 INTRODUCTION

In the recent years, LTE has become a hot topic and difficulty in wireless communications. Its key technologies are OFDM and MIMO that can achieve high spectral efficiency and high transfer rate. Both of the capacity of MIMO system and the performance of the LTE receiving algorithm greatly depend on the characteristics of the channel MIMO, especially the correlation between each antenna (9). Therefore, in order to design LTE system better, it is necessary to establish a LTE MIMO channel model to reflect effectively the spatial correlation and be suitable for the simulation of LTE system.

Methods currently used for MIMO channel modeling are mainly divided into deterministic and stochastic methods (5, 11), literature (2) gives the relevant Rayleigh envelope model, but it does not consider the correlation between the antennas; literature (8) gives a single ring method based on Jakes model, which could characterize channel case in a certain extent, but the computational complexity; literature (3, 6, 10) gives the shaping filter method, it could basically reflect the situation, but ignore the related effects of different paths and spectral leakage. The paper builds the model for MIMO multi-path fading channel of LTE system based on the statistical properties of the fading correlation modeling and provides an improved algorithm of shaping filter on the basis of the literature (3), which add random phase Gaussian spectrum producing by sum of sinusoids to eliminate the relevance of each path, and adds a low-pass filter to reduce spectral leakage, to simulate the independent fading channel better.

2 MIMO CHANNEL MODELING OF LTE SYSTEM

Building a LTE channel model aims to get the response of channel impulse, and the key is to determine the tap matrix (4). The paper uses the MATLAB software to build a wireless space-time-frequency MIMO fading channel simulation platform of LTE downlink, the workflow of this simulation platform is shown as Fig. 1.

Fig.1 shows that user need to set channel type and the number of antennas according to their demands firstly. Then according to the parameters set by the user, it can get the spatial correlation matrix of the base station and mobile station. So it can get the channel correlation matrix by calculating the Kronecker product of and. Then it calculates the Cholesky decomposition of to get the spatial correlation matrix. And finally it can get the multipath fading channel matrix H by distributing the power of each path according to the delay power spectrum of selected channel type.

Because this system is mainly used for the LTE downlink, and according to the requirement of application, so this thesis builds the model and simulation platform by using LTE 3GPP standards as a benchmark. And LTE MIMO radio channel model is based on the space-time-frequency characteristics of statistical MIMO channel model (1).

By analyzing the traditional shaping filter method, it shows that there are some correlation and spectral spurious between different paths. In order to solve this problem, this paper proposes an improved algorithm of forming-filter algorithm, which adds random phases of random Gaussian noise produced

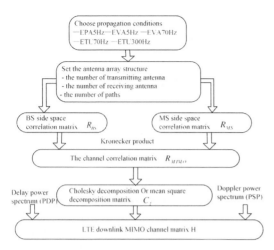

Figure 1. Simulation flow chart of space-time-frequency MIMO channel for LTE downlink.

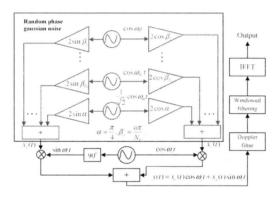

Figure 2. Improved forming filter simulation diagram of the frequency domain.

from sum of sinusoids on the basis of original Smith forming-filter algorithm in order to eliminate the correlation between different paths effectively, and it can effectively improve the spectral quality, and reduce spectral leakage by adding a frequency low-pass filter to it. The structure block diagram of the improved modeling algorithm can be seen in Fig. 2, and its mathematical model can be expressed by equ. (1).

$$y(t) = x_c(t)\sin(\omega_c t) + jx_s(t)\cos(\omega_c t) \quad (1)$$

Where, $x_s(t) = \dfrac{2}{\sqrt{N}}\sum_{n=0}^{M}\alpha_n$, $\alpha_n = 2\cos\beta_n$

$x_c(t) = \dfrac{2}{\sqrt{N}}\sum_{n=0}^{M}b_n$, $b_n = 2\cos\beta_n$

$\omega_c = \omega_d \sin\dfrac{2\pi n - \pi + \theta}{4M}$

$\beta_n = \begin{cases} \alpha = \pi/4, & n = 0 \\ (\pi n)/M, & n = 1, 2, \cdots, M \end{cases}$

Setting the number of antenna of the base station (BS) and the one of the mobile station (MS) are both

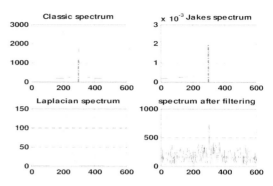

Figure 3. Doppler spectrum.

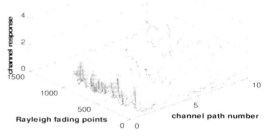

Figure 4. Curves of time domain decay.

one. The system is LTE down link, the symbol rate is 3.84 Mbps, Rayleigh fading channel is 1024, and the carrier frequency is 2.15 GHz. The channel Doppler spectrum is shown in Fig. 3, and the mathematical expression of classic spectrum is shown in equ. (2).

$$S(f) = \begin{cases} \dfrac{A}{\sqrt{1-\left(\dfrac{f}{f_d}\right)^2}} & -f_d \le f \le f_d \\ 0 & else \end{cases} \quad (2)$$

According to the expression of equ. (2), it shows that the classical spectrum shown in Fig. 3 should be a U-shaped waveform. But it makes a translation to make the spectrum below zero shifted to 300 Hz and the Jakes spectrum and Royce spectrum are drawn only half of spectrum in the figure. So actually it should be U-shaped. Fig. 4 shows the fading characteristic curve of the channel matrix H in time domain while the maximum Doppler frequency shift is 300 Hz. The numbers of the path are 7,9 and 9, which correspond to EPA, EVA and ETU respectively in LTE 3GPP standard.

3 THE ALGORITHM VERIFICATION

Seen from the above, channel modeling is the process to solve the channel matrix H. Therefore, detecting whether a channel model is correct or not is equal to the correctness of channel correlation matrix H. Three aspects are used to verify the algorithm of this paper, and they are the spatial correlation curve, the power delay profile and Doppler distribution (shown in Fig. 5 to Fig. 7 below).

3.1 The spatial correlation characteristic of channel matrix

It uses the channel matrix to calculate the spatial correlation coefficients between the various elements. Assume that:

$$R_{BS} = \begin{bmatrix} \rho_{11}^{BS} & \cdots & \rho_{1M}^{BS} \\ \vdots & \ddots & \vdots \\ \rho_{M1}^{BS} & \cdots & \rho_{MM}^{BS} \end{bmatrix}_{M \times M}, \quad R_{MS} = \begin{bmatrix} \rho_{11}^{MS} & \cdots & \rho_{1N}^{MS} \\ \vdots & \ddots & \vdots \\ \rho_{N1}^{MS} & \cdots & \rho_{NN}^{MS} \end{bmatrix}_{N \times N},$$

where $\rho_{m1m2}^{BS} = \langle \alpha_{m1n}^l, \alpha_{m2n}^l \rangle$ represents the signal correlation received transmit signals of the same MS by the antenna m1 and m2 of BS. α_{mn} represents the fading coefficients from the no. n antenna of MS to the no. m antenna of BS. As Fig. 5 shows, it's the spatial characteristic curve between each elements of spatial channel matrix for channel EVA70Hz. Wherein Fig. 5(1) is a perspective view, Fig. 5(2) is a plan view. Fig. 6 shows the spatial characteristic comparison chart of the improved and the ideal curve.

The red dotted line in Fig. 5 represents the ideal value (i.e., the element value of channel correlation matrix), and the blue solid line represents the simulated value (i.e., the spatial correlation coefficient value of each element calculated by the channel matrix H). The pink asterisk line in Fig. 6 represents the simulated values after it has been improved. $< h_{ij}^l, h_{kl}^l >$ represents the correlation coefficients between the fading coefficients from sending antenna i to receiving antenna j and the fading coefficients from sending antenna k to receiving antenna l when the path l was invested. Fig. 5 and Fig. 6 show that the correlation between actual and theoretical values has a gap, which is due to restrictions by the simulation iterations. The iterations N is bigger, it can get more sampled points of the Rayleigh fading channel. Because the fading amplitude of this points obey the Rayleigh distribution, so it needs $N \to \infty$ in order to generate Rayleigh fading distribution channel better, but the N value is impossible infinitely large and the sampling rate of the filter requirements will be higher if N is larger, and the computational complexity also increases. So there are some errors in simulation processing. Fig. 6 shows the correlation coefficient decreases after it has been improved, i.e., the correlation is reduced. The improved algorithm in this paper verified can reduce the impact of the correlation of each path.

3.2 Doppler power spectrum (PSD) of tap coefficients

Fig. 7 shows the Doppler spectrum of the various paths. It uses a classical Doppler spectrum which is completely symmetrical "U" type spectrum in this paper. And seen from Fig. 7, it shows that the simulation

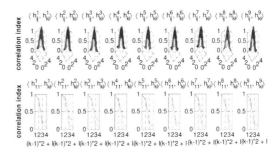

Figure 5. Curve of channel correlation coefficient before unmodified.

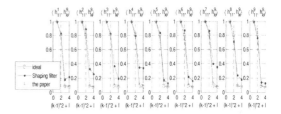

Figure 6. Curve of improved channel correlation coefficient.

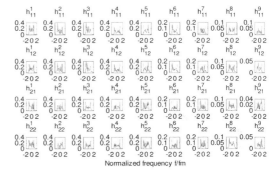

Figure 7. Channel Doppler spectrum of EVA70Hz.

Doppler spectrum of the tap is approximate to a "U" type spectrum. Then a comparison curve of the mean error values of the traditional method and the algorithm in the paper are shown in Fig. 8. The mean error value of improved algorithm is 0.032, and the mean error of traditional algorithm is 0.045. The improved algorithm can make the error value reduce 1.35 dB, and relatively close to the ideal state. To be explained that, since the error of each path is random, so the curve of Fig. 8 is not linear related.

3.3 The power delay profile (PDP) of Tap coefficients

Fig. 9 shows the tap coefficients PDP of EPA, EVA and ETU, wherein the ideal line is the red asterisk PDP curve and blue circle line is the simulation values. It can be seen from the figure that the simulation value is very close to the theoretical value.

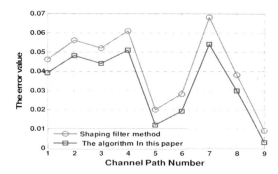

Figure 8. Error value of Spectral Doppler and the ideal value.

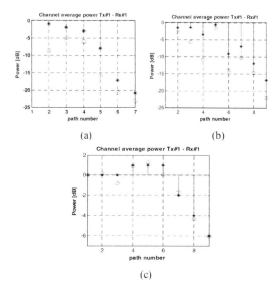

Figure 9. (a) PDP value of EPA, (b) PDP value of EVA, (c) PDP value of ETU.

Through the above analysis we can see that the algorithm in this paper is very close to the LTE channel model defined by LTE 3GPP standard, and it can be better to simulate LTE spatial channel.

4 CONCLUSION

In this paper, it uses MATLAB simulation software to build a LTE downlink space channel simulation platform. By analyzing the simulation results and the theoretical value compared, it shows that the simulation results are close to theoretical values, and it verifies the correctness of the improved algorithm.

This algorithm can not only be applied to LTE system, but also can be appropriated for the 3GPP wireless MIMO systems, and it also can be suitable for the assessment of the key technologies and the processing algorithms in LTE downlink.

REFERENCES

[1] 3GPP. MIMO discussion summary. 3GPP TSG R1-02-0181, January 2002.
[2] Foschimi G J, Gans M J. On Limits of wireless communications in fading environment when using multiple antennas. *Wireless Personal Communications (S0929-6212)*, 1998, 36(6):311–334.
[3] Fuxiang Zhou & Xiaojing Zheng. Research on Simulation Algorithm of Radio Fading Channel Using MATLAB. *Journal of Hebei North University (Natural Sience Edition)*. Vol. 26, No. 3. 2010, 6:20–24.
[4] Kermoal J P & Schumacher L. A stochastic MIMO radio channel model with experimental validation. *IEEE J Select Areas Communication*, 2002, 20(6): 1–2.
[5] K. Yu & B. Ottersten, "Models for MIMO propagation channels". *Wireless Commun. Mobile Comput, vol. 2*, pp. 653–656, Oct. 2002.
[6] Rappea Rappaport TS. Wireless Communications: *Principles and Practice (2nd ed)*. New York: Prentice Hall PTR, 2000: 450–500.
[7] Schumacher L & Pedersen K I & Mogensen P E. From antenna spacings to theoretical capacities-guidelines for simulation MIMO system. *Proc. IEEE PIMRC'2002, Lisbon, Portugal*, Sept. 2002: 587–592.
[8] Shiu Da-Shan & Foschimi G J. Fading correlation and its effect on the capacity of multielement antenna system. *IEEE Transactions on communications (s0090-6778)*, 2000, 48(3): 502–513.
[9] Weichselberger W & Herdin M & Qzcelik H & Bonek E. A stochastic MIMO channel model with joint correlation of both link ends. *IEEE Trans. on Wireless Communications*, 2006, 5(1): 1–2.
[10] William C J. Microwave Mobile Communications. *New York: John Wiley & Sons Inc*, 1994: 378–420.
[11] Yu Kai. Modeling of multiple-input multiple-output radio propagation channels. [Ph. D. dissertation]. *KTH university*, 2002: 11–28.

Study on the benefit of ethnic culture tourism industry in Yunnan based on method of SSA

Hongyu Xie
College of Travelism and Geography Sciences, Yunnan Normal University, Kunming, China

Jidong Yi & Jianhou Gan
Key Laboratory of Educational Informalization for Nationalities, Ministry of Education, YNNU, Kunming, China

Zhilin Zhao
College of Travelism and Geography Sciences, YNNU, Kunming, China

ABSTRACT: Based on the method of Shift-Share Analysis (referred to as SSA), the benefits between the ethnic culture tourism industry of Yunnan and the domestic tourism industry of China; the relationship between the benefit of ethnic culture tourism industry and the total GDP in Yunnan during the period of 2003–2012 were analyzed. This paper showed that the developing trend of ethnic culture tourism industry in Yunnan during the ten years is increasing year by year, the regional competitive advantage became stronger and stronger, which also showed a growth momentum to the benefit of the domestic tourism industry; at the same time, it also contributed to the GDP of Yunnan at the same period. Through the analysis, the conclusion is that ethnic culture tourism industry should be developed vigorously in Yunnan where it is rich in ethnical culture resources, and the local ethnic cultural tourism should be seen as a pillar industry to develop, so that it can become the new growth point for the development of regional economy in Yunnan.

Keywords: Ethnic culture tourism; Industrial benefit; Shift-Share analysis; Tourism growth

1 INTRODUCTION

1.1 Ethnic culture tourism industry

Ethnical culture tourism industry refers to the activities of producing and providing ethnical culture products to meet the industrial culture need of people. In detail, the ethnical culture tourism industry can obtain certain economic and social benefits through the investment on the ethnical culture and cultural resources. In China, the concept of ethnic culture tourism was first seen in the book "Tourism Culture and cultural tourism" written by Wei Xiaoan; he proposed that the activities of cultural tourism can be divided into four sections, that is the national culture, traditional culture, institutional culture and folk culture. And he also believed that the cultural tourism will become the focus of cultural industry, so the tourism market should be cultivated and developed. Huangfu Xiaotao thought it is a key way of changing the cultural resource to cultural capital for the integration of culture and tourism development. He also supported that the cultural tourism industry will become the focus of future cultural industry. Li Lifang discussed the construction of the ethnic culture industry which should be referred to the construction of protecting the historical and cultural heritage in the west; [1] in the article "Research on Ethnic Culture industrial in Yunnan", Li Rui pointed out the development of ethnic culture should be based on the characteristic of each region in Yunnan and take the unique path of development, due to the unbalanced development of economy and the diversity of ethnical culture resources in Yunnan, the development of ethnical culture industry should take the directions of urban cultural industrialization and the backward development ethnic cultural industrialization; [2] Fang Yunping and Wen Nanxun discussed the relationship between the regional ethnic culture and the regional economic development; they thought the regional ethnic culture impacts the regional economic development, which is really rare in the study of ethnic culture industry; [3] Li Zhongzhi and Ye Lei proposed people should solve the relationship among the cultural industry and cultural undertakings, government and the market, the allocation of cultural resources in order to let the ethnic cultural industry develop smoothly; [4] Sun Fumin thought cultural industry can promote the development of many industries in each region, because of its great tension, so people should form a strong sense of cultural industries [5].

2 BENEFIT ANALYSIS OF THE ETHNIC CULTURE INDUSTRY

How the benefit of the ethnic culture tourism in a region is mainly seen whether the economic benefits of ethnic cultural industries can play to the extreme and whether it is able to contribute to local economic development. Through the effective measure and analysis on the ethnic culture tourism industry, it can be judged whether it is reasonable to make the culture tourism industry as a mainstay industry in the process of making the policy for the economic development in Yunnan.

2.1 General introduction of study area

Yunnan province (21° 8'32"−29° 15'8"N, 97° 31'39"−106°11'47"E) is located in the southwest of China, and the Tropic of Cancer across the total south of this province. The maximum horizontal distance of the whole territory is 864.9 kilometers; the maximum longitudinal distance is 900 km, so the total area is 394,000 square kilometers, which account for 4.1% of the total land area in China, of which the mountain accounts for about 84% of the total land area in Yunnan, plateau and hill are about 10%, basins and valleys account for about 6%. The average elevation is about 2,000 meters, the highest elevation is 6740 meters, and the lowest elevation is 76.4 meters. There are numerous ethnics in Yunnan, where is living 51 ethnics. In the sixth national census of 2010, more than 5,000 people have 25 ethnics, so Yunnan is the most ethnical province in China. Because of the diversity of each ethnic culture, there are various kinds of customs and cultures, so Yunnan province is known as "the Grand Garden of ethnical customs". The rich ethnical customs has laid a good base for the development of ethnic culture tourism.

2.2 Methodology

The Shift-Share analysis is the normal method in the western regional economic research which was used in this paper; it analyzed the benefit of ethnic culture tourism industry from 2003 to 2012 in Yunnan province. The detail operation mainly made the growth rate of ethnic culture tourism industry in a certain period as a benchmark, and estimated a assumed share according to the average growth rate of the ethnic culture tourism in Yunnan, which compared with the average growth rate of domestic tourism in China; the conclusion is whether the benefit of ethnic culture tourism is good or not from the deviation situation. It can be judged whether it is reasonable to make the culture tourism industry as a mainstay industry in the process of making the policy for the economic development in Yunnan.

Specific calculations of the Shift-Share analysis is as follows:

$$i_a = \frac{p_{at} - p_{ao}}{p_{ao}} (a=1,2,\cdots n) \quad (1)$$

$$I_a = \frac{P_{at} - P_{ao}}{P_{ao}} (a=1,2\cdots n) \quad (2)$$

$$Y_a = \frac{Y_o \times P_{ao}}{P_o} \quad (3)$$

where i_a is the revenue growth rate of ethnic culture tourism in Yunnan, I_a is the revenue growth rate of the national tourism, Y_a is the revenue share of ethnic culture tourism in Yunnan in accordance with the national tourism industry. In equation (1), p_{at} is the revenue of ethnic cultural tourism in a certain year of Yunnan province; p_{ao} is the revenue of ethnic culture tourism in the base period (2002); in equation (2), P_{at} is the revenue of the national tourism industry in a certain year; P_{ao} is the revenue of the national tourism industry in the base period (2002); and in equation (3), Y_o is the total revenue of ethnic culture tourism in a certain year in Yunnan; P_o is the total revenue of the national tourism industry in a certain year.

In the period of $(0, t)$, the growth value of ethnic culture tourism industry is W_a, which can be decomposed into the sharing component N_a; structural deviation component S_a and competitive deviation component D_a, the detail is as the following:

$$W_a = p_{ao} - p_{at} = N_a + S_a + D_a$$

where $N_a = Y_a I_a$; N_a is the sharing component which refers to the variables of ethnic culture tourism revenue in Yunnan according to the growth rate of the national tourism, that is the diversification of ethnic culture tourism scale in Yunnan which is prorated among the total amount of national tourism revenue. $S_a = (P_{ao} - Y_a) \times I_a$; S_a is the structural deviation component, which refers to differences generated between the benefit of ethnic cultural tourism in Yunnan and the benefit of the national cultural tourism, and these differences lead to the deviation of ethnic cultural tourism in Yunnan compared with the standard benefits of the national culture tourism, it excludes the difference between the average benefits speed of Yunnan ethnic culture tourism and the growth rate of the national tourism benefits, assuming they are equal, so that the benefits of ethnic culture tourism in Yunnan contribute to the benefits of national tourism was analyzed. Therefore, the larger value is S_a, the more contribution to the benefit of the national tourism. $D_a = P_{ao} \times (i_a - I_a)$; D_a is the competitive deviation component, which the deviation is generated by the benefit speed of ethnic culture tourism in Yunnan and the growth rate of the national tourism; it reflects the competitive ability of the ethnic culture tourism in Yunnan. So the larger is the D_a value, the stronger is the competitive ability of the ethnic culture tourism, and the contribution to economic growth will also be bigger.

$$WD_a = W_a + D_a$$

Table 1. Benefit growth rate of national tourism and the benefit deviate share of ethnic cultural tourism from 2003 to 2012 in Yunnan.

Year	I_a	i_a	$I_a - i_a$	N_a	S_a	D_a
2003	−0.122889	0.055852	−0.178741	−0.004320	−0.640801	0.994872
2004	0.228890	−0.143312	0.372202	0.004660	0.122741	−0.207168
2005	0.380884	0.476105	−0.095221	0.011863	0.200137	0.530000
2006	0.605282	0.711458	−0.106176	0.018845	0.318055	0.590976
2007	0.968469	0.986165	−0.017696	0.028536	0.510514	0.984960
2008	1.084082	1.271351	−0.187269	0.034502	0.568898	1.042340
2009	1.263744	1.776180	−0.512436	0.045258	0.658142	2.852219
2010	1.820697	2.447812	−0.627115	0.064989	0.948411	3.490522
2011	3.042400	3.476588	−0.434188	0.097863	1.559554	2.416690
2012	3.563421	4.830080	−1.266659	0.132942	1.850458	7.050224

where WD_a represents the overall growth advantage of ethnic culture tourism industry in Yunnan.

Since then, this paper will introduce two equations, they are $K_{at} = \frac{p_{at}}{P_{at}}$ and $K_{ao} = \frac{p_{ao}}{P_{ao}}$ which can be able to get the growth rate of ethnic culture tourism over the early and final period in the proportion of the total national tourism revenue, and the regional ethnic culture tourism competitiveness index X can be drawn:

$$X = \frac{\sum_{a=1}^{n} K_{at} \times p_{at}}{\sum_{a=1}^{n} K_{ao} \times p_{at}}$$

2.3 Results and discussion

2.3.1 Comparative analysis of benefits between ethnic culture tourism in Yunnan and the national tourism industry

According to the method of Shift-Share Analysis, the benefit growth rate of national tourism and the benefit deviate share of ethnic culture tourism from 2003 to 2012 in Yunnan were respectively calculated through the basic data from the Yearbook of Yunnan Tourism Statistics, provided the data of 2002 is the base date. As shown in Table 1.

As can be seen from Table 1, the development of ethnic culture tourism industry from 2003 to 2012 in Yunnan is as the following:

1. The revenue growth speed of ethnic culture tourism of Yunnan was faster than the national growth rate in 2003, and the competitive deviation component was larger, which indicates that the ethnic culture tourism in 2003 of Yunnan has a certain advantage to the national tourism industry, but the deviation component of ethnic culture tourism is a little small that indicates the contribution of ethnic cultural tourism of Yunnan to the national tourism industry is relatively small in 2003.
2. In Table 1, the revenue growth speed of Yunnan ethnic culture tourism in 2004 was negative; the relative rate of change was large. Accordingly the competitive deviation component separation D_a was negative, which can be drawn the competitive ability of ethnic culture tourism of Yunnan in 2004 is smaller than the competitive ability in 2003, but the structural deviation component increased that indicates the momentum of ethnic cultural tourism of Yunnan was weaker in 2004, but the contribution to the revenue growth of the national tourism still exists.
3. The revenue growth speed of ethnic culture tourism in Yunnan began to become faster and faster than the speed of national tourism from 2005 to 2007, it shows a steady growth momentum, whether the competitive deviation component or sharing component were increasing year by year, which indicates the competitive ability of ethnic culture tourism began to become stronger and stronger, and the sharing component also become bigger and bigger in the national tourism, therefore the contribution to the entire national tourism revenue began to appear gradually.
4. The revenue growth speed of ethnic culture tourism in Yunnan still shows a momentum of growth from 2008 to 2010 on the basis of the previous three years, and they were higher than the national growth rate. Of which the competitiveness deviation components D_a, the sharing components N_a and structural deviation component S_a are positive, it indicates the competitive ability of ethnic culture tourism increased steadily from 2008 to 2010 in Yunnan, the sharing component accounted for a growing share of the entire national tourism, so the contribution to the growth of national tourism revenue was also growing.
5. Though the revenue growth speed of national tourism is fast from 2011 to 2012, the growth speed of ethnic culture tourism is faster than the speed of national tourism, especially the competitive deviation component D_a of Yunnan ethnic culture tourism is 7.050224 in 2012, which increased 4.633534 points compared with the D_a in 2011. It illustrates the competitive ability of ethnic culture tourism in Yunnan was getting stronger and stronger, and the contribution to the national tourism growth is greater.

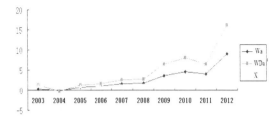

Figure 1. The overall benefit diagram of ethnic culture tourism from 2003 to 2012 in Yunnan.

Through the data in Table 1, according to the formula of the competitive index of regional ethnic culture tourism X, the overall benefit diagram of ethnic culture tourism from 2003 to 2012 in Yunnan can be calculated (see Fig. 1).

From Figure 1, it can be seen that the total revenue growth of ethnic culture industry W_a showed a gradual growth trend from 2003 to 2012, accordingly the total overall growth of the advantages of the ethnic culture tourism industry WD_a highlighted year by year, and the contribution to the entire national tourism industry become stronger and stronger. At the same time, the total regional competitive ability of ethnic culture tourism in Yunnan also enhanced, those phenomenon is very consistent to the values of Table 1. It can be inferred that though there had a small decline of the ethnic culture tourism industry in 2003 and 2004, the overall benefit was an upward trend in those ten years, both the value of national cultural tourism growth and the competitive advantage has emerged as a strong growth momentum since 2004, the contribution to the national tourism industry also grew faster and faster.

All in all, through the Shift-Share Analysis, the benefit of ethnic cultural tourism industry in Yunnan from 2003 to 2012 has been calculated, which showed a growth trend. The competitive ability is becoming stronger and strong in the development of national tourism industry, which the contribution to the national tourism industry has become bigger and bigger. It can be seen that the momentum of ethnic cultural tourism industry has shown a good development in Yunnan province, even in China.

2.4 *Comparative analysis between the benefit of ethnic cultural tourism industry and the GDP in Yunnan*

Since the Yunnan government proposed the suggestion of "construction of distinctive national culture province" in 1996, and the policy of constructing Yunnan province to become a bridgehead for the Southeast in 2009, the Yunnan government has taken full advantage of the rich ethnic culture and has developed the various culture vigorously, mined the ethnic culture resources to fully develop the ethnic cultural tourism, after ten years of development, how is the economic benefits of cultural tourism industry in Yunnan? How is the contribution to the gross national product of Yunnan? It is a key problem for the Yunnan government to make the next economic policy. So the comparative

Table 2. Comparative analysis between the benefit of ethnic culture tourism industry and the GDP from 2003 to 2012 in Yunnan.

Year	Total Revenue (Unit: 100 million RMB)	GDP (Unit: 100 million RMB)	Percentage (%)
2003	308.33	2556.02	12.06
2004	250.17	3081.91	8.12
2005	430.10	3461.73	12.42
2006	499.78	3988.14	12.53
2007	580.00	4772.52	12.15
2008	663.28	5700.00	11.64
2009	810.70	6178.25	13.12
2010	1006.83	7220.00	13.99
2011	1300.30	8750.95	14.86
2012	1702.50	10309.80	16.51

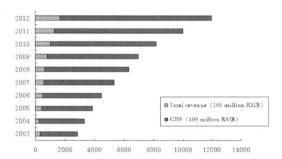

Figure 2. Analysis of the Relationship between Ethnic culture tourism industry and GDP in Yunnan.

analysis between the benefit of ethnic cultural tourism industry and the GDP from 2003 to 2012 in Yunnan can be drawn according to the data of Yunnan Bureau and Yunnan Provincial Tourism Bureau, the detail as follows (Table 2).

From Table 2, it can be seen that the economic benefits of cultural tourism industry was increasing year by year, especially from the year of 2009, the growth trend is faster and faster, the contribution to the gross national product is bigger and bigger, the values in Table 2 are consistent to the values of Table 1, which shows that the ethnic culture tourism in Yunnan is an emerging pillar industry, it will play a more and more important role in the development of the economy in Yunnan in the future.

Accordingly the relationship between ethnic culture tourism industry and GDP in Yunnan was showed in Figure 2.

As an important industry in the third part of culture tourism industry can promote the whole economic growth of GDP in Yunnan, while the GDP growth will promote the development of tourism in turn in Yunnan, so it is a positive correlation between the growth rate of ethnic cultural tourism and GDP in Yunnan. As can be seen from Figure 2, each one unit of the growth rate of the ethnic cultural tourism will pull about eight units of GDP in Yunnan, this is because, on the one hand, the ethnic culture tourism industry is an important part of the gross national product in Yunnan. On the other hand, the ethnic culture tourism has a strong leading role in the whole economic development, not only it affects the whole third industry, but also it has a stimulating effect on social economic development. For example, it can promote the development of financial services, transportation, environmental protection and other industries and sectors.

So the ethnic culture tourism industry should be developed vigorously in Yunnan which is rich in ethnical culture resources, at the same time, the government should see the local ethnic cultural tourism as a pillar industry to develop and give strong support to the ethnical culture tourism from the policy, funding so that it can become the new growth to the regional economy. Because if the government plays the leading role of the ethnic cultural tourism industry, especially emphasis on the development of the service sector of the regional tertiary industry, it will not only bring better economic benefits, but also play the advantages of labor-intensive industries to absorb surplus labor. At the same time, because of the development of the ethnic culture tourism industry, it will involve the problem of protecting and integrating the leaving idle resources, and finally evolve the attractive tourism product, which has the contribution to the ethnic culture tourism, even to the entire national tourism.

3 CONCLUSION

Based on the method of Shift-Share Analysis (referred to as SSA), the benefits between ethnic culture tourism industry of Yunnan and the domestic tourism industry of China; the relationship between benefit of ethnic culture tourism industry and the total GDP in Yunnan during the period of 2003–2012 were analyzed. The results showed that the developing trend of ethnic culture tourism industry in Yunnan during the ten years is increasing year by year, the regional competitive advantage became stronger and stronger, which also showed a growth momentum to the benefit of the domestic tourism industry; at the same time, it also contributes to the GDP of Yunnan at the same period. Through the analysis, this paper concluded that it should develop the ethnic culture tourism industry vigorously in Yunnan which is rich in ethnical culture resources, and the government should see the local ethnic cultural tourism as a pillar industry to develop and give strong support to the ethnical culture tourism from the policy, funding so that it can become the new growth point for the development of regional economy in Yunnan.

ACKNOWLEDGEMENTS

This work was supported by Soft Science Program of China (2013GXS4D149), Ministry of Education, Humanities and Social Sciences Project (12YJCZH053), Minority Native Culture of Quality Engineering Projects in Yunnan.

REFERENCES

[1] Lifang Li, 2000. Yunnan People: Another "eyes to see the world"—Implications for construction of overseas cultural Associated with Yunnan Ethnic Culture Industry, Creating, Yunnan, 12(10):04–07.
[2] Rui Li, 2002. Research on Yunnan Ethnic Culture Industry, Journal of Yunnan Administration College, Yunnan, 4(2): 59–62.
[3] Yuanping Fang & Nanxun Wen, 2000. Correlation Research between the Regional Ethnic Culture and Regional Economic Development, Yunnan Institute of Economic Management, Yunnan, 1(12):57–60.
[4] Zhongzhi Li & Lei Ye, 2003. Some Relationships in the Process of Cultural Industries, Journal of Changzhou Teachers College, Jiangsu 1(12):19–21.
[5] Fumin Sun, 1998. National Cultural Expansion and Cultural Industries Process in the Ethnical Region, Hubei Institute for Nationalities (Social Science Edition), Hebei, 5(6): 74–78.

Numerical simulation of the flow field and cavitation in centrifugal pump

Jun Wang
Jiangsu University of Science and Technology Institute of Energy and Power Engineering, Zhenjiang, China

ABSTRACT: Authors of papers is based on CFD technology, application of GAMBIT software to establish the model of centrifugal pump, internal flow of the centrifugal pump is investigated by means of FLUENT software. Analysis of the centrifugal pump impeller in the internal flow field of velocity and pressure and turbulence intensity distribution, and the conditions of centrifugal pump cavitation occurs. Saturated steam pressure inside the impeller of a centrifugal pump is less than the pressure of water at normal temperature, is one of the important causes of cavitation of centrifugal pump.

Keywords: Centrifugal Pump; Cavitation; Numerical Simulation

1 GENERAL INSTRUCTIONS

1.1 Introduction

Centrifugal pump cavitation is the main reason for the decline of the performance and efficiency of the centrifugal pump. Cavitation in pump flow, pressure head and efficiency are reduced. If produce a lot of bubbles, which will decrease sharply and even can lead to traffic, forcing the pump stop. Cavitation not only influence the fluid flow state, but also to the centrifugal pump damage, decreased so that the pump life, poor reliability. Study of centrifugal pump cavitation, the occurrence condition, to reduce the harm of centrifugal pump, is of great significance to reduce economic losses.

1.2 Cavitation

When the pump pressure decreases to a critical value, the liquid in the pump will produce vaporization phenomenon, hole mass formed by the destruction of the continuous liquid phase. The hole is filled with liquid vapor and gas precipitated from the solution. When these holes go to the lower pressure region, they will grow into larger bubbles, then, bubbles are fluid into pressure is higher than the critical value of the region, the bubble will collapse, this process is called cavitation.

Cavitation is a destructive phenomenon of centrifugal pump. Static pressure when the impeller entrance near the liquid transporting temperature equal to or lower than the saturated vapor pressure of liquid, liquid in this part of the gasification, bubble. Bubble containing liquid into the impeller pressure zone, the bubble rapidly condensation or rupture. A partial vacuum caused by air bubbles disappear, the surrounding liquid with high velocity flow to the original bubble occupy space, the local impact pressure is great. In the repeated use of this huge, causing the pump casing and impeller is damaged.

1.3 Cavitation number

The flow around the object, due to the relative motion between the object and the flow, pressure elsewhere will be different. Set the distance flow pressure as p_0, speed as v_0; surface point of discussion for the initial pressure as p_1, speed as v_1, ignoring the hydraulic loss, list the Bernoulli equation:

$$\frac{p_0}{\rho g} + \frac{v_0^2}{2g} = \frac{p_1}{\rho g} + \frac{v_1^2}{2g} \qquad (1)$$

Pressure coefficient:

$$C_p = \frac{p_1 - p_0}{\rho v_0^2 / 2} = 1 - \frac{v_1^2}{v_0^2} \qquad (2)$$

The minimum pressure on objects is $p_{\min}$, that is:

$$C_{p_{\min}} = \frac{p_{\min} - p_0}{\rho v_0^2 / 2} \qquad (3)$$

When $p_{\min}$ is reduced to a critical value, it will cause phenomenon of cavitation. A cavitation just happened, $p_{\min} = p_v$, (p_v: vaporization pressure), and define the cavitation number is:

$$C_V = \frac{p_v - p_0}{\rho v_0^2 / 2} \qquad (4)$$

Thus, $C_V = C_{p_{\min}}$. The value of $(p_0 - p_v)$ is bigger, C_V value is large, the water is not prone to cavitation; the value of v_0 is larger, the smaller the value of C_V, which is more easily lead to water cavitation.

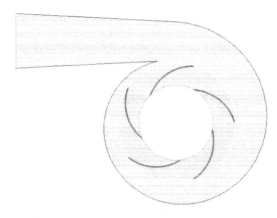

Figure 1. Mesh divided map.

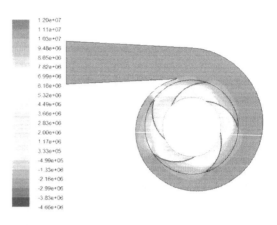

Figure 2. Static pressure contours.

2 MODELING OF CENTRIFUGAL PUMP IMPELLER AND BOUNDARY CONDITIONS

2.1 Impeller model and mesh divide

To the single stage single suction centrifugal pump, using GAMBIT to establish model of centrifugal pump. The design parameters of centrifugal pump are: velocity $= 2$ m/s, inlet radius $= 70$ mm, rotation rate $= 1200$ r/min, number of blades $= 5$, outlet radius $= 110$ mm.

Mesh the model by GAMBIT: Division of impeller flow region grid: the grid element "Elements" in "Tri", "Pave" type selection, grid spacing for the "1", mesh generation, mesh number is 50604. Divided volute flow region grid: the grid element "Elements" in "Tri", "Pave" type selection, grid spacing for the "2", mesh generation, mesh number is 32816. The results of mesh generation is shown in Figure 1.

2.2 The calculation hypothesis

Assuming that the fluid is Newton the incompressibility of the fluid, the working process of the fluid properties do not change, and in the calculation, without considering the influence of gravity on flow field. The use of the computational domain boundary conditions includes inlet boundary conditions, the outlet boundary condition, and the periodic boundary conditions.

2.3 Boundary conditions of import and export

This paper is used in the calculation of inlet boundary conditions for the flow, and assuming that the inlet velocity distribution uniformity, outlet boundary conditions is pressure and outlet pressure is uniform distribution hypothesis.

2.4 Wall conditions

Take the no slip wall boundary condition. For the wall movement, who follow the impeller rotating wall by the moving wall, and the direction and velocity of movement and the impeller rotation direction and speed, while the rest of the wall are static, the speed value is zero. In determining the slip characteristics and movement characteristics of the given after the wall roughness and other characteristics.

3 THE RESULTS OF NUMERICAL SIMULATION

3.1 The static pressure distribution

As shown in Figure 2, can be seen along the blade of centrifugal pump pressure flow direction, the first is to reduce, reaches a minimum value, static pressure began to rise again, arrive at the impeller outlet, the pressure reached the maximum. As can be seen from the diagram, static pressure minimum in the blade inlet later, centrifugal pump at the entrance, due to the inhalation of the liquid inlet, therefore requires a low pressure, because the liquid is discharged, the blade inlet later formed the vacuum or pressure, the liquid into the post due to centrifugal motion, obtain a higher speed, by the impeller discharge fluid through the chamber, most speed can be transformed into pressure energy, so in the outlet of a high pressure. Water at 25°C, which saturated vapor pressure is $3.169 \times 10^3 p_a$. While at the behind inlet blade, it formed a low pressure zone, seen from Figure 2, which is a negative pressure, and it is less than saturated vapor pressure of water. Because of the centrifugal effect, the flow velocity is larger, so, at here prone to cavitation phenomenon. Along the blade direction, the pressure reached to the lowest point, and then rises rapidly, resulting in cavitation. Pressure increase causes the bubbles quickly broken, resulting in a huge local impact, which is damage to the blade surface, which has formed a cavitation damage.

3.2 Velocity distribution

As shown in Figure 3, in the impeller, the fluid along the flow path, velocity increases gradually, reached

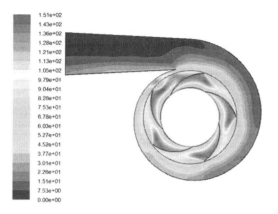

Figure 3. Velocity distribution contours.

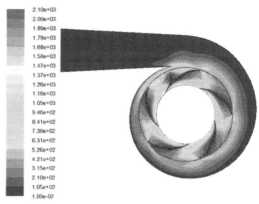

Figure 4. Distribution of turbulence.

the maximum at the impeller outlet, an impeller inlet velocity is the least, which is consistent with the theory, when the fluid enters the centrifugal pump impeller flow areas, its speed is small, is the initial velocity of fluid. When high speed centrifugation fluid through the impeller, it has a very high speed, so, along the radial direction, the fluid velocity increases gradually.

From the graph, we can also see that, when the fluid enters the volute flow area, the flow rate decreased, and it reached the lowest at the volute outlet, which is consistent with the theory. When the fluid enters the volute, the fluid must obtain a high pressure, so it can be discharged, that is to say, the fluid internal pressure is greater than the volute outlet pressure, and in order to obtain a high pressure fluid, much more kinetic energy of the fluid must be transformed into pressure energy, so the speed decreases.

3.3 Analysis of turbulence

As shown in Figure 4, this is the turbulence distribution in a centrifugal pump, it can be found that turbulence is larger at the impeller inlet, which is consistent with the theory, since the entrance flow of impeller inlet angle as a small positive impact angle, which will lead kinetic energy loss at the impeller inlet, in the design flow, the vortex phenomenon is happened, it will lead to a certain area of the negative pressure produced by cavitation.

It also can be seen from the picture that the turbulence intensity at the later place of the inlet is small, and this is consistent with the theory. The impeller passage is diffused, and the uneven pressure distribution makes the fluid suffer a force which keeps balance with the fluid inertia force in the middle of the impeller channel pointing to the opposite direction of pressure gradient. Because of the small fluid velocity in the boundary layer that makes the fluid particle move in the opposite direction to the pressure gradient and forms the secondary flow perpendicular to mainstream direction, the separation loss is produced and the condition that the turbulence intensity is smaller.

4 CONCLUSIONS

Through the results of numerical simulation, the paper analyzed the internal flow field distribution of the centrifugal pump and revealed the special rule of the internal flow and the flow mechanism of the impeller and finally obtained the following main conclusions:

(1) the pressure changing trend inside the centrifugal pump impeller: the pressure gradually reduces after entering from the entrance and then gradually rises towards export volute, and there is a low pressure area in the later place of the inlet where the pressure is lower than the saturated vapor pressure under normal temperature, which is one of the important reasons for the centrifugal pump to produce cavitation. Most of the kinetic energy is converted into pressure energy in the volute area, and there is a higher discharge pressure at the exit.

(2) due to the centrifugal movement of vanes, the fluid speed increases gradually in the impeller, and the circumferential velocity gets larger with the increase of radius and reaches the maximum at the impeller outlet. After the fluid flows into the volute, because of the gradual expanding of the volute cross-section area, the kinetic energy is reduced and translated into pressure energy of the fluid and reaches the minimum at the exit.

(3) when the inlet angle of fluid at the entrance of the impeller is a positive attack angle, the fluid produces vortex phenomenon and causes negative pressure and kinetic energy loss; the influence of the axial eddy current between adjacent vanes makes the turbulence intensity of this part small.

REFERENCES

[1] Yuan Jianping, Yuan Shouqi, He Zhixia. 2004. Research Advance in Flow of Centrifugal Pumps. *Transactions of the Chinese Society of Agricultural Machinery* 35(4): 21–26.

[2] Shang Yangfeng, Ren Shijun, Zhu Yuqin. 2009. Harm and Preventive Measures for Cavitation of Centrifugal

Pump. *Chemical Engineering Design Communications* 35(3):32–37.
[3] Hammit, F.G. 1980. *Cavitation and Multiphase Flow Phenomena*. New York: McGraw-Hill Book Co.
[4] Zhang Jian. 2011. Centrifugal Pump Cavitation and its Prevention. *Guangzhou Chemical Industry* 39(6):141–142.
[5] Wang Yong, Liu Houlin, Yuan Shouqi. 2011. CFD Simulation on Cavitation Characteristics in Centrifugal Pump. *Journal of Drainage and Irrigation Machinery Engineering* 29(2):99–103.

Development of a full-wave underground MRS receiver system based on LabVIEW

G.X. Cao, J. Lin, X.F. Yi, Q.M. Duan & L.B. Feng
College of Instrumentation and Electrical Engineering, Jilin University, Changchun, China

ABSTRACT: For forecasting the mud and water inrush disaster in underground engineering, this paper constructed a full-wave underground Magnetic Resonance Sounding (MRS) receiver system. In this paper, used the receiver coil matching technology and the receiver system timing control technology to complete the lower machine main circuit design; PC control software used "producer–consumer" programming framework developed in LabVIEW platform; using Tablet PC as a terminal data collection, and data transfer via a remote network; the overall used of water proof enclosure, safety design for underground engineering; volume and weight of the system is more lightweight and easy to carry out field work. This paper presents key system design methods and stable and reliable test results, provides full-wave underground detection of MRS method and provides instrumentation platform in underground engineering applications.

Keywords: Magnetic resonance sounding; LabVIEW; Full-wave acquisition; Underground engineering

1 INTRODUCTION

Groundwater has been the main target of geophysical surveys. With the vigorous development of the Chinese underground engineering construction in recent years, engineering construction such as traffic tunnel, water conservancy, hydropower and mining energy buried depth increasing. MRS (Magnetic Resonance Sounding) detection is a non-destructive method for direct detection of groundwater geophysical exploration. It uses the transmitter coil to emit changing electric field to make the hydrogen protons of water space produce energy level transition and then uses the receiver coil to induce the electromagnetic signal that hydrogen proton relaxation process produced.

Domestic existing JLMRS-I type of MRS instrument uses the envelope mode of the signal acquisition which used 4N times of the MRS signal frequency sampling rate and completed signal extraction. However, in the larger electromagnetic noise underground construction site, the sensitivity of the instrument fall cause the MRS signal easily drowned. Full-wave acquisition mode is a substantial increase in the sampling rate and true to acquire and collect the receiving coil sensor signal more directly as possible. This acquisition method complemented the signal and noise full-wave data of the underground water in front detection; improved the data integrity of the underground water detection; provided a richer basis for post-processing.

This paper aimed at the experimental environment for underground engineering, designed a full-wave underground MRS receiver through MRS receiver coil matching network circuit and receiving system timing control, with a "Producer–Consumer" programming architecture PC control software by LabVIEW, to receive the full-wave MRS signal.

2 DESIGN OF THE RECEIVER SYSTEM

Since the receiver is in the underground construction site work environment, the full-wave underground MRS receiver should have lightweight, high sampling rate, fast execution speed, high-precision synchronization, clearly display basic features. The full-wave underground MRS receiver consists of hardware and software: hardware system is designed and implemented mainly based on an 8-channel 24 bits signal acquisition card MPS140801; PC control software developed by LabVIEW graphical programming platform.

2.1 *Hardware architecture*

The full-wave underground MRS receiver system constructed by the touch tablet PC as a PC platform and the USB bus forms signal acquisition card MPS 140801 with 8-channel 24 bits as the main part, matching up with the high voltage relays of the signal conditioning part, the front-end matching network and MRS amplifier. Meanwhile this paper designed the overall timing of the receiver system to control the various modules. Synchronization with the transmitter section uses a wired/wireless synchronization receiver. Users can acquire MRS signal through a variety of sync mode.

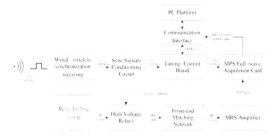

Figure 1. Receiver system hardware block diagram.

Figure 2. The software structure.

The receiver system hardware block diagram is shown in Figure 1.

The PC platform via the communication interface transmits control commands to the MCU. The timing control board is the sequential control center of the receiver system. The wired/wireless synchronization receiver can receive cable or wireless synchronization commands from the transmitter. Synchronizing signal conditioning circuit decodes the wireless signal into a wired and protects the follow-up circuit from false triggering. Take the falling edge of the sync signal as synchronization reference then it into the timing control board. Timing control board after receiving switching time, collection time and other parameters sent from the host computer calculates accurate commands based on transmitting synchronous signal to control the MPS full-wave acquisition card and high voltage relays. MRS signal received by the receiving coil follows by voltage relays, the front-end matching network and MRS amplifier input MPS full-wave acquisition card. The front-end matching network is the key technology of the system design. The MRS amplifiers magnification is adjustable between 80 dB~130 dB. After acquisition digital signals via the communication interface return the host computer to display and superposition processing. This process completed the MRS signal reception.

2.2 Receiver control software

Receiver control software is programmed by software platform LabVIEW2012. By analyzing the specific requirements of a full-wave underground MRS receiver, the software designed to be single-threaded of main line and multi-threaded of local line. The software structure is shown in Figure 2. The software is divided into three hierarchies: the user input layer, the hardware driver layer and the system processing layer.

User input layer mainly for software users to facilitate the realizations such as setting selection of signal sampling rate, acquisition time parameters (including launch time, release time, switching time and acquisition time), matching the transmitted pulse moments, selecting the data file storage path and so on. The hardware driver layer mainly for the underlying hardware to implement the whole system hardware drive in time sequence control, the main functions include testing equipment connection normal, the download of the whole sequence control, the full-wave acquisition board of drive and control and so on. The system processing layer mainly for the internal software to implement of the full-wave sampling software functions such as software initialization, queue cache of data acquired, data superposition algorithm, data storage and so on. Three layers work together constituting the full-wave underground MRS receiver operating system software.

3 SYSTEM FUNCTIONS REALIZATION

3.1 Research on the matching network circuit

The structure of the front-end matching network of the full-wave underground MRS receiver is shown in Figure 3. When the system receives the MRS signal, signal through the closed high voltage relays and the bidirectional diode port protection circuit, the receiver coil and the resonant capacitor formed an LC resonant circuit, this realizes the MRS signal frequency selection. Select the resonance capacitor and the parallel matching resistor reasonable based on the receiving coil parameters can improve the quality factor of the receiving circuit to better suppress noise, and make sure the output signal is reaching the required amplitude level of detection device. Resonant circuit output signal through the preamplifier becomes the input to the post-stage MRS amplifier circuit.

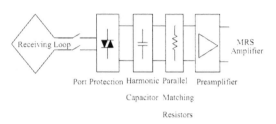

Figure 3. The front-end matching network block diagram.

The electrical equivalent model of the LC resonant circuit can be represented in Figure 4(a), where R_L is the equivalent resistance of the receiving coil, L is the equivalent inductance of the receiving coil, C is the equivalent capacitance of the harmonic resonance

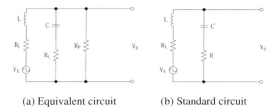

(a) Equivalent circuit   (b) Standard circuit

Figure 4. Equivalent circuit model.

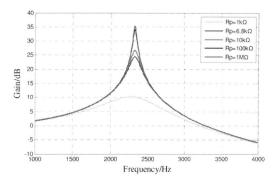

Figure 5. Amplitude-frequency characteristic curves at the different matching resistances.

circuit, R_c is the equivalent resistance of harmonic capacitor, R_P is the parallel matching resistors, V_S is the receiving coil sensed MRS signal source, V_O is the loop output MRS signal to access preamplifier. To simplify the calculation, Figure 4(b) shows the normalized into standard resonant circuit equivalent model.

According circuit equivalent transformation, Figure 4(b) translated the electrical parameters of the circuit equivalent resistance R' and equivalent capacitance C' and calculated the voltage transfer function on the MRS signal in the resonant circuit. We obtained the expression quality factor Q of the circuit

$$H(j\omega) = \frac{V_o(j\omega)}{V_s(j\omega)} = \frac{1 + j\omega C'R'}{-\omega^2 LC' + j\omega C'(R_L + R') + 1} \quad (1)$$

$$|H(j\omega)| = \left[\frac{C'^2 R'^2 \omega^2 + 1}{\omega^2 C'^2 (R' + R_L)^2 + (\omega^2 C'L - 1)^2}\right]^{\frac{1}{2}} \quad (2)$$

$$Q = \frac{\sqrt{L/C'}}{R_L + R'} \quad (3)$$

where ω is the angular frequency of the receiving resonant circuit, the C' and R' is given by

$$C' = \frac{1 + \omega^2 C^2 (R_P + R_C)^2}{\omega^2 C R_P^2} \quad (4)$$

$$R' = \frac{R_P + \omega^2 C^2 R_P R_C (R_P + R_C)}{1 + \omega^2 C^2 (R_P + R_C)^2} \quad (5)$$

This article uses 6 meters 30 turns square coil as a receiving antenna. Here L is 20.2 mH, R_L is 2.16Ω, take the local Larmor frequency as resonant frequency, here is 2326 Hz, harmonic capacitance C is 232 nF, ignore R_c. When changing the parallel matching resistors R_P, the amplitude-frequency characteristics of the receiving matching network will change. Figure 5 shows the matching networks amplitude-frequency characteristic curve when R_P take 1 k, 6.8 k, 10 k, 100 k and 1 MΩ. Within a certain range, with the increase of R_P, the gain at the center frequency of the matching network also increases, the bandwidth of the network gradually decreases and the value of the quality factor Q increases. When R_P beyond a certain range, the curve tends to be stable gradually. As shown in Figure 6 the frequency amplitude characteristic curve roughly coincides between $R_p = 1$ MΩ and $R_p = 100$ kΩ and the Q values changed little.

Since the matching network in the forefront of the received signal circuits, not only to guarantee a certain gain, but also to ensure adequate bandwidth. Comprehensive consideration, the R_p selected 6.8 kΩ in this system, when f = 2326 Hz, gain is 24.37 dB, quality factor Q is 16.

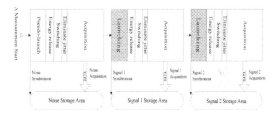

Figure 6. Measurement process of Receiver system.

3.2 Sequential control of receiver system

During underground MRS detecting, the detection process can be divided into: large current emission, energy release, switch, eliminate jitter and signal sampling. Among them, the large current emission is the process that launching an alternating current to groundwater bodies to produce MRS phenomenon; Energy release is the process that releasing the remaining energy after launching; switch is the process that using the toggle switch to connect the receiving coil into reception system; eliminate jitter is the process that to remove the switch jitter interference. The above processes connect between the receiver and transmitter. Sequentially collected once the noise and then collected two signals. Measurement process is shown in Figure 6.

In the PC control software, time settings menu reflected the timing control of full-wave underground MRS receiver system. Users running PC software, we can set the control system parameters in the menu,

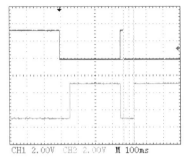

(a) Transmitting sync signal & Relay switching signal

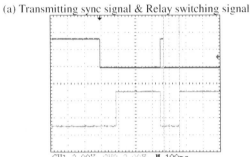

(b) Transmitting sync signal & Acquisition start signal

Figure 7. Timing control board outputs test diagram.

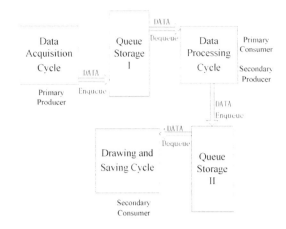

Figure 8. Block diagram of the data acquisition structure.

including the transmitting time T_S, energy release time T_N, switching time T_R and eliminate jitter time T_X. Input parameters comply upper and lower computer communication protocol via the serial port VI that VISA function module configured downloaded to the MCU.

In the hardware, timing control board reflected the timing control of full-wave underground MRS receiver system. Timing control board receives the transmitting synchronized signal generated by MRS transmitter. The board generates timing signals to control voltage relays and full-wave acquisition card according the time parameters downloaded by the PC software. We take the transmitting synchronization time as t_1, voltage relay switching time as t_2 and acquisition start time as t_3. The relationship between the three can be expressed as:

$$t_2 = t_1 + T_S + T_n \quad (6)$$

$$t_3 = t_1 + T_S + T_n + T_r + T_x \quad (7)$$

The programmable logic device calculated t_2 and t_3 by captured t_1 and then output corresponding control edges.

Figure 7(a) shows the transmitting sync signal and relay switching signal comparison. The blue curve is the transmitting sync signal and the red curve is the relay switching signal. Figure 7(b) shows the transmitting sync signal and acquisition start signal comparison. The blue curve is the transmitting sync signal and the red curve is the acquisition start signal.

Falling edge of the transmitting sync signal is effective; rising edge of relay switching signal and acquisition start signal is effective. The PC software downloaded $T_s = 40$ ms, $T_n = 20$ ms, $T_r = 30$ ms, $T_x = 1$ ms, timing control board to make precise timing control. As we can see from Figure 7, the relay switch starts switching at 60 ms after transmitting and the full-wave acquisition card starts collecting at 91 ms after transmitting.

3.3 Producer–consumer architecture

Acquisition part in control software is the producer–consumer mode architecture. In this architecture, data acquisition cycle as primary producer, data processing cycle as primary consumer and secondary producer, data drawing and saving cycle as secondary consumer. The three parts above constitute a collection framework shown in Figure 8. DATA generated by the data acquisition cycle is pushed into the queue Storage I. While Storage I not empty, data processing cycle will pop DATA from Storage I and superimpose then push DATA into the queue Storage II. Similarly, while Storage II not empty, data drawing and saving cycle will pop DATA from Storage II and draw diagrams display and save.

This architecture takes little time and realizes the data fast calculation and other functions stably and orderly. This acquisition architecture implemented multi-threaded execution, ensured the real-time monitoring for the program, saved running time in the maximum extent and improved the efficiency of full-wave underground MRS software.

4 TEST RESULTS AND ANALYSIS

4.1 Environmental noise test

Connect simulated coil to full-wave underground MRS receiver to test environmental noise in electromagnetic shielding room. Receive simulated coil inductance is

38.13 mH. After matching the measured data of the receiving system as shown in Figure 9. V_{pp} of the environmental noise is about 1000 nV, measured results accord with indoor noise range.

4.2 Simulated MRS signal test

The test of full-wave underground receiver system is to receive simulated MRS signal output by signal generator by the coils coupling way in the laboratory. Tests using AFG3021B Tektronix signal generator outputting initial 500 mV damped sine wave to simulate proton MRS signal shown in Figure 10(a). The sine wave frequency is 2326 Hz. After coils coupling the simulated MRS signal becomes nV level. After the front-end matching network and MRS amplifier, the signal is amplified 153,597 times. Collected by 128 k sampling rate, the signal is shown in Figure 10(b). The acquired waveform is not distortion through receiving system. This confirmed the authenticity of the receiving system.

5 SUMMARY

This paper designed and implemented full-wave underground MRS receiver based on the LabVIEW. By receiving matching network technology and the sequential control technology, cooperating with the independent development PC control software which uses "Producers–Consumers" programming framework realized the full-wave underground MRS signal reception. For the application of MRS method in underground engineering provides the instrument platform.

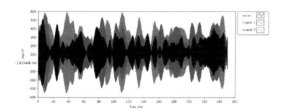

Figure 9. Environment noise measurements.

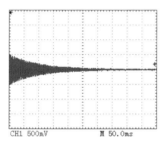

(a) Signal generator generates simulated MRS signal

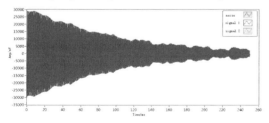

(b) Receiver test simulated MRS signal

Figure 10. Simulated MRS signal test.

REFERENCES

[1] Jonathan F. Stebbins and Ian Farnan. 1989. Nuclear magnetic resonance spectroscopy in the earth sciences; *structure and dynamics*. 245(4915):257–263.
[2] Müller-Petke M., Costabel S., Lange G. and Yaramanci U. 2009. Assessment of improved measurement technology for magnetic resonance sounding. *4th International Workshop on Magnetic Resonance Sounding, 20–23 October 2009*:153–158.
[3] Callaghan P. 2007. *Principles of Nuclear Magnetic Resonance Microscopy*, New Zealand: Oxford University Press.
[4] Wang Z X, Rong L L, Lin J, et al. 2010. FID Signal Detection Based on DLIA Sampled by Quadruple Lamor Frequency. *Journal of Data Acquisition & Processing*, 25(5): 1004–9307.
[5] David O. Walsh. 2008, Multi-channel surface NMR instrumentation and software for 1D/2D groundwater investigations. *Journal of Applied Geophysics*, 66: 140–150.
[6] Wang Y J, Lin J. 2008. Amplifier design of Surface nuclear magnetic resonance instrument for underground water invest *Chinese Journal of Scientific Instrument*. 29 (8): 1627–1632.
[7] Yi X F, Li P F, Lin J. 2013. Simulation and experiment research of MRS response based on multi-turn Loop. *Chinese Journal of Geophysics*, 56(7):2484–2493.
[8] Grunewald E, Knight R. 2012. Nonexponential decay of the surface-NMR signal and implications for water content estimation. *Geophysics*, 77, EN1–EN09.

The cooling effect and cooling energy savings potential in Beijing for metamerically color-matched cool colored coatings

J. Song, J. Qin, J. Qu, W. Zhang, Z. Song, Y. Shi, L. Jiang, J. Li, X. Xue & T. Zhang
Technical Center, China State Construction Engineering Co. Ltd., Beijing, P.R. China

ABSTRACT: Conventional and cool black, brown, red, green, yellow and blue coatings were prepared and their optical and thermal properties were systematically investigated using a UV/VIS/NIR spectrophotometer and a portable differential thermopile emissometer. The cooling effect and the cooling energy savings in Beijing resulting from these cool colored coatings were estimated. The cool colored coatings created in this work metamerically match the corresponding conventional analogies. The predicted cooling effect of the cool colored coatings relative to the corresponding conventional coatings range from 4.2°C for the cool green coating pigmented with phthalocyanine green to 17°C for the yellow coating pigmented with nickel titanium yellow. The estimated cooling energy savings resulting from the cool colored coatings in Beijing range from 1.07 to 3.23 kWhm^{-2}yr^{-1} for the buildings with moderate roof insulation; they range from 1.07 to 5.39 kWhm^{-2}yr^{-1} for the buildings with poor roof insulation.

Keywords: Solar reflectance; Thermal emittance; Near infrared reflecting pigment; Cooling effect; Metameric match; Cooling energy savings.

1 INTRODUCTION

Cool colored coatings play an important role in the building energy efficiency because they are more popular than the coolest white solar reflective coatings for pitched roofs and exterior walls due to the aesthetic and visual considerations (Song et al., 2014; Levinson et al., 2007a; Levinson et al., 2007b; Santamouris et al., 2011; Synnefa et al., 2007; Uemoto et al., 2010) and their better contamination resistance. The key point to create a cool colored coating is to improve its near infrared (NIR, 700–2500 nm) reflectance because strong ultraviolet (UV, 250–400 nm) absorptance is required for all the coatings to shield the coatings and substrates and the visible (VIS, 400–700 nm) spectral reflectance is fixed to yield a specific color (Levinson et al., 2007a; Levinson et al., 2007b; Santamouris et al., 2011; Synnefa et al., 2007).

Generally speaking, there are two methods for improving the NIR reflectance of a cool colored coating. Method 1 is the so-called two-layer technique (Levinson et al., 2007a; Levinson et al., 2005; Brady and Wake, 1992). The idea is to prepare a thin colored topcoat colored with NIR-transmitting pigments and a white basecoat with high NIR reflectance, and then spray them onto roof and/or wall substrates with low NIR reflectance (e.g. concrete substrates). Method 2 is to directly spray the NIR-transmitting colored coating onto roof and/or wall substrates with high NIR-reflectance (e.g. aluminum and wooden roofs) (Levinson et al., 2007a; Levinson et al., 2005; Brady and Wake, 1992).

In the present paper, the representative cool black, brown, red, green, yellow and blue coatings were prepared in our laboratory. Their optical properties were systematically investigated and compared with the corresponding conventional colored coatings. Furthermore, the thermal emittance values of all the conventional and cool colored coatings were also tested. In addition, the cooling effect and cool energy savings in Beijing of the cool colored coatings relative to the corresponding conventional analogies were estimated and discussed. The main goal of this paper is to provide more insight into the research area of the building energy efficiency.

2 EXPERIMENTAL

2.1 *Selection of materials*

In order to prepare the cool colored coatings, a water-based pure acrylic emulsion was selected as binder. Talcum was also purchased as extender pigment since it is transparent and non-reflective throughout the visible and near-infrared regions and thus does not affect the performance of other pigments (Brady and Wake, 1992; Bendiganavale and Malshe, 2008). The commercially available conventional and NIR-transmitting pigments selected in this work, together with the designations of the colored coatings pigmented with these colorants, are listed in Table 1.

In addition, appropriate paint additives, such as a wetting agent, a dispersant, an antifoaming agent, a

Table 1. Conventional and NIR-transmitting colored pigments selected in this paper and the designations for the colored coatings pigmented with these colorants.

Colors	Pigments	Coatings
Black	Carbon black	S_{c-b}-black
	Perylene black	S_P-black
	Dioxizine purple	S_{di}-black
Brown	Zinc iron chromite brown	S_{Fe-Cr}-brown
	Chrome antimony titanium buff rutile	S_{Ti-Cr}-brown
Red	Iron oxide red	S_{Fe}-red
	Diketopyrrolopyrrole red	S_{dik}-red
Green	Cobalt green	S_{Co}-green
	Phthalocyanine green	S_{ph}-green
Yellow	Iron oxide yellow	S_{Fe}-yellow
	Cromophtal yellow	S_{Cr}-yellow
	Titanium-chrome yellow	S_{Ti-Cr}-yellow
	Nickel titanium yellow	S_{Ti-Ni}-yellow
Blue	Iron blue	S_{Fe}-blue
	Cobalt blue	S_{Co}-blue
	Phthalocyanine blue	S_{Ph}-blue

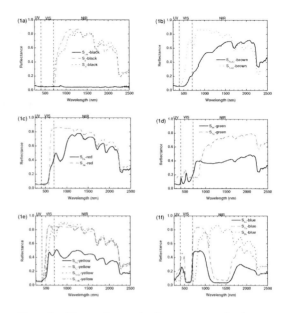

Figure 1. Measured spectral reflectance curves for the conventional and cool black coatings (a), brown coatings (b), red coatings (c), green coatings (d), yellow coatings (e) and blue coatings (f).

leveling agent and a coalescent, were also selected to improve the quality and performance of the coatings.

2.2 Preparation of cool colored coatings and samples

The preparation process of the conventional and cool colored coatings was nearly the same as that described in detail elsewhere (Song et al., 2014). The obtained colored coatings were airlessly sprayed on to the self-manufactured white basecoats, whose optical and physicochemical properties were described in detail elsewhere (Shi et al., 2013; Song et al., 2013; Zhang et al., 2013), to prepare the specimens for the measurements of their spectral reflectance and thermal emittance.

2.3 Spectral reflectance and thermal measurements

A UV/VIS/NIR spectrophotometer (Perkin Elmer Lambda 750), equipped with an integrating sphere (150-mm diameter, Labsphere RSA-PE-19), was employed to measure the spectral reflectance of colored coatings following ASTM E903-96 (Standard test method for solar absorbance, reflectance and transmittance of materials using integrating spheres). The solar reflectance was computed by integrating the measured spectral data weighted with the air mass 1.5 beam-normal solar spectral irradiance.

A portable differential thermopile emissometer AE1 (Devices & Services Co., Dallas, TX) was used to measure the thermal emittance of the colored coatings according to ASTM C 1371 (Standard test method for determining the emittance of materials near room temperature using portable emissometer).

3 RESULTS AND DISCUSSION

3.1 The spectral reflectance and thermal emittance of colored coatings

Fig. 1 shows the spectral reflectance curves of the cool colored coatings and the corresponding conventional analogies over cool white basecoats, whose solar reflectance is approximately 0.88. The corresponding computed spectral and solar reflectance values are summarized in Table 2. Several observations can be made from Fig. 1 and Table 2. First, the NIR and solar reflectances of all the cool colored coatings are higher than those of the corresponding conventional colored coatings. Second, as indicated in Fig. 1 and Table 2, although the cool brown, red, green, yellow and blue coatings have the same colors as the corresponding conventional analogies, their reflectance curves in the VIS region do not overlap and their VIS reflectance values are not equal, indicating that the created colored coatings are only metameric to the corresponding conventional colored coatings. Despite the reflectance curves in the VIS region and the computed VIS reflectances of the cool black coatings separately pigmented with perylene black and dioxazine purple colorants are quite similar to those of the standard black coating pigmented with carbon black colorant, the three coatings are still metamerically matched. Actually, the black coating pigmented with perylene black has a green shade regardless of its jet black appearance, while the black coating pigmented dioxazine purple has a violet shade, as indicated by the small overshoot in the VIS region between approximately 400–435 nm. In addition, as indicated in Fig. 1f,

Table 2. Measured spectral and solar reflectance values for the conventional and cool colored coatings.

Samples		Reflectance			
		Solar	UV	VIS	NIR
Black	S_{c-b}-black	0.049	0.053	0.050	0.047
	S_p-black	0.346	0.051	0.075	0.715
	S_{di}-black	0.334	0.054	0.107	0.644
Brown	S_{Fe-Cr}-brown	0.288	0.057	0.131	0.508
	S_{Ti-Cr}-brown	0.582	0.065	0.449	0.793
Red	S_{Fe}-red	0.368	0.054	0.216	0.582
	S_{dik}-red	0.552	0.057	0.389	0.798
Green	S_{Co}-green	0.255	0.060	0.165	0.387
	S_{ph}-green	0.341	0.172	0.183	0.557
Yellow	S_{Fe}-yellow	0.352	0.054	0.314	0.423
	S_{Cr}-yellow	0.661	0.048	0.604	0.781
	S_{Ti-Cr}-yellow	0.609	0.063	0.469	0.831
	S_{Ti-Ni}-yellow	0.685	0.060	0.632	0.802
Blue	S_{Fe}-blue	0.222	0.149	0.189	0.274
	S_{Co}-blue	0.408	0.210	0.328	0.528
	S_{Ph}-blue	0.342	0.065	0.093	0.686

Table 3. Measured thermal emittance of the conventional and cool colored coatings.

Samples		Thermal emittance
Black	S_{c-b}-black	0.90
	S_p-black	0.91
	S_{di}-black	0.89
Brown	S_{Fe-Cr}-brown	0.88
	S_{Ti-Cr}-brown	0.91
Red	S_{Fe}-red	0.90
	S_{dik}-red	0.89
Green	S_{Co}-green	0.92
	S_{ph}-green	0.89
Yellow	S_{Fe}-yellow	0.90
	S_{Cr}-yellow	0.90
	S_{Ti-Cr}-yellow	0.91
	S_{Ti-Ni}-yellow	0.92
Blue	S_{Fe}-blue	0.89
	S_{Co}-blue	0.89
	S_{Ph}-blue	0.90

in the VIS region, the spectral reflectance curve of the coating pigmented with phthalocyanine blue is similar to that of the standard black pigmented with carbon black except that there is a reflectance peak around 450 nm, which is attributed to the reflectance of blue color. As shown in Table 2, the computed VIS reflectance of the coating pigmented with phthalocyanine blue is 0.093, which is slightly higher than that of the standard black coating (0.05). The blue coating pigmented with phthalocyanine blue actually shows a black appearance with a blue shade.

The thermal emittance of different conventional and cool colored coatings are listed in Table 3. As anticipated, the thermal emittance of the cool colored coatings is approximately 0.9. It is a common knowledge that polymer-based coatings have high thermal emittance values ranging from 0.85 to 0.95 (Levinson et al., 2007a; Santamouris et al., 2011; Uemoto et al., 2010; Brady and Wake, 1992), which are not affected by cool pigments (Synnefa et al., 2007).

3.2 Estimates of the cooling effect of cool colored coatings

In the previous studies (Uemoto et al., 2010; Shi et al., 2013; Song et al., 2013; Song et al., 2012; Smith et al., 2003; Cao et al., 2011), self-developed apparatuses equipped with infrared (IR) lamps were employed to assess the cooling effect of cool coatings. However, the spectrum of IR lamps is different than that of the solar spectrum. The former has a larger infrared component and a much smaller visible component (Song et al., 2013; Zhang et al., 2013). Therefore, the measured cooling effect of the cool coatings is not the true cooling effect. The true cooling effect of a cool coating may be directly tested in the sun on a sunny calm summer day or estimated following ASTM E 1980-01 (standard practice for calculating solar reflectance index of horizontal and low sloped opaque surfaces) under the following standard conditions: insolation = 1000 W/m², sky temperature = 300 K, ambient air temperature = 310 K and convection coefficient (medium wind) = 12 W/(m²K). Because the first method is not always available and the testing conditions change over time, the cooling effect of the cool colored coatings developed in this work was estimated using the second method. The estimated cooling effect of the cool colored coatings relative to the corresponding conventional analogies are tabulated in Table 4. Compared to the corresponding conventional analogies, all the cool colored coating have lower surface temperatures. The cooling effect of the cool colored coatings, characterized by the surface temperatures reduction values relative to the conventional colored coatings, is quite pronounced for the cool black, brown and yellow coatings; it is less obvious for the cool red and blue coatings; it is least noticeable for the cool green coating.

3.3 Estimates of cooling energy savings in Beijing

The cooling energy savings in Beijing resulting from the cool colored coatings on the small sloped and/or flat roofs were estimated using DOE-2.1 E computations, which correlates the cooling energy savings in summer to annual cooling degree days (base 18°C, CDD18). The climate data for Beijing Capital Airport were obtained from the Chinese typical meteorological year (CTMY) database, which was co-developed by Zhang et al. (Zhang et al., 2002) and Lawrence Berkeley National Laboratory. The simulation represents the energy savings relative to the corresponding conventional analogies.

Assuming that the coefficient of performance (COP) of the cooling air conditioner is an average

Table 4. Estimated cooling effect of the cool colored coatings relative to the corresponding conventional analogies.

Samples		Surface temperatures (K)	Temperature reduction values (°C)
Black	S_{c-b}-black	355.6	–
	S_p-black	341.2	14.4
	S_{di}-black	341.8	13.8
Brown	S_{Fe-Cr}-brown	344.1	–
	S_{Ti-Cr}-brown	329.3	14.8
Red	S_{Fe}-red	340.1	–
	S_{dik}-red	330.8	9.3
Green	S_{Co}-green	345.7	–
	S_{ph}-green	341.5	4.2
Yellow	S_{Fe}-yellow	340.9	–
	S_{Cr}-yellow	325.1	15.8
	S_{Ti-Cr}-yellow	327.9	13.0
	S_{Ti-Ni}-yellow	323.9	17.0
Blue	S_{Fe}-blue	347.3	–
	S_{Co}-blue	338.1	9.2
	S_{Ph}-blue	341.4	6.9

Table 5. Estimated annual cooling energy savings in Beijing resulting from the cool colored coatings used on roofs with different thermal insulation levels relative to the corresponding conventional analogies.

		Annual cooling energy savings (kWhm^{-2}yr^{-1})	
Samples		RSI (1.74 m^{2}KW^{-1})	RSI (0.87 m^{2}KW^{-1})
Black	S_{c-b}-black	–	
	S_p-black	3.23	4.31
	S_{di}-black	2.15	4.30
Brown	S_{Fe-Cr}-brown	–	
	S_{Ti-Cr}-brown	3.23	4.31
Red	S_{Fe}-red	–	
	S_{dik}-red	1.07	2.15
Green	S_{Co}-green	–	
	S_{ph}-green	1.07	1.07
Yellow	S_{Fe}-yellow	–	
	S_{Cr}-yellow	3.23	5.38
	S_{Ti-Cr}-yellow	3.23	4.31
	S_{Ti-Ni}-yellow	3.23	5.39
Blue	S_{Fe}-blue	–	
	S_{Co}-blue	1.07	2.15
	S_{Ph}-blue	1.07	1.07

value of 2.0, the estimated cooling energy savings resulting from the above cool colored coatings in Beijing for the prototypical model house with roof insulation of R-10 (1.74 m^{2}KW^{-1}) and R-5 (0.87 m^{2}KW^{-1}) are tabulated in Table 5. As shown in Table 5, for the cool green and blue coating separately pigmented with phthalocyanine green and phthalocyanine blue colorants, which have the smallest increase in solar reflectance relative to the corresponding conventional analogies, the cooling energy savings are the same for the roofs with different insulation levels. For the other cool colored coatings with higher increase in solar reflectance, the cooling energy savings for the buildings with poorer roof insulation are larger than that for buildings with better roof insulation. This indicates that increasing the solar reflectance is more important than increasing roof insulation for building energy efficiency. For the buildings with moderate roof insulation, the estimated annual cooling energy savings in Beijing range from 1.07 to 3.23 kWhm^{-2}yr^{-1}; for the buildings with poor roof insulation, the estimated annual cooling energy savings in Beijing range from 1.07 to 1.39 kWhm^{-2}yr^{-1}.

4 CONCLUSIONS

The above new findings lead us to draw the following conclusions:

- All the colored coatings prepared in this work metamerically match the conventional colored coatings.
- The cooling effect of the cool black, brown and yellow coatings is most pronounced, while the cooling effect of the cool green coating is least noticeable. The cooling effect of the cool red and blue coatings is in between.
- The estimated energy savings resulting from the cool colored coatings in Beijing range from 1.07 to 3.23 kWhm^{-2}yr^{-1} for the buildings with moderate roof insulation; they range from 1.07 to 5.39 kWhm^{-2}yr^{-1} for the buildings with poor roof insulation.
- Increasing solar reflectance of roofs is more effective than improving insulation of roofs for building energy efficiency.

ACKNOWLEDGMENTS

This work was performed under the "Water-Borne Cool Coatings for Building Energy Efficiency" project with funding from the Technical Center of China State Construction Engineering Co. Ltd.

REFERENCES

Bendiganavale, A. K. & Malshe, V. C. 2008. Infrared reflective inorganic pigments. *Recent Patents on Chemical Engineering* 1(1): 67–79.

Brady, R. F. & Wake, L. V. 1992. Principles and formulations for organic coatings with tailored infrared properties. *Progress in Organic Coatings* 20(1): 1–25.

Cao, X., Tang, B., Zhu, H., Zhang, A. & Chen, S. 2011. Cooling Principle Analyses and Performance Evaluation of Heat-Reflective Coating for Asphalt Pavement. *J. Mater. Civ. Eng.* 23 (7): 1067–1075.

Levinson, R., Berdahl, P., Akbari, H., Miller, W., Joedicke, I., Reilly, J., Suzuki, Y. & Vondra, M. 2007a. Methods of creating solar-reflective nonwhite surfaces and their application to residential roofing materials. *Solar Energy Materials and Solar Cells* 91(4): 304–314.

Levinson, R., Akbaria, H. & Reilly, J. C. 2007b. Cooler tile-roofed buildings with near-infrared-reflective non-white coatings. *Building and Environment* 42(7): 2591–2605.

Levinson, R., Berdahl, P. & Akbari, H. 2005. Solar spectral optical properties of pigments—Part I: model for deriving scattering and absorption coefficients from transmittance and reflectance measurements. *Solar Energy Materials & Solar Cells* 89(4): 319–349.

Santamouris, M., Synnefa, A. & Karlessi, T. 2011. Using advanced cool materials in the urban built environment to mitigate heat islands and improve thermal comfort conditions. *Solar Energy* 85(12): 3085–3102.

Smith, G. B., Gentle, A., Swift, P. D., Earp, A. & Mronga, N. 2003. Coloured paints based on iron oxide and silicon oxide coated flakes of aluminium as the pigment, for energy efficient paint: optical and thermal experiments. *Solar Energy Materials and Solar Cells* 79(2): 179–197.

Shi, Y., Song, Z., Zhang, W., Song, J., Qu, J., Wang, Z., Li, Y., Xu L. & Lin, J. 2013. Physicochemcial properties of dirt-resistant cool white coatings for building energy efficiency. *Solar Energy Materials and Solar Cells* 110:133–139.

Song, Z., Qin, J., Qu, J., Song, J., Zhang W., Shi, Y., Zhang, T., Xue, X., Zhang, P., Zhang, H., Zhang, Z. & Wu, X. 2014. A systematic investigation of the factors affecting the optical properties of near infrared transmitting cool non-white coatings. *Solar Energy Materials and Solar Cells* 125: 206–214.

Song, Z., Zhang, W., Shi, Y., Song, J., Qu, J., Qin, J., Zhang, T., Li, Y., Zhang H. & Zhang, R. 2013. Optical properties across the solar spectrum and indoor thermal performance of cool white coatings for building energy efficiency. *Energy and Buildings* 63: 49–58.

Song, Z., Shi, Y., Zhang, W., Song, J., Qu, J., Li, Y. & Wang, Z. 2012. Development of a continuous testing apparatus for temperature reduction performance of cool coatings. *Review of Scientific Instruments* 83: 054901-1–054901-5.

Synnefa, A., Santamouris, M. & Apostolakis, K. 2007. On the development, optical properties and thermal performance of cool colored coatings for the urban environment, *Solar Energy* 81(4): 488–497.

Uemoto, K. L., Sato, N. M. N. & John, V. M. 2010. Estimating thermal performance of cool colored paints. *Energy and Buildings* 42 (1): 17–22.

Zhang, Q., Huang, J. & Lang, S. 2002. Development of typical year weather data for Chinese locations. *ASHRAE Transactions* 108(2): 1063–1075.

Zhang, W., Song, Z., Shi, Y., Song, J., Qu, J., Qin, J., Zhang, T., Li, Y., Xu, L. & Xue, X. 2013. The effects of manufacturing processes and artificial accelerated weathering on the solar reflectance and cooling effect of cool roof coatings. *Solar Energy Materials and Solar cells* 118: 61–71.

Zhang, W., Song, Z., Song, J., Shi, Y., Qu, J., Qin, J., Zhang, T., Li, Y., Zhang, H. & Zhang, R. 2013. A systematic laboratory study on an anticorrosive cool coating of oil storage tanks for evaporation loss control and energy conservation. *Energy* 58: 617–627.

Improved K-means algorithm with better clustering centers based on density and variance

G.C. Deng, J.C. Tao, M.J. Zhou & Y.C. Xu
School of Information Engineering, Nanchang University, Nanchang, China

ABSTRACT: Considering the defection that the traditional K-means algorithm has sensitivity to the initial clustering centers, an improved algorithm is presented. It computes the global density of every data object, and then adjusts the density with the variance of data object's density area, and then finds K data objects which have the high density area to be the K initial centers dynamically. On the premise of the K is given, experiments on UCI demonstrate that the improved algorithm can convergence more quickly and can get a better clustering.

Keywords: K-means algorithm; Clustering centers; Global density; Variance

1 INTRODUCTION

Clustering analysis is an important research field in data mining. Clustering aims at dividing a collection of data objects into different classes, and makes the same class of data objects as similar as possible, and between different classes of data objects as dissimilar as possible.

K-means is a kind of partition-based clustering algorithm. First, find out K data objects to be initial clustering centers, which represent the initial average value of each class. Second, for the rest of data objects, according to the distance between them and initial clustering centers, classifies them into the most similar class. Third, update clustering centers by calculating the new average distance of each class. This process is repeated until reaching convergence conditions (Han & Kamber 2006).

Traditional K-means algorithm selects initial clustering centers form all data objects randomly, which gets different clustering results. Therefore, it is of great significance for K-means algorithm to find a set of suitable initial clustering centers, which can not only reduce the instability of clustering results, but also get better clustering results.

On the whole, there are three methods to select initial clustering centers (He et al. 2004): random sampling, distance optimization and density estimation. (Qing & Zheng 2009) improves K-means algorithm by means of dividing sampling. (Tong et al. 2009, Yao & Shi 2010, Cao et al. 2009, Feng 2007) use max-min distance means to determine the initial centers. (Chen & Wang 2012, Sun & Liu 2008) find those data objects which are farthest away from each other in the highest density area to be initial centers and effectively reduce the possibility of selecting isolated objects. Besides considering the area density of data objects, (Feng et al. 2012, Li & Wang 2010) get current optimal initial centers through minimum spanning tree. (Zhang & Duan 2013) introduces individual silhouette coefficient, through running K-means algorithm several times, calculating and comparing the silhouette coefficient of final results, thus to get the optimal initial clustering centers and final clustering result.

In this paper, an improved K-means algorithm is proposed, which improves the method of selecting initial clustering centers and updates the new clustering center during the clustering process. This can make sure the clustering centers can reflect the relationship between data objects and actual distribution in dataset. Experiments show that the improved algorithm in this paper achieves better clustering results and faster convergence speed than traditional K-means algorithm.

2 DEFINITIONS OF IMPROVED K-MEANS ALGORITHM

Data objects set to be clustered: $S = \{x_1, x_2, \ldots, x_3\}$, n is the total number of data objects, $x_i = (x_{i1}, x_{i2}, \ldots, x_{ir})$, r is the property dimension of x_i

K initial clustering centers: $Z = \{c_1, c_2, \ldots, c_k\}$, k is the number of classes as well.

There are definitions as follows:

Definition 1 Euclidean distance between two data objects (Liu 2009)

$$d(x_i, x_j) = \sqrt{\sum_{p=1}^{r}(|x_{i1} - x_{jp}|)^2} \quad (1)$$

Definition 2 Density parameter of data objects.

In general, people want to choose the data objects to be centers which have big data concentration. But it is only based on distance factor, usually easy to get some noise data objects. So when choosing the initial clustering centers, consider some factors together such as distance, density and distribution of data objects.

$$density(x_i, MinPts) = R \times \frac{D_{x_i}}{D_{x_i} + R} \quad (2)$$

R is the minimum radius of a circle, which is centered on x_i, containing the specified MinPts numbers of data objects, and R is called the density area of x_i. Value of R is smaller, x_i has a higher density in its density area, and on the contrary, x_i has a lower density.

$$D_{x_i} = \frac{1}{MinPts} \sum_{j=1}^{MinPts} \left(d(x_i, x_j) - \overline{d(x_i, x_j)} \right)^2 \quad (3)$$

D_{x_i} is the distance variance of x_i and other data objects in the x_i's density area. $\overline{d(x_i, x_j)}$ is the mean distance of x_i and other data objects in the x_i's density area. The smaller the value D_{x_i} is, the more evenly distributed data objects in the x_i's density area are, on the contrary, the less evenly distributed data objects in the x_i's density area are. $\frac{D_{x_i}}{D_{x_i}+R}$ is the adjusting parameter of R. It can make the selection of initial clustering centers more reflect the distribution of data objects. Therefore, the smaller R and D_{x_i} are, the higher density and more evenly distribution of data objects in the x_i's density area are. That is the higher the density parameter of x_i.

Definition 3 Global density parameter of data objects.

Data object x_i's global density, remember to densityG(x_i), $MinPts = \lambda_1 n/k$, x_i ($x_i \in$ S'). Among them, $0 < \lambda_1 \leq 1$, n is the total number of data objects; k is the number of clustering classes. S' is the set of available data objects in S, it changes dynamically with the selection of initial clustering centers.

Definition 4 Local density parameter of data objects.

Data object x_i's local density, remember to densityL(x_i), $MinPts = \lambda_2 n_k$, x_i ($x_i \in$ cluster$_k$). Among them, $0 < \lambda_2 \leq 1$, cluster$_k$ is the set of data objects in the kth class; n_k is the number of data object in cluster$_k$.

Definition 5 Updated clustering centers

$$z_i' = (x_{maxL} + z_i)/2 \quad (4)$$

x_{maxL} is the minimum density L in the ith class, that is the data object which has the maximum local density.

z_i is the arithmetic average distance of data objects in the same class (Liu 2009),

$$z_i = \frac{1}{n_i} \left(\sum_{j=1}^{n_i} x_{j1}, \sum_{j=1}^{n_i} x_{j2}, \cdots, \sum_{j=1}^{n_i} x_{jr} \right)$$

Among them, n_i is the number of data objects in the ith class, x_{jp} is the pth property, $j = (1, 2, \ldots, n_i)$, $p = \{1, 2, \ldots, r\}$.

Definition 6 Evaluation of clustering validity (Rand 1971)

$$RI = \frac{TP + TN}{TP + FP + FN + TN} \quad (5)$$

Rand Index (RI) is used to compare the consistency of the clustering results and known optimal clustering results. Among them, TP means two data objects correctly are classified into the same class which should be classified into the same class. TN means two data objects are correctly classified into two different classes which should be classified into two different classes. FP means two data objects are wrongly classified into the same class which should be classified into two different classes. FN means two data objects are wrongly classified into two different classes which should be classified into the same class. The value of RI is within the interval [0, 1], and the closer to 1, the better clustering results are.

3 DESCRIPTION OF IMPROVED K-MEANS ALGORITHM

The description of improved initial clustering centers K-means algorithm is as follows:

Input: the number of classes k and the set containing n data objects;
Output: k clustering centers meeting the convergence condition and final clustering results.

1. Calculate d(x_i, x_j) according to the formula (1), which is the distance between any two data objects in data objects set S, and then get a set Dist, Dist = {d(x_i, x_j)|$i,j \in (1,2,\ldots,n)$}.
2. Calculate the global density of each data objects in S according to the formula (2), and then get a set DG, D_G = {densityG(x_i)|$i \in (1,2,\ldots,n)$}.
3. Set the data object p_i which has the highest global density in set DG as the ith initial clustering center c_i. Then delete the MinPts data objects which is in the circle centered on p_i from set S, and delete those data objects' global density from set DG, and delete distance value involved with those data objects from set Dist, $i = (1, 2, \ldots, k)$.
4. Repeat steps (2) and (3) until find certain k data objects.
5. Set these k data objects above to be the initial clustering centers of K-means algorithm, representing the clustering centers of k classes.
6. Calculating the distance between each data object with these clustering centers on the base of each class center, classify the data object into the class which has the minimum distance.
7. Calculate local density of data objects in each class, then get set DL, D_L = {density L(x_i)|$i \in (1,2,\ldots,n_k)$}, and update clustering center z_k' of each class.

8. Repeat steps (6) and (7) until meeting convergence condition (clustering centers no longer change) (Liu 2009).

4 EXPERIMENT RESULTS AND ANALYSIS

4.1 Experiment description

In order to study the effectiveness of the improved initial clustering centers algorithm, this paper chooses dataset Iris, Wine and Balance-scale in UCI as the experiment dataset, and the number of iterations and RI as the evaluation of clustering results. Information about these datasets is shown in table 1.

4.2 Experiment results

Comparing the results of the traditional K-means algorithm and proposed improved K-means algorithm in this paper, the former runs 10 times, gets the average iterations and RI. Because the initial clustering centers and clustering centers during clustering of the latter are fixed, it just runs once.

The experiment results are shown in table 2.

Among them, A.K is the traditional K-means algorithm. A.I.K is the improved K-means algorithm in this paper. I.C.C is the initial clustering centers. N.It is the number of iteration during running algorithm.

In table 2, the data objects to be initial clustering centers are described by their line number in the specified dataset. For example, 12, 79, 109 indicates selecting the 12th data object (4.8, 3.4, 1.6, 0.2), the 79th data object (6.0, 2.9, 4.5, 1.5) and the 109th data object (6.7, 2.5, 5.8, 1.8) in dataset Iris to be initial clustering centers.

4.3 Analysis of experiment results

The traditional K-means algorithm selects initial clustering centers randomly, resulting every time the algorithm runs, the centers are different, and the numbers of iterations are different, and clustering results are different, and the value of RI has a certain range.

From table 2, for dataset Iris, if runs traditional K-means algorithm, the value of RI can top 87.97%, the lowest is 71.43%, the average is 81.20%, which is lower than 87.97% gotten by the improved K-means algorithm, and the latter's convergence speed is much faster. For dataset Wine, if runs traditional K-means algorithm, the value of RI can top 71.87%, the lowest is 68.80%, the average is 70.72%, which is lower than 72.73% gotten by the improved K-means algorithm, and the latter's convergence speed is just a little faster. For dataset Balance-scale, if runs traditional K-means algorithm, the value of RI can top 61.85%, the lowest is 56.86%, the average is 58.38%, which is lower than 62.05% gotten by the improved K-means algorithm, and the latter's convergence speed is faster than the former. Look at the whole initial clustering centers by these two algorithms, we can find that the initial clustering centers by the improved K-means algorithm are from each class in datasets. Therefore, comparing the traditional K-means algorithm and the improved K-means algorithm, the latter has a better clustering result and a faster convergence speed.

Table 1. Information of each dataset.

Dataset	PRM			
	Number of data	Dimension of properties	Number of classes	Number of data in each class
Iris	150	4	3	50 50 50
Wine	178	13	3	59 71 48
Balance-scale	625	4	3	49 288 288

Table 2. Experiment results of two algorithms.

Dataset		Iris			Wine			Balance-scale		
Algorithm		I.C.C	N.It	RI (%)	I.C.C	N.It	RI (%)	I.C.C	N.It	RI (%)
A.K	1	13,31,122	4	71.45	95,43,14	5	69.20	415,41,161	13	58.00
	2	13,92,17	3	71.43	133,15,152	5	71.87	223,89,544	15	57.23
	3	65,103,15	6	87.37	40,13,5	4	68.80	246,475,542	20	58.73
	4	92,144,101	7	87.97	6,9,144	11	69.20	178,154,203	3	58.38
	5	90,51,40	9	87.37	132,50,163	5	71.87	148,500,604	11	61.85
	6	136,9,66	5	87.97	160,58,129	6	71.87	537,434,514	16	58.85
	7	11,63,30	3	71.43	98,58,33	3	68.81	185,331,536	19	56.93
	8	89,101,109	7	87.97	49,77,128	8	71.87	178,216,412	15	56.86
	9	29,102,36	8	71.69	50,108,136	3	71.87	397,227,548	31	59.40
	10	90,62,124	8	87.37	86,89,123	10	71.87	353,606,28	23	57.60
	AVG		6	81.20		6	70.72		17	58.38
A.I.K		12,79,109	3	87.97	108,154,47	5	72.73	313,587,65	12	62.05

5 CONCLUSION

Because the traditional K-means algorithm depends on the selected initial clustering centers, the clustering results differ in a large range with the different initial clustering centers. A new improved K-means algorithm is proposed in this paper. The algorithm gets the density concentration of data objects in specified density area, and introduces distance variance to adjust the density, and then figures out the density parameter of data objects. By ranking the density parameters to select the initial clustering centers, and then runs K-means algorithm. During the process of clustering, it updates the clustering centers on the base of the average of distance and the local density. The results of experiments on several datasets from UCI show that the improved K-means algorithm in this paper can get a better clustering result, and have a higher stability, and converge faster.

ACKNOWLEDGEMENTS

National Natural Science Foundation of China (61262049)

REFERENCES

[1] Cao, Z.Y. et al. 2009. K_means Clustering Algorithm with Fast Lookup Initial Start Center. Journal of Lanzhou Jiaotong University, 28(6):15–18.
[2] Chen, G.P. & Wang, W.P. 2012. An Improved K-means Algorithm with Meliorated Initial Center//The 7th International Conference on Computer Science & Education (ICCSE), 150:153.
[3] Feng, B. et al. 2012. Optimization to K-means Initial Cluster Centers. Computer Engineering and Applications.
[4] Feng, C. 2007. Research of K-means Clustering Algorithm. Dalian: Dalian University of Technology, 25–29.
[5] Han, J.W. & Kanmber, M. 2006. Data mining: Concepts and Techniques (Second Edition), San Francisco: Morgan Kaufmann Publishers, 402–404.
[6] He, J. et al. 2004. Initialization of Cluster Refinement Algorithms: A Review and Comparative Study, Proceedings of International Joint Conference on Neural Networks, 297–298.
[7] Li, C.S. & Wang, Y.N. 2010. New Initialization Method for Cluster Center. Control Theory & Application, 27(10):1435–1440.
[8] Liu, B. 2009. Web data mining. Beijing: Tsinghua University press, 89-94.
[9] Qing, X.P. & Zheng, S.J. 2009. A new Method for Initializing the K-means Clustering Algorithm. Second International Symposium on Knowledge Acquisition and Modeling, 41–44.
[10] Rand, W. 1971. Objective Criteria for the Evaluation of Clustering Methods. Journal of the American Statistical Association, 66(336): 846–850.
[11] Sun, X.J. & Liu, X.Y. 2008. New Genetic K-means Clustering Algorithm Based on Meliorated Initial Centers. Computer Engineering and Applications, 44(23):166–168.
[12] Tong, X.J. et al. 2011. Optimization to k-means Initial Cluster Centers. Computer Engineering and Design, 32(8):2721–2788.
[13] Yao, Y.H. & Shi, X.L. 2009. K-means Rough Clustering Algorithm Based on Optimized Initial Center. Computer Engineering and Applications, 46(34):126–128.
[14] Zhang, J. & Duan, F. 2013. Improved k-means Algorithm with Meliorated Initial Centers. Computer Engineering and Design, 34(4):1691–1699.

Kind of image defect detection algorithm based on wavelet packet transform and Blob Analysis

L.X. Ao & X.B. Zhou
School of Information Engineering, Nanchang University, Nanchang, China

ABSTRACT: By using wavelet packet, the image can be decomposed both in the low frequency part and high frequency part. Wavelet packet analysis technology has been used in image processing as a new time-frequency analysis tool. In this paper, by using wavelet packet to decompose image and proper threshold to process the coefficients of the wavelet packet tree, then reconstruct the defect image. Using blob analysis method to extract the defect object and determine its type.

Keywords: wavelet packet; Blob Analysis; printing defects

1 INTRODUCTION

Due to production equipment, production environment, operators and other factors, presswork will appear misregister, misting, stain, leakage and many other defects in the high speed printing [1]. All the difference between standard template and printing image is printing defects. The online print defect detection system based on machine vision can meet the real-time requirements of printing quality detection in industrial production. Many researchers do a lot of experiments and research on printing defects detection, and their basic thought was the same. Print defects performance in the image is the difference of the gray value between the defect image and standard image. Compare the gray value of defect image with standard image, and determine the difference is beyond the scope of preset threshold, so that decide the presswork with defects or not [2].

In this paper, by using wavelet packet algorithm to isolate defect area, then take the advantage of regional analysis method to extract defect features effectively and quickly, thereby judging defect types.

2 THE BASIC THEORY AND PRINCIPLE OF WAVELET PACKET TRANSFORM

Due to the orthogonal wavelet transform only further decomposition to the low frequency part of signal, but don't further decomposition to the high frequency part which is the details, so the wavelet transform can effectively represent a large family of low frequency information as the main component signal, but it can't well decomposition and represent the signal that contains a large number of details (small edge or texture), such as non-stationary machinery vibration signal, remote sensing images, seismic signal, biomedical signal. In contrast, the wavelet packet transform can provide a finer decomposition for the high frequency part, and the decomposition without redundancy, no omissions, so it can be better time-frequency localization analysis to the signal that contains a large number of medium and high frequency information.

2.1 The definition of wavelet packet

Assume $\{V_j\}_{j \in z}$ is the multi-resolution analysis of $L^2(R)$ [3], W_{j-1} is the orthogonal complement space of V_{j-1} in V_j, $u_0(t)$ and $u_1(t)$ are the corresponding scaling function and wavelet function, write for:

$$\begin{cases} u_0(t) = \sqrt{2} \sum_{k \in z} h_k u_0(2t-k) \\ u_1(t) = \sqrt{2} \sum_{k \in z} g_k u_0(2t-k) \end{cases} \quad (1)$$

Among them: $\{h_k\}_{k \in z}$, $\{g_k\}_{k \in z}$, $g_k = (-1)^k h_{1-k}$, then:
A set of functions define by

$$\begin{cases} u_{2n}(t) = \sqrt{2} \sum_{k \in z} h_k u_n(2t-k) \\ u_{2n+1}(t) = \sqrt{2} \sum_{k \in z} g_k u_n(2t-k) \end{cases} \quad (2)$$

Is $\{u_n(t)\}$ $(n = 0, 1, 2, \ldots)$, this is the wavelet packet determined by orthogonal scaling function $u_0(t)$ [4].

Define subspace U_j^n is the closure subspace of function $u_n(t)$, assume $g_j^n(t) \in U_j^n$ then $g_j^n(t)$ can be represented as:

$$g_j^n(t) = 2^{j/2} \sum_{k \in z} d_{jk}^n u_n(2^j t - k) \quad (3)$$

Then wavelet packet decomposition algorithm is:

$$\begin{cases} d_{jl}^{2n} = \sum_{k \in z} \tilde{h}_{k-2l} d_{l-1,k}^{n} \\ d_{jl}^{2n+1} = \sum_{k \in z} \tilde{g}_{k-2l} d_{l-1,k}^{n} \end{cases} \quad (4)$$

Wavelet packet reconstruction algorithm is:

$$d_{j+1,l}^{n} = \sum_{k \in z} (h_{l-2k} d_{jk}^{2n} + g_{l-2k} d_{jk}^{2n+1}) \quad (5)$$

Among them, j is scale index, k is translation indicators, n is frequency refining indicators, $\tilde{h}$ and $\tilde{g}$ is wavelet filter, h and g is wavelet synthesis filter, $\tilde{h}$ is the reverse order of h, namely $\tilde{h} = h_{-n}$.

2.2 The basic implementation steps of wavelet packet analysis

(1) Select the appropriate wavelet filter and determine a wavelet decomposition level N, then get the wavelet packet coefficients tree structure of a given sampling signal by wavelet packet transform.
(2) Choose the information cost function, using the best wavelet basis selection algorithm to choose a best basis.
(3) Use threshold to process the wavelet packet coefficient of the best orthogonal wavelet packet basis.
(4) Reconstruction signal by using reconstruction algorithm of wavelet to process the wavelet packet coefficient.

2.3 The applications of the wavelet packet transform in image processing

As a new time-frequency analysis tool, wavelet packet analysis is well applied in signal analysis and processing. It has made significant research results in many aspects, such as signal processing, pattern recognition, image analysis, data compression, speech recognition and synthesis. Graphic images can be seen as a two-dimensional signal [5], therefore, wavelet packet analysis is naturally applied in the field of image processing. Such as it has a good application in the image compression, image denoising, image enhancement and image fusion.

Since image is a non-stationary signal and wavelet analysis has the characteristics of multi-resolution analysis, so wavelet analysis is suitable for processing image signal. But the frequency spectrum difference is large in different types of images, so must to finer division of spectrum, and it can achieve better effect by using wavelet packet analysis to process the image at this time [6]. Since wavelet packet analysis not only divides the frequency band in multi-level, but also makes a further decomposition of the high frequency part that wavelet multi-resolution analysis has not subdivide. It overcomes the disadvantages of orthogonal wavelet transform, so the wavelet packet analysis can provide images a more detailed analysis than wavelet multi-resolution analysis method, it is of great significance in image processing.

Most of the energy concentrated in just a few coefficient when images decomposed by wavelet packet, that is, in the transform domain, the numerical of the low frequency part of the image is bigger. The energy of the high frequency part and noise is relatively scattered on the transform, wavelet packet coefficient is smaller. Therefore, by using wavelet packet transform to process is a certain processing on coefficient, then reconstruction of wavelet packet.

3 BINARY IMAGE

Binary image is to select a suitable threshold to set the pixel on the grayscale level 256 gray value contain only 0 or 255. So the whole picture is showing a clear black and white effect, and this makes a clear distinction between target and background, highlights the interested target outline, simplifies the image and reduces the amount of data that image processing, make image analysis easier. Binarization is an important means of image segmentation. In the gray or color image, a simple method to represent a target or object is to use a binary image [7], where 255 represents the point on the target and 0 means other points. After the object is separated from the background, calculate the geometric and topological characteristics from its binary image. The key of the binary image is the selection of threshold binarization. The target information will be missing if the value of threshold is too small, and the background information will be mixed with the target information while the threshold is too large.

Gray histogram is a function of the grayscale distribution, which represents the number of pixels of each gray level in the image, reflect the frequency of each gray level in the image. The abscissa of the gray histogram is grayscale and the ordinate is the frequency of gray levels. It's the basic statistical characteristics of image. The histogram of presswork has a clear bimodal nature, the trough between the two peaks can be selected as the separate, so the threshold for binarization is:

$$\text{level} = N/255 \quad (6)$$

When binaring gray image the transformation of each pixel gray value is as the following expression:

$$f(x, y) = \begin{cases} 255 & f(x, y) \geq N \\ 0 & f(x, y) < N \end{cases} \quad (7)$$

In the process of image processing, binarization is effectively reducing the amount of data that need processing, and it's an important way to improve the real-time processing algorithm, which is very important in the design of online system of presswork defect

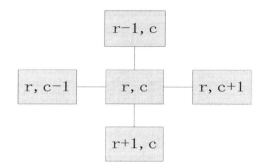

Figure 1. 4-connected

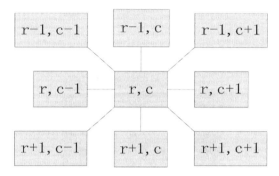

Figure 2. 8-connected

detect. At the time of the analysis of the defect image, select the appropriate threshold to binary the defects result image, guarantee the real-time requirement [8].

4 BLOB ANALYSIS

Blob in the image [9] refers to a set of interconnected pixels with similar properties. Blob analysis is the method that analysis the connected domain with similar pixels in the image. Thus, blob analysis is mainly to the shape, size, area, location and other geometric features of the closed blob.

There are two kinds of image region partition method. One is the region-based approach, the other is the contour estimation by using edge detection method. In the method based on region, all the pixels corresponding to an object together and labeled to indicate that they belong to a region. Under a certain criteria, the pixels assigned to a particular area, so that those pixels can be separated from the rest pixels on the image, this process can be referred to as the image segmentation. Connected component marking can be the way of image segmentation. For most operations, the binary image object recognition depends on the way the image is determined when the agreed connection adjacent pixels. There are two kinds of connection mode, that 4-connected and 8-connected.

An image I is divided into n regions, which are represented as R_i. The consistency predicate of the sub-region is $P()$, so:

(1) The consistency predicate and region division meet the condition that following: $P(R_i) = \textit{True}$
(2) Any two regions can't be combined into on region: $P(R_i \cup R_j) = \textit{False}$

The regional analysis of binary image is to label the connected component of the image tag. That is, extracting the different parts from the binary image, and analyzing the characteristics of the region, so as to achieve the purpose of detecting or identifying objects.

Regional analysis can also be described as the calculation of one or more geometric characteristics of all regions of the divided image. The geometric characteristics these areas contain are: area, bounding, centroid, orientation, perimeter, convex image, filled image, and so on.

5 PRINT DEFECT RECOGNITION

Print defect recognition system is not only able to detect the print defect is existed or not, but also requires to accurately identify and determine the types of defects, such as bite, scratches, ink, etc. The type of defects can explain the cause of the defect, so it can promptly issue a warning which can be the basis of adjusting printing system timely, and realize real-time control of the printing process, improving the printing quality. According to the requirement for the quality of printed matter, as well as the factors affecting the quality of printed matter, summed up in the print quality inspection can be basically divided into the following several actual classification of defects:

The point defects that small but differ from that of pixel values, such as ink dots, white dots;
The linear defects that small but wide distribution, such as scratches.
The large defects, such as ink, and white leakage.

The defect feature extraction is a crucial step in the classification of defects. The analysis of block is based on area, centroid and image. Obtain the image size r and c of the defect area, and max width represent as wid, max height represent as hei. In this paper, mainly the identification of the common defects that is dots, white spots, scratches, ink, trapping identification. R1, C1, W1, W2, H1, H2 is the threshold for determining the defect type. Generally the experience value is:

Scratch: Defects object characteristics must meet one of the following conditions:

(1) wid $<=$ W1 || hei $<=$ H1 and r $>$ R1 || c $>$ C1;
(2) wid $<=$ W1 && hei $>$ H2 and wid $>$ W2 && hei $<=$ H1;

White dots: Defects object characteristics must meet one of the following conditions:

(1) ink $=0$ and wid $<=$ W1 || hei $<=$ H1 and r $<=$ R1 && c $<=$ C1;

Figure 3. Standard image.

Figure 4. Inspected images.

Figure 5. ink1 = 1 defect image.

Figure 6. ink0 = 0 defect image.

(2) ink = 0 and wid <= W2 && hei <= H2 and r <= R1 && c <= C1;

Ink dots: Defects object characteristics must meet one of the following conditions:

(1) ink = 1 and wid <= W1 || hei <= H1 and r <= R1 && c <= C1;
(2) ink = 1 and wid <= W2 && hei <= H2 and r <= R1 && c <= C1;

White leakage: Defects object characteristics must meet one of the following conditions:

(1) ink = 0 and wid <= W2 && hei <= H2 and r > R1 || c > C1;
(2) ink = 0 and wid > W2 && hei > H2;

Ink leakage: Defects object characteristics must meet one of the following conditions:

(1) ink = 1 and wid <= W2 && hei <= H2 and r > R1 || c > C1;
(2) ink = 1 and wid > W2 && hei > H2;

6 EXPERIMENT AND ANALYSIS

The above proposed the defect recognition that binary image analysis and area-based approach based on wavelet packet theory. It uses MATLAB R2008a to simulate experiments.

6.1 Defect image

Select db2 2 layer wavelet packet decomposition, and calculate the best wavelet packet tree with Shannon entropy standard [10]. In this paper, calculated the wavelet packet coefficient of the standard image and inspected image to reconstruct the defect images. Considering the need to determine defects is ink dots, white dots or others, defect image is divided into two defect images to label depending on the features of defect ink (ink1 = 1 represents ink is too much, ink0 = 0 represents there is lack of ink).

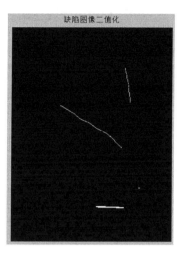

Figure 7. ink1 = 1 binary defect image.

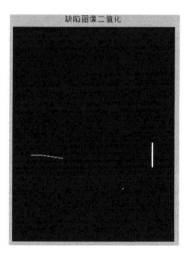

Figure 8. ink0 = 0 binary defect image.

Figure 9. ink1 = 1 label defect.

Figure 10. ink0 = 0 Label defect.

Figure 11. The experimental results.

6.2 Binary the defect image

In this paper, choose level = 0.1 as the threshold. The following white area is the printing defects detected.

6.3 Defect object notation

In the binary image processing, the main application of the morphological is to describe and express the shape useful image feature extraction. In the binary image, the object is the collection of pixels that value is 1 and link together. In this paper, select the 8-connected method to extract defects.

6.4 Defect type recognition

The following characteristic parameters of defects can be obtained in the experiments: area, size (r, c), wid and hei. According to these defects features can identify the type of defect.

7 CONCLUSION

This paper introduces the decomposition and reconstruction algorithm based on wavelet packet and the method of regional analysis, analyzes the significance of the threshold of binarization in defect detection. Combined with morphology area analysis method to extract specific features of defects, determine the type of defect. Through a lot of simulation test the feasibility of the types of defects recognition scheme, accurate identification of defects to improve production efficiency and quality control.

ACKNOWLEDGEMENTS

Research and implementation of online quality control in print process and production control system. Nanchang Technology Project.

REFERENCES

[1] Zhang, L.F. Research of printing defect detection based on machine vision. Beijing printing institute, 2010, 2–3.
[2] Tsai, D.M. & Hsieh, C.Y. Automated surface inspection for directional textures. Image and Vision Computing, 1999, (18): 49–62.
[3] Cheng, W.B. & Wang, H.J. Image denoising methods based on wavelet packet of the improved threshold. Journal of hunan university of science and technology, 2009.
[4] Liu, M.C. The wavelet analysis and application. Tsinghua university press, 2005.
[5] Li, D.Q. & Liu, C.S. Wavelet packet transform technology application in image processing. Journal of Zhengzhou institute of light industry 2011, 26 (3): 33–37.
[6] Zhang, N. & Wu, X. Lossless compression of color mosaic images. IEEE Trans on Image Proc, 2006, 15(16):1379.
[7] Xu, L., Zeng, Z., Liu, J.Z. & Wang, X.J. Application of machine vision in the online printing defect detection and research. The computer system application, 2013, 22(3):186–190.
[8] Chen, Y.J. Print defect detection system based on machine vision research. Xian university of science and technology, 2006.
[9] Xu, M., Tang, W.Y., Ma, Q.L. & Hao, J.Q. Research of online printing defect detection system based on Blob algorithm. Packaging engineering, 2011, 32(9):20–23.
[10] Du, L.Z., Yin, K. & Zhang, X.P. Based on the optimal wavelet BaoJi noise reduction method and its application. Journal of engineering geophysics, 2008, 5(2):25–29.

A resolution test method for high precision accelerometer based on two-axis turntable

Rendong Ma, Gongliu Yang, Kui Zhang & Weizhen Zheng
School of Instrument Science and Opto-electronics Engineering, Beihang University, Beijing, China

ABSTRACT: Microgravity measurement system and high precision navigation system require a high precision resolution of the accelerometers, but it is difficult to determine the resolution of the high precision accelerometers by traditional methods. In this regard, a new resolution test method based on two-axis turntable was presented. In this method, microgravity can be aroused by dividing the gravitation twice. The resolution can be obtained by comparing the actual output increment of the accelerometer and the theoretical value. The precision of this test method was given by theory analysis. Lastly, the new resolution test method was validated by the simulation and experiment. The simulation result showed this method could measure the resolution of 3×10^{-8} g and the experiment proved this method is effective.

Keywords: high precision accelerometers; resolution test; twice dividing; two-axis turntable

1 INTRODUCTION

High precision accelerometer is not only the key component of high precision inertial navigation equipment, but also an important part of the microgravity measurement system (Bogue 1984). The resolution of the accelerometers refers to the minimum variation of the acceleration that the accelerometer aroused away from 0 g. So far, the resolution of the accelerometer was generally measured by using the gravity dip method. The gravity acceleration was subdivided by using high precision optical dividing head. The subdivided acceleration was used as the accelerometer input. In order to improve the measuring accuracy, the test instruments must be settled in isolation foundation. The precision digital voltmeter or the data acquisition card can be used to measure the accelerometer output. The measurement accuracy of the test instruments should be at least one order of magnitude higher than that of the accelerometer. A dividing head of 2″ accuracy could provide an minimum acceleration variation of 10^{-5} g, which was unable to meet the need of the high precision resolution test (Mei 1990).

This paper presents a new method to measure the high precision resolution by using the two-axis position turntable with which can solve the dilemma.

2 TRADITIONAL RESOLUTION TEST METHOD

Resolution test is designed to measure the minimum acceleration variation that the accelerometer can detect, when the acceleration input is not near 0 g. The precision optical dividing head is generally used in the traditional resolution test. By rotating the dividing head, the accelerometer input can get an increment. The corresponding relationship between acceleration increment and angle increment for dividing head is

$$\Delta a = g[\sin(\theta_0 + \Delta\theta) - \sin\theta_0] \qquad (1)$$

The increment of the ideal output voltages is:

$$\Delta U = K_1 \Delta a \qquad (2)$$

The dividing head should be turned to the position where the output of the accelerometer is not near 0g. For instance, the dividing head can be turned to the position where $\theta = 30°$. The angle $\Delta\theta$ corresponding to the resolution of the accelerometer can be calculated by using the formula above. When $\Delta\theta = 24$, theoretical resolution value is 1×10^{-4} g. Record the output at this position, and then turn the dividing head again. The angle increment is $\Delta\theta = 24$. Record the output voltage and repeat the procedure three times (Dong 1990).

According to the experimental data, calculate the average voltage increment ΔU_P of each position. The data processing method: compare the average output increment value ΔU_P and the ideal output increment value ΔU. If

$$\frac{\Delta U_p}{\Delta U} \times 100\% \geq 50\% \qquad (3)$$

then the resolution is qualified.

3 RESOLUTION TEST METHOD BASED ON TWO-AXIS TURNTABLE

3.1 The mathematical model and the input of the accelerometer

In the traditional resolution test, the accelerometer input is a gravitational acceleration component. A more accurate acceleration input is required to measure higher resolution. In the gravitational field experiments, two low precision angles can be combined into a high precision angle by the two-axis turntable, with which method the high-precision acceleration input can be obtained (Li et al. 2012).

There are two install statuses of the accelerometer: gate status and pendulum status. Gate status: at the initial position, set the pendulum axis of the accelerometer parallel to the horizontal axis of the turntable. The input axis will be in a horizontal state while the output axis is in a vertical state. Pendulum status: at the initial position, set the output axis of the accelerometer parallel to the horizontal axis. The input axis will be a horizontal state while the pendulum axis is in a vertical state. In this paper, the accelerometer will be installed in pendulum state. The testing principle of the pendulum status is shown in Figure 1. In the XYZ coordinate system, X, Y axes are on the horizontal plane, Z axis is on the gravitational direction. IA, OA, PA axes of the accelerometer are parallel to axes X, Y, Z axes on the initial installation (Park 2005).

Rotate the horizontal axis of the turntable by an angle α, and rotate the rotary axis of the turntable by an angle β. Then the gravitational component on the input axis, output axis, and that on the pendulum axis of the accelerometer, respectively:

$$\begin{bmatrix} g_{IA} \\ g_{OA} \\ g_{PA} \end{bmatrix} = \begin{bmatrix} g\sin\alpha\cos\beta \\ g\sin\alpha\sin\beta \\ g\cos\alpha \end{bmatrix} \quad (4)$$

The gravitational component on the input axis should be away from 0 g (0.5 g is generally used). Rotate the turntable by $\Delta\alpha$, $\Delta\beta$, then we get

$$\Delta g_{IA} = g\sin(\alpha+\Delta\alpha)\cos(\beta+\Delta\beta) - g\sin\alpha\cos\beta \quad (5)$$

The high-precision acceleration input, which is a high angle positioning, can be got by the combination of two low precision deviation angle positioning of the two-axis turntable.

3.2 The principle of the resolution test based on two-axis turntable

Two-axis turntable has two angular position errors (Bai & Zhao 2006). Considering the turntable angular position errors, we get

$$\Delta g_{IA}' = g\sin(\alpha+\Delta\alpha+\gamma)\cos(\beta+\Delta\beta+\mu) - g\sin\alpha\cos\beta \quad (6)$$

γ and μ are two angular position errors of the turntable.

$$\begin{aligned}\Delta g_{IA}' &= g[\sin\alpha\cos(\Delta\alpha+\gamma) + \cos\alpha\sin(\Delta\alpha+\gamma)] \\ &\quad \cdot [\cos\beta\cos(\Delta\beta+\mu) - \sin\beta\sin(\Delta\beta+\mu)] \\ &\quad - g\sin\alpha\cos\beta \\ &= g[\sin\alpha\cos\beta\cos(\Delta\alpha+\gamma)\cos(\Delta\beta+\mu) \\ &\quad - \sin\alpha\sin\beta\cos(\Delta\alpha+\gamma)\sin(\Delta\beta+\mu) \\ &\quad + \cos\alpha\cos\beta\sin(\Delta\alpha+\gamma)\cos(\Delta\beta+\mu) \\ &\quad - \cos\alpha\sin\beta\sin(\Delta\alpha+\gamma)\sin(\Delta\beta+\mu)] \\ &\quad - g\sin\alpha\cos\beta \\ &= g\sin\alpha\cos\beta[\cos(\Delta\alpha+\gamma)\cos(\Delta\beta+\mu)-1] \\ &\quad - g\sin\alpha\sin\beta\cos(\Delta\alpha+\gamma)\sin(\Delta\beta+\mu) \\ &\quad + g\cos\alpha\cos\beta\sin(\Delta\alpha+\gamma)\cos(\Delta\beta+\mu)\end{aligned} \quad (7)$$

Ignore the high order quantities, then we get

$$\begin{aligned}\Delta g_{IA}' &= g\cos\alpha\cos\beta\sin(\Delta\alpha+\gamma)\cos(\Delta\beta+\mu) \\ &\quad - g\sin\alpha\sin\beta\cos(\Delta\alpha+\gamma)\sin(\Delta\beta+\mu)\end{aligned} \quad (8)$$

To simplify the operation, $\alpha=30°, \beta=0°$ can be chosen as the initial position which can ensure the gravitational component on the input axis is 0.5 g. Then we get

$$\Delta g_{IA}' = \frac{\sqrt{3}}{2}g\sin(\Delta\alpha+\gamma)\cos(\Delta\beta+\mu) \quad (9)$$

When $\Delta\alpha=0$, the acceleration will change with $\Delta\beta$ in a small range.

$$\Delta g_{IA}' = \frac{\sqrt{3}}{2}g\sin\gamma\cos(\Delta\beta+\mu) \quad (10)$$

Through the formula derivation, we can summarize a resolution test method based on the two-axis turntable. First, choose $\alpha=30°, \beta=0°$ as the initial position (position 0). Then rotate the rotary axis of the turntable by an angle increment of $\Delta\beta$ to reach position 1. Position 2~5 can be reached by repeating the last step. Go back to the initial position (position 6), then rotate the rotary axis of the turntable by an angle increment of $\Delta\beta$ each time in the opposite direction to reach position 7~11. Record the output of the accelerometer at each position to calculate the resolution.

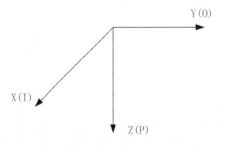

Figure 1. The schematic diagram of the pendulum status.

The error rate of the acceleration output is

$$g_{Error} = \frac{(\Delta g_{IA}' - \Delta g_{IA})}{\Delta g_{IA}} \cdot 100\% \quad (11)$$

with which we can determine the effectiveness of the method. Under the condition that the angular position error is a random value less than $2''$ of Gaussian distribution, we can obtain the error rate of the acceleration output by matlab simulation (Yan & Qin 2007). The simulation results of the error rates under different acceleration input are shown in Figures 2 and 3.

According to the simulation results, when the turntable angle position increment is $2''$, the acceleration input error rate caused by the turntable angular position error is up to 100%. When the turntable angle position increment increases to $24''$, the acceleration input error rate reduced to 10%. That is to say, the acceleration input error caused by the turntable angle position error is an order of magnitude lower than the ideal input, which has proved that the method can be used to measure the resolution of the accelerometer when the turntable angular position increment is 24. The minimum resolution that the method can measure is

$$\Delta g_{IA} = g\sin\alpha\cos(\beta + 24'') - g\sin\alpha\cos\beta$$
$$= 3.3169 \times 10^{-8} g \quad (12)$$

We can draw a conclusion that, in consideration of the turntable angular position error, the resolution test method based on the two-axis turntable can measure a minimum resolution of 3.3169×10^{-8} g.

3.3 Experimental verification

A high precision quartz flexible accelerometer has been used to verify the resolution test method. The experimental apparatus includes an electronic level, high precision digital multimeter, high precision accelerometer acquisition circuit, high precision turntable with attitude accuracy $2''$, computer etc.

The accelerometer was installed in pendulum state after the turntable was leveled. Rotate the horizontal axis of the turntable by an angle $30°$ as the initial position. Rotate the rotary axis of the turntable by an angle $24''$ each time. The high precision digital multimeter was used to record the output of the accelerometer. The output voltages are shown in Table 1.

The theoretical voltage increment can be calculate with formula (2),

$$\Delta U = K_1 \Delta a = 3.28332 \times 10^{-6} V \quad (13)$$

Use formula (3) to determine the effectiveness of the test method. The calculation results are shown in Table 2.

As is shown in Table 2, all of the calculation results are above 50%. That is to say that the resolution of the high precision accelerometer is above 3×10^{-8} g.

4 CONCLUSIONS

This paper focus on the question that traditional resolution test method is not suitable for high precision accelerometer, and we proposed a new resolution test method based on the two-axis turntable to deal with

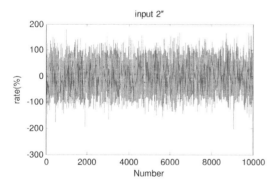

Figure 2. Error rate with the accelerometer input $2''$.

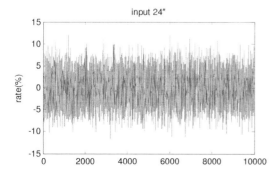

Figure 3. Error rate with the accelerometer input $24''$.

Table 1. Output voltage of the accelerometer.

Position	Voltage (mV)	Position	Voltage (mV)
0	0.51358628	6	0.51358694
1	0.51358910	7	0.51358323
2	0.51359143	8	0.51358057
3	0.51359367	9	0.51357698
4	0.51359536	10	0.51357334
5	0.51359824	11	0.51356984

Table 2. Calculation results of $\Delta U_P/\Delta U$.

Position	$\Delta U_P/\Delta U$ (%)	Position	$\Delta U_P/\Delta U$ (%)
0	85.9%	6	113.0%
1	71.0%	7	81.0%
2	68.2%	8	109.3%
3	51.5%	9	110.9%
4	87.7%	10	106.6%
5		11	

this problem. We obtained the high precision acceleration by dividing the gravitation twice. At the same time, considering the angle position error of the two-axis turntable, we derived the test precision of the method. The matlab simulation and the experimental verification had shown that a minimum resolution of 3.3169×10^{-8} g could be measured by using the method in this paper which could improve the precision of resolution test greatly.

REFERENCES

Bai Xue-feng & Zhao Yan. 2006. Errors propagation analysis of single-axis rate and three-axis position turntable. Aerospace Control: 24(2): 26–29.
Bogue, R. 1984. Developing Science of Accelerometers. Control and Instrumentation: 69–71.
Dong Jing-xin. 1990. Micro inertial instrument-micro mechanical accelerometer. Tsinghua University Press: 246–247.
Hunter, J.S. Lance Q-Flex Accelerometer Qualification Test Program. AD-A121264: 2.0–2.10.
Lee, I. et al. 2005. Development and Analysis of the Vertical Accelerometer. Sensors and Actuators A: Physical: 119(1): 8–18
Li Hai-bing et al. 2012. Dynamic estimation method for resolution of high precision accelerometer. Journal of Chinese Inertial Technology: 20(4): 496–500.
Love, W.J. & Scarr, A.J. 1973. Determination of Multi-Axis Machine. MTDR Conf. Proc: (14): 307–315.
Mei Shuo-ji. 1990. Test and data analysis of inertial instrument. Northwestern Polytechnical University Press: 93–98.
Shestakov, A.L. 1996. Dynamic Error Correction Method. IEEE Tran. On Instrument and Measurement. Vol. 45. No.1, Feb.
Tan, C. W. Park, S. 2005. Design of Accelerometer based Inertial Navigation Systems. IEEE Transactions on Instrumentation and Measurement: 54(6): 2520–2530.
Yan Gong-min & Qin Yong-yuan. 2007. Calibration simulation for laser strapdown IMU with two-axis turntable. Journal of Chinese Inertial Technology: 15(1): 123–127.
Zhou Qi et al. 2008. Precision calibration techniques research for laser strap-down inertial measurement unit. Measurement & Control Technology: 27(9): 95–98.

Pressure fluctuation model of hydraulic turbine based on LS-SVM

X.L. An & F. Zhang
China Institute of Water Resources and Hydropower Research, Beijing, China

ABSTRACT: The influence of active power and working head on pressure fluctuation model of hydraulic turbine is investigated based on the field test data, and Least Square Support Vector Machine (LS-SVM) method is applied to establish pressure fluctuation model of hydraulic turbine. Practical examples demonstrate that the proposed LS-SVM model has good accuracy. Using this model, the normal value of pressure fluctuation can be computed quickly based on current parameters of operation condition. This model can be applied to online anomaly detection system of hydropower units, there is good practicality.

Keywords: hydraulic turbine; pressure fluctuation; least square support vector machine; test data

1 INTRODUCTION

With the increase of a hydropower unit's capacity and size, the effect of the unit's running condition on the hydropower station and its interconnected grid hydropower becomes increasingly great (An et al. 2013). The pressure fluctuation of hydraulic turbine is one of the main factors which affect the unit's stability (Zhang et al. 2011). Pressure fluctuation will cause unit vibration, output swing, blade crack and tear of draft tube wall. When the pressure fluctuation is severe, the resonance of workshop or adjacent buildings may occur. The resonance has a direct threat to the safe operation of the entire hydropower station. Therefore, in order to ensure the safe and stable operation of the unit, deeply mining the field test data is need. It can better grasp the true state of the turbine's pressure fluctuation, and early warn the possible anomalies.

In this paper, based on the field test data of hydraulic turbine's pressure fluctuation, the Least Squares Support Vector Machine (LS-SVM) is introduced to build a pressure fluctuation model for hydraulic turbine. In this model, the influence of active power and working head on pressure fluctuation is considered. The proposed model provides a new way for the online assessment and fault warning of hydropower unit.

2 PRINCIPLE OF LEAST SQUARES SUPPORT VECTOR MACHINE REGRESSION

Least Squares Support Vector Machine (LS-SVM) is an extension of support vector machine, it converts the quadratic programming problem to solve linear equations. It has faster solving velocity, is widely used in the regression analysis, pattern recognition and many other areas (An et al. 2011a, Zhou et al. 2011). The regression principle of LS-SVM is as follows:

Let the sample sets $s = \{(x_1, y_1), (x_2, y_2), \ldots, (x_l, y_l)\} \in R^n \times R$, the high-dimensional feature space linear function, equation (1), is applied to fit the sample sets.

$$f(x) = w^T \varphi(x) + b \quad (1)$$

where $\varphi(x)$ is a nonlinear mapping from input space into a high dimensional feature space, w is the weight vector, b is a bias constant. The regression problem of LS-SVM algorithm is based on structural risk minimization principle to solve constrained optimization problem:

$$\min \frac{1}{2} w^T w + \frac{\gamma}{2} \sum_{i=1}^{l} e_i^2 \quad (2)$$

The constraint condition is:

$$y_i = w^T \phi(x_i) + b + e_i \quad (i=1 \sim l) \quad (3)$$

The Lagrangian function is introduced to transform the optimization problem of the formula (2) into the dual space, then

$$L = \frac{1}{2} w^T w + \frac{\gamma}{2} \sum_{i=1}^{l} e_i^2 - \sum_{i=1}^{l} \alpha_i [w^T \phi(x_i) + b + e_i - y_i] \quad (4)$$

where α_i is Lagrange multiplier, γ is a constant. The method in reference (Zhou et al. 2011) is used to solve equation (4) finally get α and b. Finally, LS-SVM regression function is gotten, the detailed steps can be seen in reference (Zhou et al. 2011).

3 PRESSURE FLUCTUATION MODEL OF HYDRAULIC TURBINE

A large number of field data analysis showed that active power and working head are the main factors which affect the hydraulic turbine's pressure fluctuation. Based on LS-SVM method, the pressure fluctuation model of hydraulic turbine $y = f(P, H)$ is founded, where P is the unit's active power, H is the working head, y is the pressure fluctuation parameters. The proposed model considers the effect of active power and working head on pressure fluctuation.

The concrete steps to build the pressure fluctuation model of hydraulic turbine are as follows:

(1) Collecting the field test data of hydraulic turbine in different conditions.
(2) Selecting the test data of hydraulic turbine operates well as learning samples.
(3) The selected samples are input LS-SVM to train, and the pressure fluctuation model $y = f(P, H)$ of hydraulic turbine is established, where P, H is the input vector, y is the corresponding target value.

4 APPLICATION EXAMPLESRES

In this paper, the pressure fluctuation field test data of a large Francis hydropower unit (rated power: 700 MW, rated speed: 75 r/min) in good operation conditions are used as standard state data of hydraulic turbine's pressure fluctuation. The data is at the upper water level 147 m ~ 170 m. The selected data is used to verify the effectiveness of the pressure fluctuation model of turbine based on LS-SVM.

From the test data, 1065 sets data in different active power and different working head are selected as standard samples to found the pressure fluctuation standard model $y = f(P, H)$. The pressure fluctuations y include pressure fluctuation at the inlet of spiral case, pressure fluctuation at blade free sections, pressure fluctuation at the head cover, pressure fluctuation at the upstream side of draft tube and pressure fluctuation at the downstream side of draft tube. Figure 1 shows five real waveforms of five pressure fluctuations in the same operating condition, where the active power of hydropower unit is 400 MW, the upstream water level 170.09 m. In order to make the model has good performance and can accurately identify the true state of the hydraulic turbine, the selected field test data should cover the unit's possible changes in working head and active power range. 300 sets data are selected as test samples to verify the standard pressure fluctuation model based on LS-SVM.

The real values and LS-SVM model calculated values of pressure fluctuation at the inlet of spiral case, pressure fluctuation at blade free sections, pressure fluctuation at the head cover, pressure fluctuation at the upstream side of draft tube and pressure fluctuation at the downstream side of draft tube are compared, the results shown in Figure 2. The horizontal ordinate in Figure 2 represents the real values of pressure

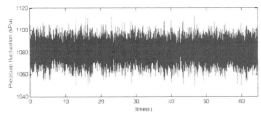
(a) Pressure fluctuation at the inlet of spiral case.

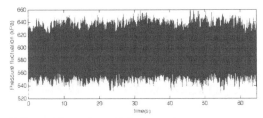
(b) Pressure fluctuation at blade free sections.

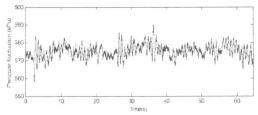
(c) Pressure fluctuation at the head cover.

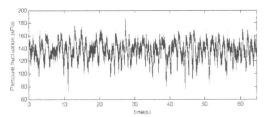
(d) Pressure fluctuation at the upstream side of draft tube.

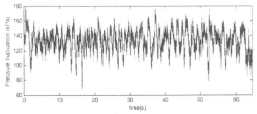
(e) Pressure fluctuation at the downstream side of draft tube.

Figure 1. Real waveforms of five pressure fluctuations in the same operating condition.

fluctuation, the longitudinal coordinate represents the calculated values of pressure fluctuation. The slash is a straight line which connects the horizontal and longitudinal coordinates at an angle of 45°. If the point in the slash indicates that the real values and the LS-SVM model calculated values of pressure fluctuation are equal. If all the points concentrate near the slash, show that this LS-SVM model has better performance.

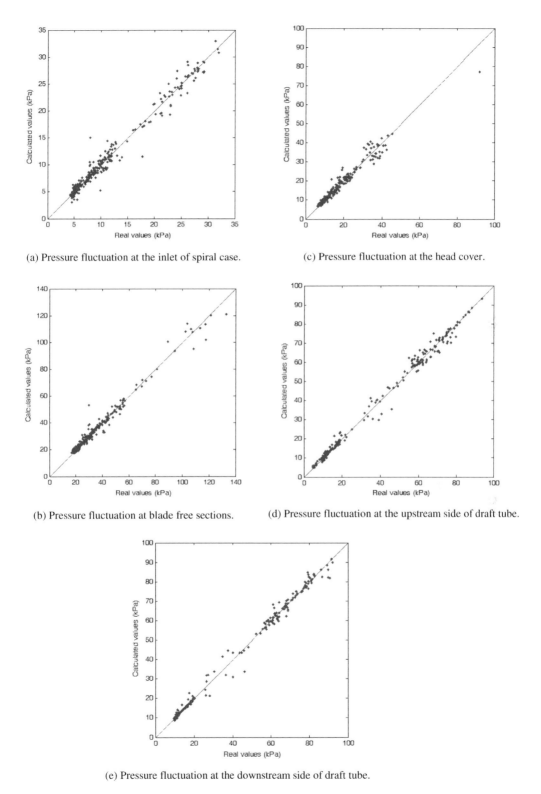

(a) Pressure fluctuation at the inlet of spiral case.

(c) Pressure fluctuation at the head cover.

(b) Pressure fluctuation at blade free sections.

(d) Pressure fluctuation at the upstream side of draft tube.

(e) Pressure fluctuation at the downstream side of draft tube.

Figure 2. Comparison of pressure fluctuation between real values and calculated values.

It can be seen from Figure 2 that the proposed model can well reflect the actual condition of the turbine's pressure fluctuation. The results show that the pressure fluctuation model of hydraulic turbine based on the LS-SVM method, which considers the effect of active power and working head on the turbine's pressure fluctuation, has a higher accuracy. The average relative error of pressure fluctuation at the inlet of spiral case, pressure fluctuation at blade free sections, pressure fluctuation at the head cover, pressure fluctuation at the upstream side of draft tube and pressure fluctuation at the downstream side of draft tube are 7.18%, 4.91%, 6.77%, 4.70% and 3.24%, respectively. The error formulas can be seen in reference (An et al. 2011b).

5 CONCLUSIONS

Based on the field test data of hydraulic turbine, the LS-SVM method is used to establish the pressure fluctuation model of hydraulic turbine. In the model, the influence of active power and working head on pressure fluctuation model of hydraulic turbine is considered. The normal value in the current working condition of the hydraulic turbine's pressure fluctuation can be quickly calculated, when the real-time operational data of the hydropower unit are input into the model. The normal value is used to determine that the hydraulic turbine's operating condition is normal or abnormal. The model can achieve early warning of failure, there is a good prospect.

ACKNOWLEDGEMENT

This work was supported by the National Natural Science Foundation of China (grant number 51309258) and the Special Foundation for Excellent Young Scientists of China Institute of Water Resources and Hydropower Research (grant number 1421).

REFERENCES

An, X., Jiang, D. & Chen, J. 2011a. Bearing fault diagnosis based on ITD and LS-SVM for wind turbine. *Electric Power Automation Equipment* 31(9):10–13.

An, X., Jiang, D. & Liu, C. 2011b. Wind farm power prediction based on wavelet decomposition and chaotic time series. *Expert Systems with Applications* 38(9):11280–11285.

An, X., Pan, L. & Zhang, F. 2013. Condition degradation assessment and nonlinear prediction of hydropower unit. *Power System Technology* 37(5):1378–1383.

Zhang, F., Gao, Z. & Pan, L. 2011. Study on pressure fluctuation in Francis turbine draft tubes during partial load. *Journal of Hydraulic Engineering* 42(10):1234–1238.

Zhou, Q., Sun, Wei. & Ren, H. 2011. Spatial load forecasting of distribution network based on least squares support vector machine and load density index system. *Power System Technology* 35(1):66–71.

The application of wireless network on quadrotor temperature observe based on NRF905

Weicheng Wang & Yang Zhu
School of Electrical Engineering, Guizhou University, guiyang guizhou, China

Ting Yu
Changchun railway vehicle equipment co., LTD, the Guokai company

Minhui Wang*
School of Electrical Engineering, Guizhou University, guiyang guizhou, China

ABSTRACT: In order to solve the problem of quadrotor temperature observe such as temperature compensation of aircraft flying height, difficulties on networking and maintaining as well as the high cost between the upper machine and lower machine, the design has changed the signal transmission into the wireless network approach, thus making the whole system easy to maintain, easy to organize network, and low cost. The use of multi-sensor data fusion technology in alarm system ensures the reliability of the entire flight control system. The classic PID control strategy used in flight control makes the aircraft flying height a better response speed and control precision. Tests show that the entire system can work reliably at the range of 500 meters.

Keywords: temperature observe; flying height ; data fusion; PID; NFR905

1 INTRODUCTION

Flying height is an important flight parameter for aircraft, and is the key to assure the safety of aircraft flight and ensure the ground commanders and operators to correctly guide and complete the flight mission. While the temperature will influence the pressure altitude in the flying height, after collecting temperature data at the flight atmosphere and making the altitude compensation, the temperature should be sent to the control chip for displaying and calculating the temperature of the atmosphere in real-time. When connecting the upper machine and the lower machine, if using the RS-485 bus or CAN bus , then the whole system will use a large number of signal cables, making the system's maintainability poor, and arousing network problems, also, a large number of communication cables increase the production costs. Moreover, due to the nature of the aircraft itself, the design changes the communication approach into a wireless network, then its communication rate is increasing, and each point has a unique address with stable system, simple circuit, and low power consumption, also, the convenient node network eliminates the need to set up a large number of signal cables, and reduces the cost, thus improving the maintainability of the system.

2 SYSTEM ARCHITECTURE

The system mainly consists of the upper machine and lower machine, and the system block diagram is shown in Figure 1. The lower machine is the aircraft exactly, consisting of wireless communication chip NRF905, STM32F103 Micro-controller, DS18B20 temperature sensor. The main task is to regulate, capture, display and transmitting the temperature. The DS18B20 detects the temperature in the flight atmosphere. The upper machine is installed in the control room for wireless monitoring, whose main task is to accept the temperature and display the temperature in real-time, then calculate the temperature data for altitude compensation; when the flying height is out of control, an alarm is generated, thus establishing the center point to multi-point wireless star network centering the monitoring station, thus achieving the measurement and control functions of temperature. When sending data, NRF905 will monitor whether the environment has the same carrier frequency, if it does, there will be no data sent, thus solving the data conflict during the communication process. The aircraft is assigned with a unique address, the upper machine sends an address signal firstly, which will be received by the lower machine, only when lower machine's address matches the code

Figure 1. System Block Diagram.

Table 1. NRF905 mode.

PWR_UP	TRX_CE	TX_EN	Operating Mode
0	X	X	Power down and SPI programming
1	0	X	Standby and SPI programming
1	1	0	Shock Burst RX
1	1	1	Shock Burst TX

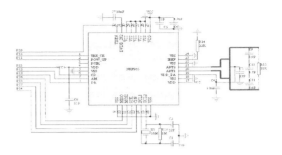

Figure 2. NRF905 hardware circuit.

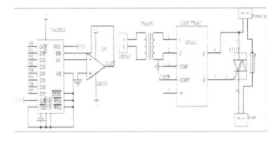

Figure 3. SCR circuit applications.

sent by the upper machine, it will send the temperature data to the upper machine, and the lower machine will be in the receiving state after transmitting the data. The upper machine will send the next address after receiving the data. Just like this, the lower machine will transmit the temperature data to the monitoring station sequentially.

3 HARDWARE DESIGN

3.1 NRF905 hardware circuit

The NRF905 chip wireless transceiver is operated in 433/868/915MHz which consists of a fully integrated frequency modulation, a receiver with a demodulator, a power amplifier, a crystal oscillator and a modulator. NRF905 has four working ways, such as Power down and SPI program, Standby and SPI programming, ShockBurst RX and ShockBurst TX mode. The working mode is mainly determined by the state of PWR_UP, TRX_CE and TX_EN pin which is shown in Table 1. The features of Shock Burst mode is automatically generated preamble and addresses CRC.32 which can easily achieve multi- machine communication. NRF905 can be programmed easily via SPI interface configuration, and the current consumption is low, when the transmit power is −10dB, the emission current is 11mA and the receive current is 12.3mA and the power-down mode can easily conserve power. NRF905 communicates with the micro-controller based on the SPI protocol. The hardware circuit diagram is shown in Figure 2.

3.2 Design on driving the heating and cooling circuit

As for the AC power regulation circuit, the most commonly used devices are bidirectional triode thyristor and the triac trigger thyristor phase-shift, which can be considered as the integration of a general anti-connected thyristors; it has two main electrodes T1 and T2, and a gate G, the gate can trigger conduction in both forward and reverse directions of the main electrode, so when the bidirectional triode thyristor triggers the current, the entire cycle of the alternating current will be conducting, thus making the full use of power. The triac trigger thyristor phase-shift is based on the strength of the control signal (usually the voltage is 0–5 V, 0–10 V, 1–5 V, and the current is 4–20 mA, 0–10 mA, etc.), the output generates double grid frequency synchronous with the grid voltage ranges from 0° to 180°, the pulse width could be used to drive the SCR. Then the voltage of the AC load is adjustable from 0 to the maximum linear regulator so as to achieve the purpose of the phase shift and change the heating or cooling power of the system. Thyristor control signal can be adjusted manually by potentiometer, as well as being automatically adjusted by DAC converters; the control is very simple and flexible. The application circuit is shown in Figure 3.

The working principle of Figure 3 is based on that the P0 port of MCU converts into the corresponding analog (0V-5V) through DAC0832, after the CONT pin of SCR phase trigger receiving this signal, the output generates double grid frequency synchronous with the grid voltage ranges from 0° to 180°, the pulse width could be used to drive the SCR. Then the voltage of the AC load is adjustable from 0 to the maximum linear

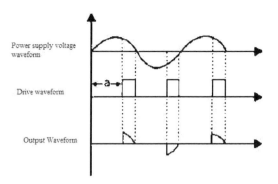

Figure 4. Thyristor phase-shift trigger output pulse waveform of the supply voltage and load voltage waveforms.

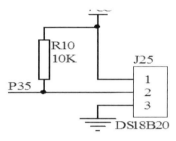

Figure 5. DS18B20 application circuit.

regulator so as to achieve the purpose of the phase shift so as to achieve the purpose of regulating to drive the heating and cooling fan, and the purpose of temperature control. The output pulse of SCR phase trigger and the waveform of the supply voltage and the waveform of the load voltage are shown in Figure 4.

3.3 Temperature detection circuit

The temperature sensor adopts the DS18B20 digital temperature sensor, with the measurement ranging from $-55°C$ to $+125°C$, and the accuracy of $+0.5°C$, thus meeting the temperature measurement of pressure altitude and control system. The adopt of DS18B20 digital temperature sensor makes the entire detection circuit simple, because the digital sensor uses a single bus technology, and port of I/O can connect multiple sensors, which have saved the I/O port resources for the micro-controller. The specific circuit application is shown in Figure 5.

4 SOFTWARE DESIGN

4.1 Design control law

The present system adopts the classic PID control law for ensuring the quality control. As for the computer system, a digital approximation method should be used to achieve PID control law. When the sampling period is quite short, a summation should replace the integration, and the rear differential should replace the differential, thus dispersing analog PID into difference equations. This design uses a digital incremental PID algorithm, and the numeric PID is calculated by the following formula (1):

$$u_k = K_p [e_k + \frac{T}{T_i}\sum_{j=0}^{k} e_j + \frac{T_D}{T}(e_k - e_{k-1})] + u_0 \quad (1)$$

$$u_{k-1} = K_p [e_{k-1} + \frac{T}{T_i}\sum_{j=0}^{k-1} e_j + \frac{T_D}{T}(e_{k-1} - e_{k-2})] + u_0 \quad (2)$$

Formula (1) minus formula (2) = formula (3)

$$\Delta u(k) = u(k) - u(k-1) \quad (3)$$

$$\Delta u(k) = K_p [e_k - e_{k-1} + \frac{T}{T_i} e_k + \frac{T_D}{T}(e_k - 2e_{k-1} + e_{k-2})] \quad (4)$$

To facilitate programming, of the formula (4) finishing the following form

$$\Delta u(k) = q_0 e(k) + q_1 e(k-1) + q_2 e(k-2)$$

$$q_0 = K_p (1 + \frac{T}{T_i} + \frac{T_D}{T})$$

$$q_1 = -K_p (1 + \frac{2T_D}{T})$$

Where: K_p is a proportionality factor
T_d is differential time constant
T The sampling period
T_i The integral time constant

4.2 NRF905 program design diagram

NRF905 adopts the technique of VLSI and Shock Burst from Nordic, which enables NRF905 to provide high-speed data transfer without the expensive high-speed MCU for data processing. The high-speed signal processing related to RF protocol will be placed within the chip, and the NRF905 will provide an SPI interface to the micro-controller application, and the rate will be determined by the interface rate set by the micro-controller. NRF905 connects by the maximum rate in RF through Shock Burst mode, and the reduction of the speed of digital applications is to reduce the portion of the average current consumption. During the Shock Burst receiving mode, the address match (AM) and Data Ready (DR) inform the MCU a valid address and each packet have been received, and during the sending mode of Shock Burst, NRF905 automatically generates preamble and CRC checksum, the signal of data ready (DR) notices MCU that the data transfer has been completed, in short, this means that reducing the memory requirements of the MCU is reducing the cost of MCU, and shortening the software development cycle as well. The receiving and transmitting data flowchart of NRF905 are shown in Figure 6.

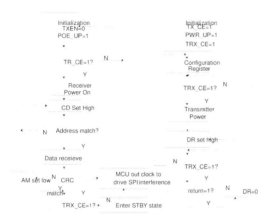

Figure 6. Data Reception and Datatransmission Flowchart.

4.3 Temperature data processing

In the aircraft control system, if flying height calculated temperature compensation is out of control, an alarm signal will be generated, the temperature data processing will use fusion

$$u = F(u_1, u_2 \cdots u_i) = \begin{cases} 1 & a_0 + a_1 u_1 + a_2 u_2 + \cdots + a_i u_i > 0 \\ 0 & other \end{cases}$$

algorithm for the accuracy of the flying altitude correction, and the temperature information will be dealt on the multi-sensor fusion principle, and u = 1 means high temperature, and u = 0 is a normal temperature. In formula U_i is the first temperature sensor collect

$$a_0 = \log(P_1 / P_0)$$

When the collection point outside temperature is higher than the set temperature

$$U_i = 1, \ a_i = \log(P_{Di} / P_{Fi})$$

When the temperature at the collection point within the range of the set temperature

$$U_i = -1, \ a_i = \log\left(\frac{1 - P_{Di}}{1 - P_{Fi}}\right)$$

PD and PF give each sensor site different probabilities of detection and false alarm probabilities for experiments and experts. In accordance with the above algorithm, it is derived u=1 or u=0 within the microcontroller, thus determining correctly whether the flying height is out of control.

5 CONCLUSIONS

This paper presents a wireless network on Quadrotor Temperature Observe and flying altitude control system based on NRF905, and adopts the wireless networks to communicate between the upper machine and the lower machine, the installation, maintenance and networking is very simple, and systematically uses the modular design, which posses good portability and scalability. Test results based on laboratory tests showed that: the data transmission through NRF905 has the advantages of Faster transfer rate, strong anti-interference, and good reliability as well, meanwhile, the system can also be used in other monitoring systems.

ACKNOWLEDGEMENT

Projects fund: Guizhou university graduate student innovation fund (Title: The new four axis in search of an aircraft flight research and design. Fund number: Research institute of technology 2014043)

REFERENCES

[1] Deng Rong. Microcontroller-based beer fermentation temperature control system. Practical World in November 2007, Volume 26, No. 11.
[2] Guo Qinghua. Greenhouses Temperature Intelligent Control System Research and Application. Anhui Agricultural Sciences 2008, 36 (11):4487–4488.
[3] Huang Zhiwei. RFIC chip circuit design theory and application. Beijing: Electronic Industry Press, 2004.
[4] Liu Hui, Xie Jian, Qin Xiaozhen, Wang Binwen. A multipoint remote monitoring system design and implementation Computer Technology and Development, 2006, 16 (8):
[5] Yu Haisheng. Micro-computer control technology. Beijing: Tsinghua University Press, 2008
[6] Zheng Jungang. NRF905 based intelligent wireless fire monitoring system design. Modern building electrical, 2009 Zhou Xiao.NRF905 based Smart Home System Design. 2012,03-0011-02

ABOUT THE AUTHOR

Weicheng Wang (1988-), Graduate, School of Electrical Engineering, Guizhou University;
   Yang Zhu (1990-), Graduate, School of Electrical Engineering, Guizhou University;
   Ting Yu (1989-), Bachelor, Changchun railway vehicle equipment co., LTD, Guokai company, Jilin Changchun
   Minhui Wang* (1963-), Assistant Professor, School of Electrical Engineering, Guizhou University.

Hypterball batch key update method based on members' behavior

Shuying Liu, Yong Xu & Fan Yang
Department of Mathematics and Computer Science, Anhui Normal University, Wuhu, Anhui, China

ABSTRACT: The core issue of secure multicast is the safe management of the group key. Through the analyzing of the multicast rekeying method of hypterball, and combining the batch key update and member's behavior, the hypterball multicast key management method based on members' behavior is proposed in this paper. The scheme breaks through the traditional method of LKH, and it not only makes full use of independence and cooperation between members, but also considers the members' behavior, and then improves the communication's efficiency. Simulation experiment results show that when members join in or leave the group, especially when the probability of the members' change is largely different, this scheme will greatly reduce the cost of key update and improve the efficiency of the key update, and at the same time the forward security and backward security are ensured.

Keywords: Multicast Secure; Batching Rekey; Members' Behavior; Hypterball; Probability

1 INTRODUCTION

As a kind of efficient communication technology, multicast [1–9] is more and more widely applied in many areas, such as video conference, financial market data transmission, remote teaching, multiplayer real-time games and so on. The network bandwidth can be significantly reduced by multicast, but the traditional multicast technology didn't consider the secure requirement of multicast applications. Therefore, how to realize multicast security while maintaining its advantages becomes a hot research point in the field. Similar to security implementation of unicast, cipher algorithm is still an effective method to realize secure multicast, but group scale determines that the effective method is to use the group key in multicast applications, that is, all members use the same encryption and decryption keys in the data communication, so the safety of the group key's distribution becomes the starting point in the secure multicast. Whenever there is a membership change, the group key must also be changed (called rekeying) to prevent a newly joining user from reading past communications, called backward access control, and a leaving member from reading future communications, called forward access control. Thus group key update is the key to implement security multicast.

At present, there are two main ways that are real-time key update [2] and batch key update [3, 7–9] in the research field of group key update. In literature [2], whenever there is a membership change, the key server must change all the keys along the path and distribute them securely among respective users to provide backward and forward access control. In order to distribute the new keys securely during the join operation, the key server encrypts the new set of keys with the corresponding previous keys, but during the leave operation new keys are encrypted with appropriate auxiliary keys and private keys of the users. However, with the scale enlargement of multicast members and the increase of fluctuation frequency, the key server must perform rekeying operations quite frequently, which increases the load on the key server and leads to low efficiency, asynchronous, resource wasting and so on. The price of the key server update is $O(\lg N)$ in the real-time update. Literature [3] addresses a kind of batch key update method, which is when members request to leave or join in the multicast group, the key server waits for a period of time to update the key instead of updating the key immediately, and then collects all the join and leave requests during the interval, generates new keys, constructs a rekey message and multicasts the rekey message. Literature [4] addresses a kind of multicast key updating method based on the spherical, which makes full use of the child node key fatherhood and independence. As a result, it reduces the load on the key server from $O(\lg N)$ to $(4\log_4 N - 1)/3$ for a group of n users with the degree of the tree being 4. To some degree, this method not only solves the above problem, but also reduces the cost of the overall key server update. And it concludes that the quad-tree is superior to the other degree key tree. On the basis of literature [4], literature [5] proposes a kind of M-dimensional spherical batch key update scheme, and this method is applied to the super ball, and then reduces the price of group key update.

Although the literature [5] reduces the price of key update to some extent, it does not consider the behavior of the members. In the practical application environment, the behavior of the group's change is often different. Based on the above theory, this paper takes the member' activity behavior into consideration, and proposes a super spherical batch key update method based on members' behavior.

2 SPHERICAL KEY TREE STRUCTURE BASED ON THE BEHAVIOR OF MEMBERS

Assuming that multicast group has N members which are considered as N nodes. According to the probability of members' behavior change, these nodes are distributed in three-dimensional space. The members are divided into groups of four, and each group's four point coordinates decide a sphere whose centre is their parent node's coordinates. In the same way, all the centers of the sphere are also divided into a group of four, so a number of spheres can be constituted until the root node. The change of any one node inevitably affects its core and its associated nodes, and the other centre of the sphere is not affected.

The paper applies the above features to multicast key update, in which N nodes are regarded as N members of multicast, and point's deletion and increase map group members' leaving and joining, and super sphere of N point is abstracted into quad-tree form. As shown in figure 1, the probability of members' change decreases from left to right in turn. And the closer to the left is, the higher the members' activity is; and the closer to the right is, the more stable the members' activity is.

2.1 Probability of the group membership change

Definition 1 U_i is defined as the member U_i in the spherical quad tree, and is also considered as the leaf node.

Definition 2 w_i is defined as the weight of the member U_i.

Definition 3 P_i is defined as:

$$P_i = \frac{w_i}{W}$$

Definition 4 W is defined as:

$$W = \sum_i w_i$$

Definition 5 L_{i-j} is defined as a sub-tree of the spherical quad tree.

Understanding of the members' change probability accurately in the multicast is impossible. Therefore, a kind of method [4] that uses the approximate weight to calculate members' change probability is proposed. The method assigns a weight w_i to each member U_i, and that the sum of P_i is one that can be concluded from the above definition 1, definition 2 and definition 3. The size of the weight can be decided by the reciprocal of members' residence time T_i in the multicast group. After understanding the behavior of the members' change, the members can be arranged according to the probability of changes from big to small, then structure the key tree by a group of four.

Literature [1] has shown that when the probability of group members' change in multicast is greater than 0.3, the structure of the star key tree is better than that of a key tree whose height is greater than 1. Structurally, the key tree based on spherical structure is also a LKH tree, so this paper discusses that the multicast group whose members' change probability is less than 0.3. Through the further research, the probability' organization of the multicast key tree is addressed, minimizing the computation and the overhead of the communication. In the probability optimization tree which is a kind of super ball key tree combined with members' behavior, the multicast members' change probability P can be divided into the following three categories:

(1) This paper considers the members as the large probability members when the members' probability P is more than 0.093 and less than 0.3.

(2) This paper considers the members as the transition period members when the members' probability P is more than 0.055 and less than 0.093 or the probability of members' change is uncertainty.

(3) This paper considers the members as the relatively stable members when the members' probability is more than 0 and less than 0.55.

According to the above idea, the multicast members can be divided into large probability members, the transition period members, and the more stable members by the probability of members' change. Then they are assigned to the corresponding position of the multicast key tree.

2.2 Initialization of the spherical key tree

If the probability of a member's change can be determined, it can be inserted into L_{1-1} and L_{1-4} at the initialization of the super spherical key tree. If their probability cannot be certain, they would be considered as a member of the transition period member. Then these members are randomly inserted into L_{1-2} and L_{1-3} of the super ball. In this way, the members of large probability are located in some areas on the surface of the super ball. As shown in figure 1, the members of large probability change are put in L_{1-1} as soon as possible. The members whose behavior is relatively stable are put in L_{1-4}, and the members who are in the transition period are put in L_{1-2} and L_{1-3}. Eventually a super ball key tree based on members' behavior is formed.

In L_{1-1}, the members' behavior activities are more frequent. If real-time key update is used, the numbers of update will increase significantly. As a result, the inefficiency and disorderly problem are prone to appearing. So a kind of batch key update strategy based on members' behavior is adopted in this paper. The members whose activities are more frequent are mostly located in the same sub tree of super ball. When members join in or leave the multicast group, we would

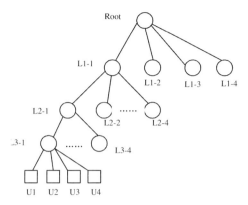

Figure 1. Spherical quad tree.

slightly increase the leaving or joining members' waiting time rather than update the key immediately. In this way, the primary two times key update are reduced to one key update, so as to achieve the aim of reducing the price of key update.

3 SPHERICAL BATCH KEY UPDATE ALGORITHM BASED ON THE MEMBERS' BEHAVIOR

The scheme is designed as follows. In order to reduce the cost which is caused by the same auxiliary key updating again and again, the auxiliary keys are delayed without affecting the security of multicast communication. The structure of the super spherical key tree is constructed by the members' probability. Therefore, the closer the probability of members' change is, the more common auxiliary keys they have. When group members join in or leave the multicast group, the key isn't updated until the cycle of batch key update reaches, and then the group controller starts to update the keys. So the common auxiliary keys that are owned by changed members only need to be updated once instead of many times, then the key update overhead can be reduced.

3.1 Members to join

Assuming that U_3 and U_4 join in the multicast group in a period of time (rekey interval), as shown in figure 1, the following are update steps of group key.

Definition 6 P_i is defined as the node in the spherical quad tree.

Step 1: U_3 and U_4 are joined in the group in this step. First of all, the group manager confirms the identity of U_3 and U_4. Then U_3 and U_4 are randomly generated a point coordinate respectively after authentication, and then the coordinates are secretly sent to the group manager. After receiving the coordinates, the manager adds U_3 and U_4 to L_{3-1}. According to U_3 and U_4, the manager calculates P_0 and then sends the coordinate of P_0 point to U_3 and U_4 openly. After receiving the messages, U_3 and U_4 calculate the distance R between the two points which are respective P_0 and the coordinates itself, and R is the group key. Then the manager needs to update the node of L_{1-1} and L_{2-1} and L_{3-1}.

Definition 7 $|P_1U_1|$ is defined as the distance between P_1 and U_1.

Definition 8 $|P_1U_2|$ is defined as the distance between P_1 and U_2.

Step 2: L_{3-1} is updated in this step. The group manager calculates the point P_1 according to the member U_1 and U_2 and makes sure that $|P_1U_1|$ is equal to R and $|P_1U_2|$ is also equal to R. P_1 and a new node L'_{3-1} point which is the new center sphere's coordinate of U_1, U_2, U_3, U_4, are encrypted by R, and then they are sent to everyone in the group publicly. After U_1 and U_2 receive the messages, they use the coordinate of P_1 and their own coordinate to calculate R, and then they figure out L'_{3-1} through R. But U_3 and U_4 use the known R to calculate L'_{3-1} directly after receiving the messages.

Definition 9 $|P_2L_{3-2}|$ is defined as the distance between P_2 and L_{3-2}.

Definition 10 $|P_2L_{3-3}|$ is defined as the distance between P_2 and L_{3-3}.

Definition 11 $|P_2L_{3-4}|$ is defined as the distance between P_2 and L_{3-4}.

Step 3: L_{2-1} is updated in this step. According to L_{3-2}, L_{3-3} and L_{3-4}, the manager calculates P_2, making sure that $|P_2L_{3-2}|$ is equal to R, $|P_2L_{3-3}|$ is equal to R and $|P_2L_{3-4}|$ is equal to R. Then P_2 and the new node L'_{2-1} which is the new center sphere coordinate of L'_{3-1}, L_{3-2}, L_{3-3} and L_{3-4}, are encrypted by R and sent openly. After receiving the messages, L_{3-2}, L_{3-3} and L_{3-4} calculate R through the coordinate of P_2 and its own coordinate, then they use R to figure out L_{2-1}', and the members in L_{3-1}' directly use the known R to figure out L_{2-1}' after receiving the messages.

Definition 12 $|P_3L_{2-2}|$ is defined as the distance between P_3 and L_{2-2}.

Definition 13 $|P_3L_{2-3}|$ is defined as the distance between P_3 and L_{2-3}.

Definition 14 $|P_3L_{2-4}|$ is defined as the distance between P_3 and L_{2-4}.

Step 4: L_{1-1} is updated in this step. The manager calculates P_3 according to L_{2-2}, L_{2-3} and L_{2-4}, making sure that $|P_3L_{2-2}|$ is equal to R, $|P_3L_{2-3}|$ is equal to R and $|P_3L_{2-4}|$ is equal to R. Then P3 and a new node L_{1-1}', which is the new center sphere's coordinate of L_{2-1}', L_{2-2}, L_{2-3} and L_{2-4}, are encrypted by R and are sent publicly. After they receive the messages, L_{2-2}, L_{2-3} and L_{2-4} use P_3 point coordinate to figure out R, and then R is used to figure out L_{1-1}'. The members in L_{2-1}' can directly use the known R to calculate L_{1-1}' after receiving the messages.

Definition 15 $|P_4L_{1-2}|$ is defined as the distance between P_4 and L_{1-2}.

Definition 16 $|P_4L_{1-3}|$ is defined as the distance between P_4 and L_{1-3}.

Definition 17 $|P_4L_{1-4}|$ is defined as the distance between P_4 and L_{1-4}.

Step 5: The root node is updated in this step. The manager calculates P_4 according to L_{1-2}, L_{1-3} and L_{1-4}, making sure that $|P_4L_{1-2}|$ is equal to R, $|P_4L_{1-3}|$ is equal to R and $|P_4L_{1-4}|$ is equal to R. Then P_4 is sent publicly. After they receive the messages, L_{1-2}, L_{1-3} and L_{1-4} use the coordinate of P_4 and its own coordinate to figure out R, then R is used to figure out L'_{1-1}. Although the root is the center sphere of the four child nodes, it doesn't work. So it doesn't need to be sent.

3.2 Members to leave

The members who are going to leave the group also need to update the auxiliary keys. Its algorithm is similar to the joined algorithm.

4 SIMULATION AND PERFORMANCE ANALYSIS

The multicast group size of 1 k, 4 k, 16 k and 32 k are chose respectively to conduct our research. 300 members of the change join in and leave the multicast group randomly, and the coordinate (X, Y) is set, in which X represents the number of members to leave, and Y represents the number of members to join in. The 10 interval periods of (240,60), (230,70), (220,80), (210,90), (200,100), (190,110), (180,120), (170,130), (160,140), (150,150) are discussed. At the same time, we repeat the calculation for each interval period, and then take the arithmetic means as the final result.

From figure 2 to figure 5 are the experimental simulation results. This paper simply presents the analysis of the worst case and mainly analyses the change of common members through the simulation experiments.

(1) The worst condition: When the joining in or leaving members are more dispersed with members' change probability more than zero and less than 0.55, the price of the batch key update method based on the members' behavior is almost the same with the batch key update way in literature [4].

(2) General situation: This scheme adopts the batch key update method based on the behavior of members, so as to make the members whose change probability is close to each other locate on the partial surface of the spherical, one sub tree of quad-tree. Therefore, the members who vary frequently are often in a sub tree, meaning that the changes of group membership concentrated relatively. If the members of the change are mostly in a sub tree in a key update cycle, they have the most common auxiliary keys [4], which are only needed to update once when the group key server update keys. Thus, it is concluded that the scheme has high efficiency.

From the simulation results of figure 2 to figure 5, that this scheme reduces the overhead of group key server greatly can be concluded, compared with the literature [4]. And with the increase of the members in multicast group, especially the closer the number of joining members and leaving members are, the more

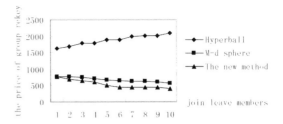

Figure 2. Group membership for 1 k.

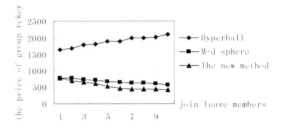

Figure 3. Group membership for 4 k.

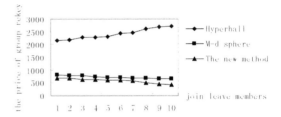

Figure 4. Group membership for 16 k.

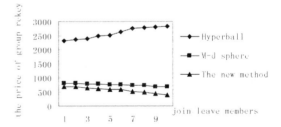

Figure 5. Group membership for 32 k.

obvious the solution' advantages are, compared with literature [6].

5 CONCLUSION

As a new model, the secure multicast key update scheme based on super ball is applied in secure multicast, and it reduces the total key update number of group manager in the multicast management. This paper proposes a kind of batch key update method based on the members' behavior of the super spherical, in which the similar members whose probability of behavior changes are close are put into a sub tree, in order that they have more common auxiliary keys. At the same time the method of batch key update is adopted, so the price of group key update is further reduced. Simulation results show that the scheme is effective to reduce the group key update cost, and improve the efficiency of the key update.

REFERENCES

[1] Zhu S, Setia S, Jajodia S. Performance optimizations for group key management schemes for secure multicast //Proceedings of the 23rd IEEE International Conference on Distributed Computing Systems. Rhode Island, USA. 2003:163–173.
[2] Wong C K, Gouda M, Lam SS. Secure group communications using key graphs. Networking, IEEE/ACM Transactions on. 2000, 8(1): 16–30.
[3] Xiaozhou Steve Li, Yang Richard Yang, Mohamed G. Gouda, Simon S. Lam. Batch rekeying for Secure Group Communications. 2001, 525–534.
[4] Shaojun Chen, Nianfeng Rong, XuYong, LiJie. multicast key updating method based on the spherical. 2007, 22(3):48–30.
[5] Haitao Xie, Yuming Wang. A kind of M-d spherical multicast key batch update scheme. Journal of Chinese Computer Systems. 2010, 31(2):254–258.
[6] Zhu F, Chan A, Noubir G. Optimal tree structure for key management of simultaneous join/leave in secure multicast //Military Communications Conference. 2003. MILCOM'03. 2003 IEEE. IEEE, 2003, 2:773–778.
[7] Li D, Aung Z, Sampalli S, et al. Privacy Preservation Scheme for Multicast Communications in Smart Buildings of the Smart Grid. Smart Grid & Renewable Energy. 2013, 4(4).
[8] Li M, Feng Z, Zang N, et al. Approximately optimal trees for group key management with batch updates. Theoretical Computer Science. 2009, 410(11):1013–1021.
[9] Soni J, Benakatti P T, Rao M V P, et al. New Approach for Secure Source Routing Protocol in MANET. Ad-hoc networks, 2013, 1:2.

Waterproof time-dependent analysis of polymer modified cement mortar

D.X. MA, Y. Liu, Y. Lai & Z.G. Luo
*China Airport Construction Group Corporation of CACC,
Beijing Super-Creative Technology Co., LTD, Beijing, China*

ABSTRACT: This paper adopting three kinds of polymers studied the crack resistance (maximum crack width and cracking weighted value) and deformation resistance (shrinkage rate, relative dynamic elastic modulus, lateral deformation and tensile bond strength) of cement mortar. The results show that with the increase of polymer content, the maximum crack width and cracking weighted value significantly decrease, which means polymer can effectively improve the crack resistance of cement mortar. In addition, polymer declines the shrinkage rate and relative dynamic elastic modulus of mortar, improves the elasticity and flexibility of mortar, increases the cohesive force between the cement and aggregate, raises the cohesiveness of various base materials, resulting in that polymer modified mortar can withstand the base material deformation, impact of external force, thermal and freeze-thaw deformation, and remarkably improve the capacity of deformation resistance. From the two aspects it can be illustrated that the long-term effectiveness of waterproof performance is based on the polymer effectively inhibiting the formation and development of cracks and significantly improving deformation resistance of mortar.

Keywords: Polymer; Mortar; Modification; Waterproof; Aging analysis

1 INTRODUCTION

Polymer has an excellent capability of improving waterproof performance of building mortar, and a great deal of work about polymer modified mortar has been done. The present studies mainly focus on the preparation of polymer-modified waterproof mortar, the effect of types, compositions, dosage of polymer on the waterproof properties, how to reduce cost and pollution with various mineral admixture etc. However, the study of time-dependent waterproofing effectiveness after adding polymer is still few literatures involved. This paper adopts three kinds of polymers researching the effect of polymer on crack resistance and deformation resistance of mortar, so as to explore the waterproof time-dependent effect of polymer modified cement mortar from the perspective of crack and deformation.

2 EXPERIMENT

2.1 Raw materials

The cement used was PII52.5 Portland cement according to Chinese Standards (GB175-2007). The aggregate was standard sand meeting the ISO679-89. The tap water was used as the mixing water. Styrene-acrylic emulsion (SAE) powder with a bulk density of $350 \sim 550$ g/L and minimum film forming temperature of $0°C$ was used, its residual moisture was less than 2%, and the pH value was $7.5 \sim 9.5$. SAE dispersion with the pH of $7.0 \sim 9.0$ was used in the mortar, the solid content of which was $55 \pm 2\%$, its glass transition temperature (Tg) and the viscosity were $-20°C$ and $500 \sim 1500$ cps respectively. The product parameter of Styrene Butadiene Rubber (SBR) dispersion was the pH (ISO976) of $7.8 \sim 10$, solid content of $50 \sim 52\%$, Tg of $-13°C$, viscosity of $35 \sim 150$ cps.

2.2 Sample preparation and testing methods

The cement-sand ratio (by mass) was 1:3 and polymer was added into mortar with the polymer-cement ratio (m_p/m_c, by mass) of 0%, 1%, 3%, 5%, 8%, 10%, 12%, 15%, 18%, 20%. The fluidity of mortar was measured according to GB/T2419-2005 and controlled at 170 ± 5 mm. The mortar specimens were molded in accordance with GB/T17671-1999. The flexibility test of modified mortar was carried out according to JC/T1004-2006 Appendix B. The water absorption was tested in accordance with DL/T5126-2001 Appendix B. After the specimen conservation the bonding strength was measured by the JC/T907-2002. All other experiments were referring to JCT984-2011.

Conservation methods [30] of 28d age mortar specimens: after the specimens being molded they were placed at $20°C$ and above 90% relative humidity conditions under curing of 24h. Afterwards the form stripping is in progress, and it's essential to cure them at $20°C$ water conservation in first 6 days followed by 21 days at 23°C and relative humidity of 60% air conservation.

3 RESULTS AND DISCUSSION

3.1 Crack resistance

This paper by using ring restraint device tests the crack of mortar. The method is mainly used for the cracking weighted value and the maximum crack width as cracking evaluation index. Assigned weights for different crack width are shown in Table 1. According to the difference of crack width the segment of crack length L_i is measured differently, which is defined as the product of the crack width and the length. And the cracking weighted value is the sum of the area of each segment according to Equation 1:

$$W = \sum A_i \times L_i \qquad (1)$$

where W is the cracking weighted value; A_i is the cracks value of different width; and L_i is the crack length of different crack width.

Table 2 shows that without mixing any polymer into mortar the maximum crack width is 0.15 mm and the cracking weighted value is quite large. While SBR dispersion content mixed into waterproof mortar is 10%, the maximum crack width decreases by 67% compared with that of not mixing the dispersion and the cracking weighted value is only 35.7. With SAE emulsion at a dosage of 8% to 12%, the maximum crack width decreases quite significantly, the falling range reaches by 71%, there is no significant decrease for the maximum crack width in the range of 15% to 20%. With SAE powder modified waterproof mortar in the range of 5% ~ 8%, the maximum crack width of which reduces from 0.10 mm to 0.06 mm, that is, about 40% reduction. In the range of 18% to 20% there is no obvious effect on the improvement of the maximum crack width for all three kinds of polymer. The data shows that regardless of what kind of polymer, when its dosage reaches a certain amount, the maximum crack width and cracking weighted values which characterizes the degree of cracking of mortar have a pretty significant decline, which proves polymer has a quite positive impact on preventing the cracking of mortar.

3.2 Deformation resistance of polymers

Mortar is a rigid waterproof material, and its waterproof performance mainly depends on their own compactness, but in the practical application it is found that mortar due to crack results in a poor waterproof performance. Except the crack caused by improper construction, another important reason of crack is essentially attributed to the different strain rate and different volume deformation resulting from different ingredients of mortar.

3.2.1 Relative dynamic elastic modulus

The changing trend of relative dynamic elastic modulus of polymer modified mortars is described in Figure 1. It is shown that with the increase of m_p/m_c the relative dynamic elastic modulus of polymer modified mortars shows a downward trend. And when the addition of polymer is relatively small the relative dynamic elastic modulus of all modified mortars has a significant decline.

When the m_p/m_c is 5% the relative dynamic elastic modulus of SBR dispersion modified mortar is 35.5 GPa, and contrasting with the control mortar the relative dynamic elastic modulus decreases by about 13%. For the relative dynamic elastic modulus of SBR dispersion-modified mortar there is a large decline in the dosage of 5% ~ 8%. When the m_p/m_c is more than 12% the relative dynamic elastic modulus of SBR dispersion modified mortar doesn't have much decrease. For the SAE dispersion with its dosage in the range of 10% ~ 12% there is a quite great decrease. The relative dynamic elastic modulus of mortar which added the SAE powder of less than 8% decreases very significantly, especially in the m_p/m_c range from 5% to 8%, the relative dynamic elastic modulus decreases from 34.5 GPa to 28.5 GPa.

As can be seen from Figure 1, when the dosage is less than 5%, the relative dynamic elastic modulus of the SBR dispersion and SAE powder is greater than that of the SAE dispersion. The falling range of

Table 2. Maximum crack width and the cracking weighted value of polymer-modified mortar

m_p/m_c, %	maximum crack width (mm)			cracking weighted value		
	A	B	C	A	B	C
0	0.15	0.15	0.15	94.5	94.5	94.5
1	0.15	0.15	0.13	95.2	93.6	90.2
3	0.12	0.17	0.12	87.8	98.4	85.4
5	0.11	0.14	0.1	79.3	90.5	76.9
8	0.08	0.1	0.06	55.4	75.8	41.3
10	0.05	0.08	0.04	35.7	60.3	32.6
12	0.05	0.04	0.03	36.3	34.7	26.3
15	0.03	0.02	0.02	27.9	25.9	22.9
18	0.01	0.01	0.01	18.7	15.2	17.4
20	0.01	0.01	0.01	17.9	10.6	13.8

* A: SAE dispersion; B: SAE dispersion; SAE powder

Table 1. Cracking weighted value.

Crack width (d × 10^{-2} mm)	d ≤ 1	1 < d ≤ 2	2 < d ≤ 3	……	14 < d ≤ 15
Weight value, A_i	1	2	3	……	15

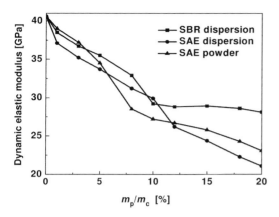

Figure 1. Variation of 28d relative elastic modulus with addition of polymer.

the dynamic elastic modulus of SAE powder modified mortar is the biggest in the amount of 5%~10%. While the dosage of polymer is 12%~20%, the dynamic elastic modulus of the mortar is that: SBR dispersion > SAE powder > SAE dispersion.

3.2.2 *Lateral deformation capacity*

The relationship between the polymer dosage and lateral deformation of polymer modified cement mortar can be seen from Figure 2. The lateral deformation value of three polymer modified cement-based waterproof mortars increases as the polymer dosage increases gradually, and when the polymer dosage is quite great the lateral deformation value of modified mortars will speed up with the increase of their dosage. When the SBR dispersion dosage does not exceed 10%, the increase of lateral deformation capacity is not pretty obvious, but after the dosage is more than 10%, the lateral deformation capacity begins to be enhanced. Compared with SBR dispersion, after SAE powder dosage reaches 8% the lateral deformation value shows a larger increase, and the increase amplitude of lateral deformation is 154% in the range of 8% to 20%. SAE dispersion has a similar lateral deformation trend with SBR dispersion. SAE dispersion increases small in the range of 0% to 10%, but after that as the dosage increases, the lateral deformation value increases sharply. The reason is when the dosage is more than 10%, the internal structure of modified cementitious waterproof mortar is a continuous polymer network structure intertwined with the formed cement paste interpenetrating network structure, resulting in the lateral deformation capacity of modified cement mortar rapidly increasing.

3.2.3 *Shrinkage rate*

Figure 3 presents that when the addition of polymer is small the shrinkage rate of three kinds of modified mortars will decrease significantly with the incorporation of polymer increasing. For SBR dispersion modified mortar of 28d age, when the addition of polymer is 0~12%, with the increase of m_p/m_c, the shrinkage rate decreased remarkably. The shrinkage

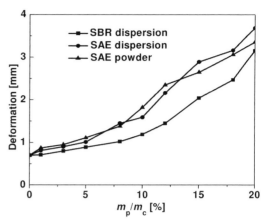

Figure 2. Relationship between lateral deformation and polymer dosage of polymer-modified mortars.

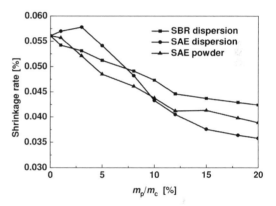

Figure 3. Relationship between shrinkage rate and dosage of 28d polymer-modified mortar.

rate of 28d age SAE dispersion modified mortar with the polymer content of 3%~8% declines observably. The changing trend of SAE powder modified mortar is pretty similar to that of SBR dispersion modified mortar in range of 0~12%. Hence, from the perspective of lowering shrinkage rate, Adding polymer into mortar is conducive to reducing the deformation of mortar.

3.2.4 *Tensile bond strength*

Mortar is generally used for the surface of construction, only the mortar itself has certain cohesive force, the effective bond with the substrate material and its long-term stability can be achieved. Hence, the majority of mortar has a strong requirement for tensile bond strength index. When the tensile bond strength of mortar is large the deformation resistance between the mortar and substrate material will be pretty great. It is discovered from Figure 4 that regardless of what kind of polymer is being incorporated, the tensile bond strength of polymer modified mortar is greater than that of control mortar, and with the increase of polymer content, the tensile bond strength

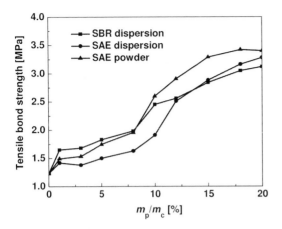

Figure 4. Relationship between tensile bond strength polymer and polymer dosage of 28d polymer-modified mortar.

improves significantly. With the polymer content in the range of 0~8%, when the same dosage of three kinds of polymers is incorporated into mortar, the tensile bond strength of mortar is as follows: SBR dispersion > SAE powder > SAE dispersion. When the polymer dosage is 1%, comparing with the control mortar, the tensile bond strength of SBR dispersion, SAE powder and SAE dispersion modified mortar increased by 34%, 21%, 15%, respectively. And with polymer dosage of 8%~20%, the tensile bond strength of SAE dispersion modified mortar is greater than that of SAE powder and SBR dispersion modified mortar.

To sum up, due to polymer decreasing the shrinkage rate and relative dynamic modulus of elasticity, raising the lateral deformation capacity and improving the tensile bond strength of mortar with all kinds of substrate materials, resulting in that mortar is able to withstand the deformation effect of substrate material, the impact of external force, the temperature deformation and freeze-thaw deformation, remarkably improving the capacity of deformation resistance, effectively inhibiting the formation and development of cracks of mortar.

4 CONCLUSIONS

With the increase of polymer content, the maximum crack width and cracking weighted value significantly decrease, which means the polymer can effectively improve the cracking resistance of cement mortar.

The polymer declines the shrinkage rate and relative dynamic elastic modulus of mortar, improves the elasticity and flexibility of mortar, increases the cohesive force between the cement and aggregate, improve the cohesiveness of various substrate materials, resulting in that the mortar can withstand the deformation effect of substrate material, the impact of external force, the temperature deformation and freeze-thaw deformation, improve the capacity of deformation resistance.

It can be illustrated that the long-term effectiveness of waterproof effect is based on significantly improving the capacity of deformation resistance, effectively inhibiting the formation and development of cracks of mortar so as to enhance the time-dependent waterproof property and durability of mortar.

REFERENCES

[1] Saija, L.M. 1995. Waterproofing of Portland cement mortars with a specially designed poly-acrylic latex. *Cement and Concrete Research* 3(25): 503–9.
[2] Zhao F.Q & Li H. 2011. Preparation and properties of an environment friendly polymer-modified waterproof mortar. *Construction and Building Materials* (25): 2635–2638.
[3] Zhang J.X. & Jin S.S. 2009. The study on basic properties of polymer latex modified cement mortar. *Journal of Beijing University of Technology* 35(8): 1062–1068.
[4] Ohama Y. 1995. *Handbook of Polymer-Modified Concrete and Mortars.* New Jersey: Noyes Publications.
[5] Joachim, S. & Otmar K. 2001. Long-term performance of re-dispersible powders in mortars. *Cement and Concrete Research* (31): 357–362.

Comparative study of highway coarse aggregate gradation used in airport pavement concrete

D.X. Ma, Y. Liu, Y. Lai & Z.G. Luo
China Airport Construction Group Corporation of CACC,
Beijing Super-Creative Technology Co., LTD, Beijing, China

ABSTRACT: This paper chose two kinds of coarse aggregate gradation (5~31.5 mm single-graded coarse aggregate, 5~20 mm, 20~40 mm double-graded coarse aggregate) to comparatively study the influence of single-graded and double-graded coarse aggregate on the performance of airport pavement concrete. The results show that the workability of airport pavement concrete prepared with double-graded coarse aggregate is superior to that of single graded coarse aggregate and both of them meet the design requirement, the flexural strength and durability of concrete prepared with single-graded coarse aggregate are better than the corresponding parameters of the concrete prepared with double-graded coarse aggregate.

Keywords: Coarse aggregate; Gradation; Airport pavement concrete; Mechanical properties; Frost resistance durability

1 GENERAL INSTRUCTIONS

Airport pavement concrete is a platform of aircraft taking off and landing which is not only affected by such harsh external environment as humidity variation, temperature change, water erosion, but mainly exposed to the instant impact and fatigue load cycles from the landing process of aircraft, so the structural layer is susceptible to damage, hence, it's pretty significant to ensure the quality stability of airport pavement concrete.

For the coarse aggregate used in airport pavement concrete, its production process is that the quarried stones are firstly crushed into the relatively small stones of 5~20 mm, 20~40 mm and then remix them according to practical application; for highways and other infrastructure projects they prefer to directly choosing the coarse aggregate of being crushed into 5~31.5 mm to design concrete. Compared with the coarse aggregate used in highway, apparently the coarse aggregate used in airport pavement concrete presents the energy consumption of production process increasing.

According to the analysis above mentioned, both of the two coarse aggregate gradations will be applied to the mix design of airport pavement concrete in this paper, comparatively studying the workability, mechanical property and durability of airport pavement concrete, further providing a certain basis to optimize the specification of airport pavement concrete.

2 EXPERIMENT

2.1 *Raw materials*

The cement used was P·O 42.5 Portland cement, the coarse aggregate were the gravel of 5~31.5 mm single-graded coarse aggregate and 5~20 mm, 20~40 mm double-graded coarse aggregate, the fine aggregate was artificial mechanism sand with fineness modulus of 2.70. The PCA-L1 superplasticizer and tap water were used as admixture and mixing water, respectively. The mix proportions of concrete were shown in Table 1.

3 RESULTS AND DISCUSSION

3.1 *Workability*

Workability is a quite important construction performance to ensure the quality of airport concrete, and the favorable workability is capable of making concrete pavement as far as possible to achieve an excellent structure in the course of construction, as a result that the hardened concrete gets an optimum mechanical properties and exceptional durability.

The influence of water cement ratio on the vebe consistency of fresh concrete with different coarse aggregate graduation is presented in Figure 1. For the single- and double-graded coarse aggregate, regardless of what the value of water cement ratio, the greater the cement content is, the lower the vebe consistency

Table 1. Mix proportions of concrete.

No.	Cement [kg/m³]	Water cement ratio	sand [%] A	sand [%] B	Water reducer [%] A	Water reducer [%] B
1	320	0.38	28	27	1.7	1.8
2	320	0.40	28	27	1.8	1.6
3	320	0.42	28	27	1.9	1.6
4	325	0.38	28	27	1.7	1.7
5	325	0.40	28	27	1.8	1.6
6	325	0.42	28	27	1.7	1.8
7	330	0.38	28	27	1.6	1.7
8	330	0.40	28	27	1.7	1.6
9	330	0.42	28	27	1.6	1.5

No.	Mix proportions A	Mix proportions B
1	1:0.38:1.80:4.63	1:0.38:1.74:4.70
2	1:0.40:1.80:4.62	1:0.40:1.73:4.68
3	1:0.42:1.79:4.60	1:0.42:1.73:4.67
4	1:0.38:1.77:4.54	1:0.38:1.70:4.61
5	1:0.40:1.76:4.53	1:0.40:1.70:4.59
6	1:0.42:1.76:4.52	1:0.42:1.69:4.58
7	1:0.38:1.73:4.46	1:0.38:1.67:4.52
8	1:0.40:1.73:4.45	1:0.40:1.67:4.51
9	1:0.42:1.72:4.43	1:0.42:1.66:4.49

*A: concrete prepared with highway coarse aggregate (5∼31.5 mm);
B: concrete prepared with airport pavement coarse aggregate (5∼20 mm, 20∼40 mm).

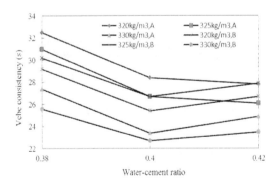

Figure 1. Influence of water cement ratio on the vebe consistency of fresh concrete.

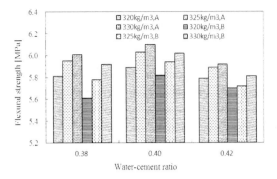

Figure 2. The relationship between flexural strength and different dosage of cement, water cement ratio.

is, and with the cement content of 330 kg/m3 there is the minimum value of vebe consistency. No matter what the amount of cement is, with the water cement ratio increasing, the vebe consistency grows to some extent after rapid decline firstly, and the lowest vebe consistency occurs in water-cement ratio of 0.40.

The vebe consistency of double-graded coarse aggregate concrete is slightly higher than the vebe consistency of single-graded coarse aggregate concrete. The main reason is that the maximum grain size of single-graded coarse aggregate which is used in highways is relatively smaller, resulting in that the specific surface area of single-graded coarse aggregate is bigger than that of double-graded coarse aggregate under the same weight condition and if the same fluidity wants to be achieved the more volume of mortar will be required for single-graded coarse aggregate concrete. Hence, the vebe consistency of single-graded highway coarse aggregate concrete appears to be a relatively large value. But both of two kinds of concretes are in range of 20 ∼ 40 s.

3.2 *Mechanical properties*

The mechanical property requirements of airport pavement concrete mainly focus on the performance of flexural strength. The variation trend of flexural strength of single-graded and double-graded coarse aggregate concretes at 28 days age is shown in Figure 2. No matter what the value of water cement ratio is, with the growth of cement content, the flexural strength increases sharply. Regardless of what amount of cement, with the water cement ratio increasing, the flexural strength firstly increases and then declines, and with the water cement ratio of 0.40 the flexural strength occurs the maximum value. Under the same conditions the flexural strength of single-graded coarse aggregate concrete is more than the corresponding values of double-graded coarse aggregate concrete. Therefore, in terms of flexural strength scope the airport pavement concrete produced with single-graded coarse aggregate is superior to the airport pavement concrete prepared with double-graded coarse aggregate.

3.3 *Durability*

Currently the durability requirement in airport domain is achieved mainly through measuring the index of frost resistance. In this paper, the weight loss and loss of dynamic elastic modulus of concrete specimen are as the characterization parameters of frost resistance. Figure 3 shows with the freeze-thaw cycles being carried out, the weight loss of these two concretes firstly decreases and then increases, and the loss

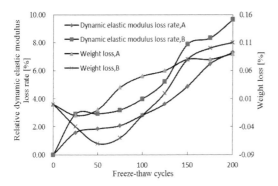

Figure 3. The influence of freeze-thaw cycles on the frost resistance durability.

rate of relative dynamic elastic modulus persistently increases.

At the beginning of the freeze-thaw cycles the weight loss of both concrete is less than zero, the water absorption is greater than the mass of spalling concrete from frost crack, being characterized by the quantity increase of concrete specimen, which is mainly caused by the porosity of concrete being relatively large. The weight loss of concrete prepared with highway single-graded coarse aggregate is quite low, showing that its porosity is comparatively small and more compact. The weight loss of both concretes starts to increases significantly after reaching the minimum value, thereinto, the weight loss of concrete prepared with double-graded coarse aggregate rises more sharply, and its erosion becomes more serious, which proves that after the freeze-thaw cycles running up to a certain number the structure of concrete prepared with double-graded coarse aggregate damages more prominently, and its frost resistance falls dramatically.

Since the freeze-thaw damage of concrete firstly starts from the capillary damage, then with the increase of freeze-thaw cycles, as the liquid in pores is transferred into solid, its volume expansion leads to the crack propagating and the relative dynamic elastic modulus decreasing gradually. During the entire process of freeze-thaw cycles, for the loss rate of relative dynamic elastic modulus the concrete prepared with single-graded coarse aggregate is less than the concrete prepared with double-graded coarse aggregate.

Hence, from the weight loss and loss rate of relative dynamic elastic modulus point of view, the frost resistance durability of concrete specimens prepared with single-graded coarse aggregate prevails over that prepared with double-graded coarse aggregate.

4 CONCLUSIONS

The vebe consistency of double-graded coarse aggregate concrete is slightly higher than the vebe consistency of single-graded highway coarse aggregate concrete, but both of them are in range of $20 \sim 40$ s.

In terms of flexural strength scope the airport pavement concrete prepared with single-graded highway coarse aggregate is superior to the airport pavement concrete prepared with double-graded conventional airport coarse aggregate.

With the freeze-thaw cycles being carried out, the weight loss of both concretes firstly decreases and then increases, the loss rate of relative dynamic elastic modulus persistently increases, but at the end of freeze-thaw cycles, the weight loss and loss rate of relative dynamic elastic modulus of concrete prepared with single-graded coarse aggregate reach the minimum value, i.e. the frost resistance durability of concrete specimens prepared with single-graded coarse aggregate prevails over that prepared with double-graded coarse aggregate .

ACKNOWLEDGEMENTS

This work was financially supported by National Natural Science Foundation of China (Grant No. 51208018), Science and Technology Project of CAAC (MHRD201225), Science and Technology Project of CAAC (MHRD20130109) and Beijing Nova Program (Z121110002512090).

REFERENCES

[1] Yi, H.L. & Cheng, L.G. 2008. The quality control of airport pavement concrete. Municipal Engineering Technology 26(6):475–478.
[2] Guo, X. & Chen, P.G. 2010. The freeze-thaw damage evaluation index of airport pavement concrete. Journal of Traffic and Transportation Engineering 10(1):13–18.
[3] Morteza, H.A. & Beygi, M.T. 2014. Evaluation of the effect of maximum aggregate size on fracture behavior of self-compacting concrete. Construction and Building Materials (55): 202–211.
[4] Basheer, L. & Basheer, P.A. & Long, A.E. 2005. Influence of coarse aggregate on the permeation, durability and the microstructure characteristics of ordinary Portland cement concrete. Construction and Building Materials (19):682–690.
[5] Zhou, F.P. & Lydon, F.D. 1995. Effect of coarse aggregate on elastic modulus and comprehensive strength of high performance concrete. Cement and Concrete Research 25(1):177–186.

Design of rapid pesticide residue detection system based on embedded technology

Shenggao Gong, Qing Liu, Zhiqin He & Pengfei Bian
School of Electrical Engineering, Guizhou University, Guiyang, China

ABSTRACT: Pesticide residue is the hidden danger of agricultural products safety, the market demand for pesticide residue detector is also increasingly urgent. The topic is to put forward to solve the detection pesticide residues in agricultural products, it uses the technology of enzyme inhibition, photochemical analysis and Lambert Beer's law to calculate the inhibition rate of pesticide residues in agricultural products the design using embedded system in terms of overall development, using visible spectrophotometer as photoelectric detecting part, using photochemical method, enzyme inhibition technology, realize the multi-channel rapid detection of pesticide residue.

Keywords: Embedded system; Visible Spectrophotometer; Detection of pesticide residue

1 INTRODUCTION

With the development of China's agricultural production level, the problem caused by extensive use of the pesticide has been widely known, traditional test method of pesticide residues include pieces of enzyme inhibition assay, biosensor method, gas chromatography, but the above methods are complex, the cost is too high, also can't adapt to the requirement of pesticide residues rapid detection. Photoelectric detection technology has the characteristic of high detection sensitivity, fast, simple and advantages of continuous tests, it has become one of the hotspots in the field of pesticide residues detection. This topic is a combination of photochemical method, technology of enzyme inhibition, using visible spectrophotometer as photoelectric detecting part, controlled by the embedded system.

2 THE METHOD OF MULTICHANNEL PESTICIDE RESIDUE RAPID DETECTION SYSTEM

Visible photometer is detector photoelectric of detection system, it consists of photoelectric signal conversion circuit, signal amplification circuit, filter circuit and complete optical signal acquisition, photoelectric signal conversion, signal amplification, and provides a basic data for calculating pesticide residues inhibition rate.

Lambert Beer's law is the basic principle of pesticide residues inhibition rate calculation. When light through a uniform and transparent solution, set the intensity of incident light for I0, transmission light intensity for I, Lambert Beer's law is to discuss solution absorbance with relation between solution concentration and thickness of solution layer, the theoretical basis of spectroscopic analysis. Its expression is: $A = \varepsilon * L * C$

Enzyme inhibition technology is based on the analysis of the degree of enzyme catalytic reaction, the degree to which the reaction of enzyme inhibition to quantitatively. Because the activity of enzyme inhibition by pesticides, and cannot catalytic or completely chromogenic agent, chromogenic reaction substrates. Visible spectrophotometer can measure color changes caused by the difference between the absorbance, then calculate the enzyme inhibition rate through the change of absorbance. The higher the enzyme inhibition rate is, the higher the content of pesticide residue in the sample.

3 THE DESIGN OF MULTICHANNEL PESTICIDE RESIDUE RAPID DETECTION SYSTEM

The system using STM-32 in terms of overall development, STM-32 based on designed for high performance, low cost, low power consumption embedded application designer ARM Cortex-M3 architecture, it has powerful function, fast processing speed, high reliability. The system is mainly composed of STM-32 control chip, spectrophotometer, multi-channel data acquisition circuit, etc. The system block diagram is shown in Figure 1.

Figure 1. Block diagram of collision avoidance system.

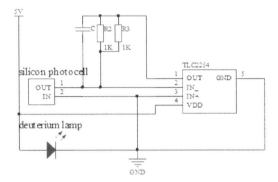

Figure 2. Circuit diagram of spectrophotometer.

3.1 The design of the visible spectrophotometer

This design uses the deuterium lamp as light source, it has continuous spectrum and the wavelength is 165 nm to 375 nm. Uses the silicon photocell as photoelectric receiving device, it has the characteristic of good portability, high sensitivity and low cost. Choose the TLC2254 chip of TI Company as the Op-amp chip. It has the characteristic of low consumption, high open-loop gain. The circuit diagram is shown in Figure 2.

3.2 Software design

The thought of software design is carried out on initialization in each part of the system, follow the instructions of the keyboard. This design needs to provide a simple and reliable operating system. The user completes function selection, parameter setting, channels selection though the screen. It can provide a well performance man-machine interface and follow the rules of the calculation. The system block diagram of software is shown in Figure 3.

According to the Lambert-Beer's law, we will complete the detect of incident light intensity, the output light intensity of standard solution and the output light intensity of sample solution inhibition rate formula is: $R = (\Delta A_{std} - \Delta A_{asm}) * 100\%$, the ΔA_{std}, ΔA_{asm} is obtained by least squares linear fitting after a certain time interval, $A_{std} = \log V_{in} - \log V_{outstd}$, $A_{asm} = \log V_{in} - \log V_{outsasm}$, R can be suppressed after calculation. Acquisition program flow chart is shown in Figure 4.

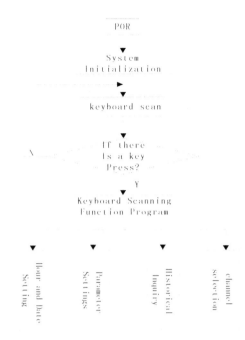

Figure 3. System block diagram of software.

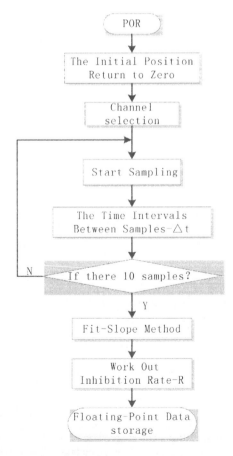

Figure 4. Acquisition program flow chart.

4 THE EXPERIMENT AND ANALYSIS OF PESTICIDE RESIDUE RAPID DETECTION SYSTEM

Use the design of pesticide residue detection system to detect pesticide samples in the actual experiment, to establish the mathematical model of pesticide residue concentration and light intensity, verify the feasibility of the designed pesticide residue detection system.

In order to guarantee the reliability of the experiment and the precision, experimental materials use mature detection reagent on the market, use the AG209 type speed test agent pesticide residues made by the Guangzhou oasis biochemical technology co., LTD.

Reaction principle

Substrate+Color Developing Agent $\xrightarrow[PH=8.0]{AchE}$ Colored Matter(Faint Yellow Precipitate)

4.1 Experiment process

Open the power supply to the system, to eliminate the influence of dark current photoelectric diode, after the completion of the system voltage to zero. According to the type AG209 speed pesticide residues test agent operating instructions carried out the operation. In the experiment, using the phosphate buffer, inspection of pesticide residues was dimethoate, the concentration are 4 mg/kg, 3 mg/kg, 2.5 mg/kg. The test results are shown as Figure 5.

According to the test results shown in Figure 5 national inspection qualification of dimethoate is 3 mg/kg, based on Figure 5, when the concentration of pesticide residues is 4.0 mg/kg, its inhibition rate of pesticide residues is 65.64%. When the concentration of pesticide residues is 2.5 mg/kg, its inhibition rate of pesticide residues is 31.59%. When the concentration of pesticide residues is 3 mg/kg, its inhibition rate of pesticide residues is 51.46%.

The density and absorbance data as shown below in Table 1.

According to Table 1, under the different concentrations of pesticide residues, curve fitting the reaction time and the absorbance, can be obtained fitting curve slope will decreased as the rising of the concentrations.

4.2 The source of system error

1) The error of light source

Any light source is not absolutely stable, then its intensity will be tiny variations, in addition, feedback resistance value of amplifying circuit in this system is 0.5k, even though tiny variations current changes will be reflected in the voltage output. The A/D circuit to collect results will produce little deviation.

2) The temperature and time conditions

If the indoor temperature below 37 C, enzyme reaction speed slows down, then add enzyme liquid and chromogenic agent after measuring easy to cause the change rate reduced, the other cool choline enzyme solution to place time is too long

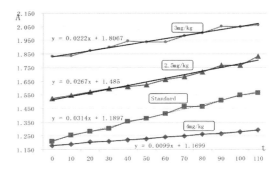

Figure 5. Trend Chart of Density and Voltage1.

Table 1. Density and absorbance.

T/s	0.	10.	20.	30.	40.	50.	60.	70.	80.	90.	100.	110.	R.
Standard	1.21	1.25	1.28	1.30	1.36	1.38	1.41	1.46	1.46	1.51	1.55	1.57	0.
4.0mg/kg	1.18	1.19	1.21	1.21	1.22	1.23	1.24	1.25	1.26	1.26	1.28	1.29	65.64%
2.5mg/kg	1.52	1.55	1.57	1.60	1.61	1.62	1.66	1.68	1.72	1.77	1.77	1.83	31.59%
3.0mg/kg	1.83	1.84	1.87	1.90	1.91	1.94	1.94	1.98	2.00	2.05	2.05	2.06	51.46%

or the temperature is not close, it can reduce the activity of enzyme, can also cause low rate.

3) The instrument error

In experimental operation, if there are external noise of light will produce certain error, with the use of other measurement instruments also will produce the error, such as balance error, the use of liquid receiver error, solution mixing uniformity and the error of time.

4) Detector error

Sensitivity to violet light source, that current will be small, the useful signals are faint, vulnerable to the interference of noise signal.

5) Solution error

Load than the quantity of each color dish solution will have small differences, according to the Lambert Beer's law, measured the absorbance of size will also have the difference, the other in the process of serving reagents have time difference, taking reaction initial point in time there will be a small difference, causes the variation of the absorbance error.

ACKNOWLEDGEMENT

Project funds: Guizhou University graduate student innovation fund (research institute of technology 2014045). Project funds: Guizhou province science and technology fund QianKeHe J Zi (2012) 2158.

REFERENCES

[1] Chunxiao Jia. Modern instrumental analysis technology and its application in food. Beijing. China light industry press, 2005.1:57–58

[2] Yikui Liao. The architecture of the M3 STM32 embedded system design. Beijing. China electric power press, 2012.2:103–105

[3] JunShen, Serge Castan. An Optimal Linear Operator for Step Edge Detection. Graphical Models and Image Proeessing.VOl.54.No.2.1992

[4] Yu Bai. Multi-channel pesticide residue detection system design. 2008:60-61

[5] Lili Qian. The rapid detection of pesticide residue and the former processing technology research. 2007

ABOUT THE AUTHOR

Shenggao Gong male born in May 1990, Guizhou University Graduate School of Electrical Engineering, Research: computer control technology

Qing Liu male born in September 1989, Guizhou University Graduate School of Electrical Engineering, Research: computer control technology

Application of water electrolysis oxy-hydrogen generator in the continuous cast products cutting

Zhenhua Nie, Ying Wang, Zhijie Gao, Chuan Wang, Di Zhang, Shipeng Liu, Yi Ren & Yuan Wang
Central Research Institute Of Building and Construction CO, LTD., MCC Group Beijing Haidian district, China

ABSTRACT: This article describes the work principle, technical performance and application of the water electrolysis hydrogen generator equipment. Compared with the use of fossil fuel gas, using hydrogen flame cutting off the fire billet cutting technology has better economic and social benefits. The technology is benefit for building a resource-saving and environment-friendly society, taking the road of sustainable development.

Keywords: Electrolyzing water oxy-hydrogen generator; Hydrogen-oxygen mixed gas; Billet; Oxy-hydrogen Flame Cutting; Propane gas

1 FOREWARD

Energy is one of the material bases for the existence and development of human society since the industrial revolution, such as coal, oil, natural gas and other fossil fuel energy systems, which greatly promotes the development of human society. Economic development is faced with severe challenges, for example, limit of fossil fuel resources and environmental protection requirements. Improving energy efficiency, optimizing energy structure, strengthening environmental protection and energy security, developing and using renewable energy are necessary conditions for the sustained development of society.

Fossil fuel gas for flame processing, with a calorific value of large, high flame temperature of advantages, but it has a lot of shortcomings, such as high cost, non-safety, polluting the environment and other issues. Hydrogen mixed gas is produced by electrolysis water, with low cost and without any contamination. Preparation of electrolytic hydrogen gas mixture as a green energy medium can meet the resource requirements of the environment and sustainable development, with good economic and social benefits. [1]

2 PRINCIPLE OF WATER ELECTROLYSIS HYDROGEN PRODUCTION

The principle of water electrolysis hydrogen production is introducing into the direct current in an electrolytic solution, then electrolysis water into hydrogen and oxygen (Figure 1).

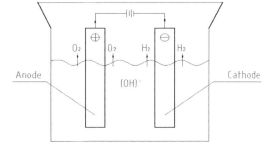

Figure 1. The electrolysis principle.

The reaction is:
Cathode $2H_2O + 2e \rightarrow H_2 \uparrow + 2OH-$
Anode $4OH - 2e \rightarrow 2H_2O + O_2 \uparrow$
The overall reaction:
$2H_2O \rightarrow 2H_2 \uparrow + O_2 \uparrow$

3 HYDROGEN GAS MIXTURE (BROWN GAS)

Brown Gas (Browns Gas) is first proposed by Australian scientists Yull Brown, who defined it in accordance with the water (H_2O) formula hydroxide molar equivalent ratios.

Hydrogen gas mixture production process consumes only water and electricity; 1 liter of water can produce 1860 liters standard cubic meters of brown gas with the consumption of about 6kW·h. Brown gas is colorless, odorless, non-toxic, and the only combustion is water. No CO, CO_2, NOX and any other toxic

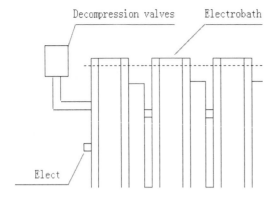

Figure 2. Structure sketch map of electrobath.

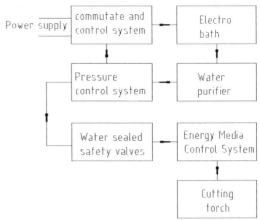

Figure 3. Hydroxide flame cutting billet process.

and harmful substances, nor the effect of greenhouse gases are produced, so it is called "water fuel."

4 WATER ELECTROLYSIS HYDROGEN GENERATOR

Conventional water electrolysis to produce hydrogen and oxygen are in the professional factory through strict management of production, and high-pressure hydrogen supply to the user and bottled form. The user must obey strict rules and production management used in order to prevent accidents, so under workplace conditions preparing hydrogen gas is not suitable for general use.

Water electrolysis hydrogen generator is based on water electrolysis hydrogen gas mixture preparation and including storage, purification devices operating unit consisting of process systems. Water electrolysis hydrogen generator for on-site production of hydrogen gas, field use, depending on the user's conditions and requirements designed to gas production of different products, you can work less than a minute after power off, gas does not need store, use pressure is low (less than 0.1 MPa) and a dedicated anti-tempering and pressure relief devices, so its use is safe and reliable.

Water electrolysis hydrogen generator works as follows: First, the alternating current through the step-down transformer and rectify into a stable adjustable low voltage DC to the electrolyzer. Electrolytic cell (Figure 2) uses a hollow trough structure, the forced cooling of the cooling fan. The temperature of the electrolyte in the cell is always kept below 85°C, required to ensure continuous operation of the generator. With a concentration of 20 to 30% KOH (or NaOH) solution in the electrolytic cell, the water is ionized into hydrogen ions and hydroxide ions, hydrogen ions in the electrolytic cell the cathode becomes hydrogen hydroxide ions in the electrolytic cell anode releasing electrons becomes oxygen and water. Mixing hydrogen and oxygen to form into the gas collection system, gas output to the device.

5 BILLET HYDROXIDE FLAME CUTTING TECHNOLOGY

Hydrogen gas in the 1990s successfully applied billet flame cutting after nearly a thousand streams outside China billet, rectangular billet, slab applications, and achieved good economic and social benefits [2]. With the continuous casting technology and equipment to improve the level of iron and steel enterprises, billet with large section long length become main products. Billet hydroxide flame cutting technology uses a new technology cutting off the fire. Hydroxide flame cutting process billet is shown in Figure 3.

5.1 *Billet hydroxide flame cutting economic*

With an annual output of 180,000 tons of carbon steel casting production lines, 150 mm × 150 mm billet, casting speed 3.2 m/min, length 6 m, for example, the use of hydrogen flame cutting off the fire, hydrogen generator is interrupted working conditions (electricity by an average of 0.6¥/kW·h, hydrogen generator actual power consumption 30 kW), the actual cutting time only 31 seconds. Considering the gas ahead of 9 seconds per cut, cutting each hydrogen generator total decreases working time 40 seconds [3].

The total length of 180,000 tons of slab ÷ 7.8 ton/m^3 ÷ 0.15 m ÷ 0.15 m ≈ 1025641 m

Each cut energy consumption: 30 kW × 0.6 ¥/kW·h × 40 s ÷ 3600 s/h = 0.2¥

Cutting times: 1025641 m ÷ 6 m ≈ 170940 times

Power consumption: 170,940 times × 0.2¥/time = 34,188¥

Cost per ton: 34,188¥ ÷ 180000 ton ≈ 0.19¥/ton

6 m length using hydrogen flame cutting slab Direct costs about 34,200¥, 9 m, 12 m length lower total cost. Using oxygen – propane flame cutting, its direct costs is 1.59¥/ton [4]. With an annual output of 180,000 tons of carbon steel casting production line, annual

direct cost is 286,200Y. Using hydrogen flame cutting, annual saving is 252,000Y, significant economic benefits.

Meanwhile, the fossil fuel gas contains carbon, high temperature $Fe + C \rightarrow FeC$, billet cutting dross after many and sticky and difficult to clean up, before rolling the need for specialized finishing. The hydrogen flame straight is good, smooth cutting section, billet less dross, and because hydrogen embrittlement dross crisp and easy to clean.

5.2 Social effect results benefit

Hydrogen gas, cheap, non-polluting, environmentally friendly, non-toxic hydrogen, will not endanger the health of the operator. Water electrolysis hydrogen generator working pressure is low and does not need store. With the production with the use of hydrogen and a small proportion of highly dispersible, even if the leak will not accumulate, gas station explosions and other accidents will not happen. CO_2 emission-free hydrogen flame, reducing greenhouse gas emissions and energy conservation, has a positive meaning for global climate change.

6 CONCLUSION

In the 21st century, with China's rapid economic growth, China has made great achievements in construction, but paid great resource and environmental costs. The contradiction between economic development and environmental resources has become increasingly acute. Meanwhile, the greenhouse gas emissions cause global warming and much of the international community attention.

REFERENCES

[1] Nie Zhenhua "Research and Development of Production Electrolyzing Water Oxy-Hydrogen Generator" Metallurgical Equipment 2011(S2):20–26.
[2] Zhou Weimin "Application and Prospect of Billet Cutting Using Oxy-hydrogen Flame" Continuous Casting 1998(4):27–29.
[3] Wang Chuan "Cutting Technology Ox hydrogen Flame Arresting for the Continuous Casting Billet" Metallurgical Equipment 2012(S2):122–123
[4] Nie Zhenhua "Economic Analysis of The Ox hydrogen Flame Cutting Continuous Cast Bloom" Metallurgical Equipment 2007(S):83–85.

Piezoelectric energy harvesting from vibration induced by jet-resonator

Huajie Zou, Hejuan Chen & Xiaoguang Zhu
School of Mechanical Engineering, Nanjing University of Science & Technology, Jiangsu Nanjing, China

ABSTRACT: Energy of airflow caused by the flight of aircrafts could be converted into electrical energy to operate requirement. An on-board physical power supply has been developed, which is essentially a Piezoelectric Energy Generator (PEG) from vibration induced by jet-resonator. And an acoustic excitation device has also been designed. CFD method is applied to analyze the pressure field inside the resonator. Meanwhile, combining with the piezoelectric effect, the output voltage of the generator is analyzed by both simulation and experiment. The result shows that the acoustic excitation device can convert the kinetic energy of moving fluids into periodic oscillation pressure during the flight. The output voltage of simulation analysis and experimental verification are consistent, which proves the feasibility of the acoustic excitation device. The conclusions provide references for the future design and experimental study of PEG from vibration induced by jet-resonator.

Keywords: Piezoelectric generator; Acoustic excitation device; Simulation; Experimental verification

1 INTRODUCTION

Physical power supply is a device converting mechanical, chemical or other forms of energy into electrical energy to operate the aircraft. It is a special power supply making use of the environment (such as airflow, heat) to generate electrical energy either in the launching or flight phase of the aircraft. Therefore, physical power supply is actually an on-board generator. For the past few years, piezoelectric energy generator (PEG) has been paid more attention on small physical power supply and vibration energy harvesting. However, PEG can only meet the electrical demand of low energy consumption products and low power lightings. How to convert the energy of moving fluid and vibration energy, which is abundant in nature, into electrical energy effectively and further improving the output power are one of the challenges in the research of PEG.

For a traditional high power piezoelectric transducer, it can adapt different demands of input and output by changing the driver and power generation mechanism. In this paper, a novel airflow induced vibration PEG based on changing the traditional driving mechanism has been developed. During the flight of aircraft, airflow passing through an annular nozzle to drive PEG produces electricity by acoustically exciting piezoelectric transducer vibration. Without increasing the complexity of the piezoelectric transducer, it is applying simple mechanical technology to solve the flow control problem and driven vibration energy and efficiency of energy conversion. Among them, airflow induced vibration technology is the key for energy conversion and the mechanical excitation system.

It is possible by the acoustic excitation device to convert the kinetic energy of moving fluid into periodic oscillation pressure load applied to piezoelectric transducer. Many scholars have studied PEG based on many acoustic excitation devices theoretically and experimentally. Allen and Smits placed a piezoelectric membrane or "eel" in the wake of a bluff body and using the von Karman vortex street forming behind the bluff body to induce oscillations in the membrane to generate electricity [1]. Li Shuguang, Yuan Jianping & Hod Lipson have studied ambient wind energy harvesting using cross-flow fluttering [2]. They proposed and test a bioinspired piezo-leaf architecture which converts wind energy into electrical energy by wind-induced fluttering motion. R Hernandez, S Jung & K I Matveev have studied acoustic energy harvesting from vortex induced tonal sound in a baffled pipe [3]. Generated sound energy was partially converted into electrical energy by a piezoelement. Wang D-A & H-H Ko have developed a new piezoelectric energy harvester for harnessing energy from flow-induced vibration [4]. It converts flow energy into energy by piezoelectric conversion with oscillation of a piezoelectric film. Sun Daming & Xu Ya developed a mean flow acoustic engine (MFAE) based on the mean flow induced acoustic oscillation effect [5, 6]. It converts wind energy and fluid energy in pipeline into acoustic energy which can be used to drive generators without any mechanical moving parts. Li Yingping proposed an on-board airflow piezoelectric generator with small scale by changing the way of conversion based on the fluidic generator [7]. However, there is no further analysis for the generator. The acoustic-excited vibration devices in the literatures are all focused on the use

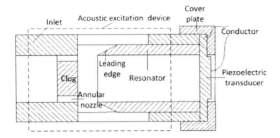

Figure 1. Schematic of airflow-induced vibration PEG.

Figure 2. Simulation model of the acoustic excitation device.

of Karman Vortex Street or vortex shearing induced cavity oscillation, and nor there are studies about the acoustic excitation devices based on jet-resonator.

Based on references 7 and 8, a piezoelectric energy generator (PEG) from vibration induced by jet-resonator applied to physical power supply of fuze has been developed by the group of Professor CHEN, shown in Fig. 1. And an acoustic excitation device containing annular nozzle and resonator is proposed. It can convert the kinetic energy of moving fluid into periodic oscillation pressure load applied to piezoelectric material during the flight. In this paper, CFD method is employed to simulate the pressure field inside the resonator. Combining with the piezoelectric effect, simulation prediction and experimental verification of the output voltage of PEG is carried out.

2 OPERATIONAL PRINCIPLE

Our design of the piezoelectric generator is based on acoustic oscillation induced by jet. The acoustic excitation device of the generator mainly consists of an annular nozzle, which is comprised of the inlet and clog, and an open-closed resonator, which is comprised of the resonator and piezoelectric transducer. A schematic of the device is shown in Fig. 1. As can be seen, during the flight, the relatively high speed airflow is guided into the inlet, and annular jet stream is formed by the annular nozzle. The annular jet stream issuing from the orifice impinges on the leading edge of the resonator, creating an acoustic perturbation which triggers air inside the resonator into resonant oscillation. Stable standing wave resonance is formed inside the resonator. And the maximum oscillation pressure is produced at the closed end of resonator. As a result of the resonator feedback system, the vocal efficiency can be greatly improved. And it can produce stable sound mainly determined by the resonator.

When the aforementioned oscillation pressure is applied to the surface of the piezoelectric transducer, which is installed at the closed end of the resonator, the piezoelectric transducer can be driven into vibration. Strain can be produced in piezoelectric material. By normal strain, charges are accumulated on both sides of piezoelectric film. Finally, voltage is formed in thickness direction.

According to the resonant principle, when the frequency of oscillation pressure is close to the resonant frequency of the piezoelectric transducer, the vibration amplitude of the piezoelectric transducer can be maximized, followed by the output power of the piezoelectric transducer being maximized.

3 NUMERICAL METHOD

CFD method is applied to simulate flow field in the resonator of acoustic excitation device. Turbulent model SST k-w is employed. It can ignore the vibration of the piezoelectric transducer. Therefore, in the simulation, the piezoelectric transducer is treated as wall.

The structure, shown in Fig. 1, has been simplified. The geometry of computational domain is shown in Fig. 2. The inlet and the resonator have the same diameter, 30 mm; the diameter of clog is 26 mm; the length of resonator is 45 mm; the gap between the inlet and resonator is 9 mm. In order to reduce computational and time cost, a two-dimensional and axial symmetric model is adopted.

In our model, ideal air is employed as the working fluid. Initial temperature and pressure are 300 K and 101.325 kPa, respectively. For compressible gas, pressure inlet is adopted as the inlet boundary, and far field pressure is adopted as the outlet boundary. No slip and thermal insulation boundary condition are applied to the wall. Wall vibration is neglected.

Governing equations are solved by commercial software FLUENT. Finite volume method is employed to discretize governing equations. The momentum, density and energy equations are both discretized by Second Order Up Wind method. Fractional step algorithm is applied to solve the discretized equations. The density of air is calculated by the ideal gas equation of state. Time step size of iteration is 1×10^{-5} s.

4 FLOW FIELD ANALYSIS

During the flight, pressure field is formed by air, which is the working pressure for the boundary condition of the inlet. In the experiment, certain pressure has been applied to the inlet. Therefore, in the simulation, the pressure-inlet has been set up as boundary condition.

When the inlet pressure is 8 kPa, the oscillation pressure variation with time at the closed end of the

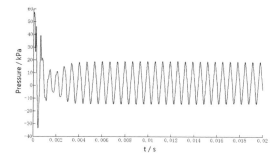

Figure 3. Oscillation pressure variation with time.

Figure 5. The finite element model of piezoelectric transducer.

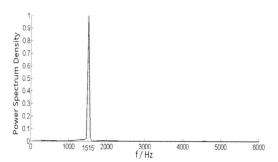

Figure 4. The power spectrum density.

Table 1. Amplitude and frequency of oscillation pressure.

Inlet pressure (kPa)	4	8	10	20	30
Amplitude (kPa)	25	33.6	38	40	50
Frequency (Hz)	1513	1515	1526	1528	1524

resonator is shown in Fig. 3. As can be seen, the oscillation pressure follows a sinusoidal pattern when it reaches to stable state. Its maximum and minimum values are 19.1 kPa and −14.5 kPa, respectively. In addition, Fig. 4 shows the fast Fourier transform (FFT) of the oscillation pressure. As can be seen, the main peak of power spectrum is very sharp and appears at f = 1515 Hz. Therefore, it can be seen that, when it reaches to stable state, the expression of the oscillation pressure can be given by

$$p(t) = 2300 + 16800 \sin(2\pi f t) \quad (1)$$

The relationship between the inlet pressure and stable oscillation pressure at the closed end of the resonator is shown in Table 1. From the table, it can be seen that the peak-peak amplitude of the oscillation pressure is 25 kPa with the inlet pressure 4 kPa; when the inlet pressure reaches 10 kPa, the peak-peak amplitude of the oscillation pressure is 38 kPa; and the inlet pressure reaches 30 kPa, the peak-peak amplitude of the oscillation pressure increases to 50 kPa; the peak-peak amplitude of the oscillation pressure increases as the inlet pressure increasing. Its frequency varies within 1%, which is beneficial to frequency matching and stable output voltage of piezoelectric transducer.

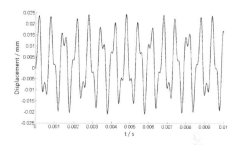

Figure 6. Displacement at the center of piezoelectric transducer.

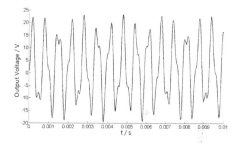

Figure 7. Output voltage of piezoelectric transducer.

5 OUTPUT VOLTAGE

5.1 Simulation analysis of output voltage

The finite element model of the piezoelectric transducer is established to simulate its vibration displacement and output voltage. The piezoelectric transducer is a laminated film including a copper layer and a PZT-5H film. The diameter of copper and PZT-5H is 35 mm and 25 mm, respectively. The thickness of them is 0.1 mm and 0.2 mm, respectively. In ANSYS, solid 45 and solid 5 is employed to define the copper and PZT-5H, respectively. The finite element model is shown in Fig. 5.

The oscillation pressure, shown in Fig. 3, is applied to the surface of the piezoelectric transducer. And piezoelectric transient analysis is solved. Fig. 6 and Fig. 7 show the vibration displacement at the center point and the output voltage of the piezoelectric transducer, respectively. As can be seen, the peak-peak amplitude of the vibration displacement and the output voltage are 0.045 mm and 40 V, respectively.

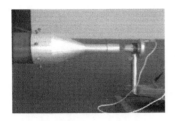

Figure 8. Experiment picture.

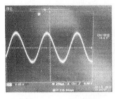 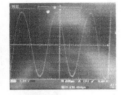

(a) Inlet pressure 4 kPa (b) Inlet pressure 8 kPa

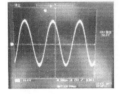

(c) Inlet pressure 10 kPa

Figure 9. Output voltage in experiment.

Table 2. Amplitude and frequency of output voltage.

Inlet pressure (kPa)	4	8	10
Output voltage (V)	16	36	52
Frequency (Hz)	1470	1470	1480

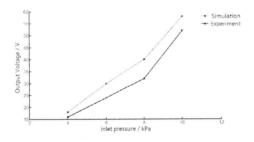

Figure 10. The relationship between inlet pressure and output voltage.

5.2 Experiment of output voltage

Fig. 8 shows the experiment picture. The inlet pressure is supplied by gas tank and the output voltage is displayed by oscilloscope. The output voltage under experiment is shown in Fig. 9 and Table 2. As can be seen, when the inlet pressure is 4 kPa, the peak-peak amplitude of the output voltage is 16 V; when the inlet pressure is 8 kPa, the peak-peak amplitude of the output voltage is 36 V; when the inlet pressure is 10 kPa, the peak-peak amplitude of the output voltage is 52 V.

Compared to the frequency of oscillation pressure, its frequency varies within 3%.

Fig. 10 shows the relationship between the inlet pressure and the output voltage. And the result of simulation and experiment are compared. The output voltage increases with the inlet pressure increasing. The result of simulation and experiment are consistent, which proves the feasibility of the acoustic excitation device.

6 CONCLUSIONS

A piezoelectric energy harvesting from vibration induced by jet-resonator is developed. The energy is harvested from airflow induced vibration during the flight of aircrafts. The oscillation pressure at the closed end of the resonator results in periodical deflection of the piezoelectric transducer and therefore voltage generation. CFD method is applied to simulate flow field inside the resonator. The output voltage of the generator is analyzed by both simulation and experiment. It is shown that the acoustic excitation device can convert the kinetic energy of moving fluids into periodic oscillation pressure. The output voltage of simulation and experiment are consistent, which proves the feasibility of the piezoelectric energy harvesting device. The conclusions provide references for the future design and experimental study of PEG.

ACKNOWLEDGEMENTS

The research is financially supported by the National Natural Science Foundation of China under contract No. 51377084.

REFERENCES

[1] J. J. Allen & A. J. Smits. 2001. Energy harvesting eel, Journal of Fluid and Structures:629–640.
[2] Li Shuguang, Yuan Jianping & Hod Lipson. 2011. Ambient wind energy harvesting using cross-flow fluttering, Journal of Applied Physics, 109:026104.
[3] R Hernandez, S Jung & K I Matveev. 2011. Acoustic energy harvesting from vortex-induced tonal sound in a baffled pipe. Journal of Mechanical Engineering, 225:1847–1850.
[4] Wang D-A & H-H Ko. 2010. Piezoelectric energy harvesting from flow-induced vibration. Journal of Micromechanics and Microengineering, 20:025019.
[5] Yu Yan & Sun Daming. 2011. CFD study on mean flow engine for wind power exploitation. Energy Conversion and Management, 52:2355–2359.
[6] Sun Daming & Xu Ya. 2012. A mean flow acoustic engine capable of wind energy harvesting. Energy Conversion and Management, 63:101–105.
[7] Li Yingping. 2006. The principle of fuze piezoelectric power supply and experimental investigations. Nanjing: Nanjing University of Science and Technology.
[8] Sun Jiacun. 2003. The study of high piezoelectric power supply for electromechanical fuze with small caliber. Nanjing: Nanjing University of Science and Technology.

Load characteristic and strategy of power quality improvement of railways in Jiangsu

Yubao Ning
Southeast University, Nanjing, China

ABSTRACT: Single-phase power supply, impact load of electrified railway will inevitably impact on the grid. In order to correctly analyze the influence of the grid after electrified railway connecting to the electricity grid, according to the current status analysis of the operation features of electrified railway in Jiangsu, in term of traction transformer connection type and traction power supply system, this paper expounds the influence of electrified railway about power grid harmonic, unbalanced influence and impact load, etc. Comparing with the method at abroad and home and the situation of Jiangsu, reasonable management measures are put forward, which can be divided into two categories. Improving starts with optimization of power supply mode: improve the electric locomotive, furnish compensation device, etc.

Keywords: Electric railways; Power quality; Management measures

1 INTRODUCTION

Electrification is not only the main direction of modernization of railway transportation, but also the important symbol of modernization of the world railway. Electric locomotive is not charged, it gets power through the external power supply and traction power supply system. However the operation of electrified railway has its own characteristics, and single-phase traction load of electrified railway destroys the symmetry of power system operation conditions. There has been growing recognition that the problems of traction power system, which include low power factor, high harmonics, negative sequence currents, and so on, are serious. Many researchers try to find the best solution to the problems in past years. Owing to the feature of different places, traditional running characteristics of electrified railway and high speed railway in Jiangsu is analyzed respectively.

2 THE RECENT SITUATION IN JIANGSU

According to the electric railway in Jiangsu, this paper carried out the corresponding analysis. As shown in Table 1 and Table 2, it introduced the main railways in Jiangsu recently[1].

The Jiangsu section of the Beijing-Shanghai electrified railway was completed on July 1, 2006, where set Xuzhou north (2 × 31.5 MVA), Yongning (2 × 50 MVA), Nanjing east (2 × 40 MVA), Liubaidu (2 × 31.5 MVA), Danyang (2 × 40 MVA), Xinzha (2 × 31.5 MVA), Shitang bay (2 × 40 MVA),

Table 1. The main traditional railways in Jiangsu.

Type User	The traditional electric railways				
	Beijing-Shanghai railway	Longhair railway	Najing-Hefei railway	Naning-Qidong railway	Shanghai-Nantong railway
Supply voltage	110 kV	110 kV	220 kV	110 kV	110 kV
Transformer connection	Impedance-matching balance	Impedance-matching balance r	Impedance-matching balance	Impedance-matching balance	Impedance-matching balance
Traction network power	Back to streamline way of direct	Back to streamline way of direct	Back to streamline way of direct	Back to Streamline way of direct	Back to Streamline way of direct
Locomotive	SS6B, SS9	SS6B, SS9	SS4, SS6B, SS9	SS9, HXD3 SS4, SS9	HXD3, SS9

Suzhou west (2 × 40 MVA), Kunshan (2 × 31.5 MVA) 9 traction station. And Xuzhou north traction station is for Beijing-Shanghai line and Long-Hai line share.

Longhai electrified railway from Xuzhou to Lianyungang, a total of five traction substations, respectively Daxujai (2 × 31.5 MVA), Pizhou (2 × 25 MVA), Xinyi (2 × 31.5 MVA), the east China sea (2 × 31.5 MVA), Yuntai mountain (2 × 31.5 MVA).

The Jiangsu section of the Beijing-Shanghai high-speed railway under section contains Xuzhou east, Nanjing south, Xiashu, Danyang, Zhenglu, Wuxi, Kunshan seven traction substations.

Shanghai-Nanjing intercity railway goes through Nanjing, Zhenjiang, Changzhou, Wuxi, Suzhou, eventually to Shanghai. The trunk line stretches approximately 303 km.

The Jiangsu section of Nanjing-Hangzhou intercity high-speed railway contains Hushu, Shangxing, Yixing 3 traction substations.

At present the electrified railway of Jiangsu adopts impedance matching balance transformer, adopts the direct power supply back to streamline method, and uses electric locomotive including shaoshan, shaoshan 6B, shaoshan9 and CRH3, CRH2. It installs fixed reactive power compensation in power supply arm.

The Zhengxu passenger dedicated line and Xulian passenger dedicated line will be opened this year, Xu belongs to the main part of the passenger dedicated line network of our country "four vertical and four horizontal".

And "three vertical and horizontal" and eight foreign channels are also in grasping the construction of the railway transportation system. Three longitudinal: Xuzhou-Nanjing railway has been put into operation; Nanjing-Shanghai railway has been put into operation; Coastal railway; Xinyi-Huaian-Yangzho-Zhenjiang. Six horizontal: Xulian railway; Nanjing-Qidong railway; West.Ningqi railway; Nanjing-Changzhou, Zhenjiang-Jiangyin-Zhangjiagang-Changshu-Taicang-Nanxiang.

Eight foreign channels: in addition to Beijing-Shanghai railway, Longhai railway, Nanjing-Wuhu railway, Xinchang railway, Nanjing-Hangzhou railway, the coastal railway, Suhuai railway and Hefei-Nanjing railway is under construction.

3 OPERATION CHARACTERISTICS OF ELECTRIFIED RAILWAY

Power supply voltage of power system is 110 kV, 220 kV and 330 kV and gives priority to with 110 kV and 220 kV. Multiple traction substations along the railway are built. Catenary system includes BT and AT system. BT system uses 27.5 kV, AT system adopts 55 kV, now general AT system.

Our country mainly adopts single-phase 25 kV voltage. AC-DC electric locomotive go through the full-wave rectifier, AC-DC-AC electric locomotive adopts PWM technology, and drives traction motor to make it drive between rail and overhead contact wire.

3.1 Traction transformer connection type

Traction transformer connection type mainly has the following kinds at home and abroad, as follows: the single-phase transformer, V/V transformer, Scott transformer, ordinary Y/d three-phase transformer connection, balance transformer.

In Jiangsu, the common railway mainly adopts the impedance matching balance wiring, and mainly high-speed railway chooses single phase V/V transformer connection form.

(1) single phase V/V transformer produces negative sequence

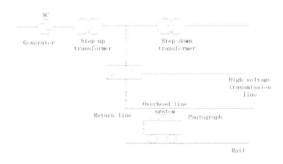

Figure 1. The traction power supply system of electrified railway.

Table 2. The main high-speed railways in Jiangsu.

Type User	High-speed electric railways			
	Nanjing-Hangzhou high-speed railway	Shanghai-Nanjing high-speed railway	Beijing-Shanghai high-speed railway	Nanjing-Anqing high-speed railway
Supply voltage	220 kV	220 kV	220 kV	220 kV
Transformer connection	Single phase V/V	Single phase V/V	Single phase V/V	Single phase V/V
Traction network power supply way	AT power supply mode	AT power supply mode	AT power supply mode	AT power supply mode
Type Of Locomotive	CRH3	CRH2, CRH3	CRH2, CRH3	

Single phase V/V connection transformer as shown in Figure 2:
Primary and secondary side:

$$\dot{I}_A = \dot{I}''/K$$
$$\dot{I}_B = \dot{I}'/K \quad (1)$$
$$\dot{I}_C = -\left(\dot{I}' + \dot{I}''\right)/K$$

Based on $\dot{I}'$, and $\dot{I}' = I'$, $\dot{I}'' = I'' e^{-j120°} = a^2 I''$.
According to the symmetrical component method,

$$\dot{I}_{A+} = \frac{1}{3}\left(\dot{I}_A + a\dot{I}_B + a^2 \dot{I}_C\right) = \frac{1}{\sqrt{3}K}(I' + I'')e^{j90°}$$
$$\dot{I}_{A-} = \frac{1}{3}\left(\dot{I}_A + a^2 \dot{I}_B + a\dot{I}_C\right) = \frac{1}{\sqrt{3}K} \quad (2)$$
$$\sqrt{I'^2 + I''^2 - I'I''}e^{j(270° + \varphi)}$$
$$\dot{I}_{A0} = \frac{1}{3}\left(\dot{I}_A + \dot{I}_B + \dot{I}_C\right) = 0$$

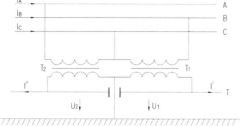

Figure 2. Single phase V/V.

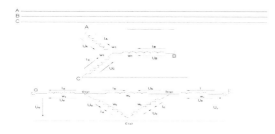

Figure 3. Impedance-matching balance transformer.

Same as other two negative sequence component.
(2) Impedance matching balance
Connection mode as shown in Figure 3:
Primary side and secondary side current relationship as shown in (3):

$$\begin{bmatrix} \dot{I}_A \\ \dot{I}_B \\ \dot{I}_C \end{bmatrix} = \frac{1}{2\sqrt{3}K}\begin{bmatrix} \sqrt{3}+1 & -(\sqrt{3}-1) \\ -2 & -2 \\ -(\sqrt{3}-1) & \sqrt{3}+1 \end{bmatrix}\begin{bmatrix} I_\alpha \\ I_\beta \end{bmatrix} \quad (3)$$

Based on $\dot{I}_\alpha$, so $\dot{I}_\alpha = I_\alpha$, $\dot{I}_\beta = \frac{a-a^2}{\sqrt{3}}I_\beta$.
According to the symmetrical component method,

$$\dot{I}_{A+} = \frac{1}{3}\left(\dot{I}_A + a\dot{I}_B + a^2 \dot{I}_C\right) = \frac{1}{\sqrt{3}K_U}(I_\alpha + I_\beta)e^{-j15°}$$
$$\dot{I}_{A-} = \frac{1}{3}\left(\dot{I}_A + a^2 \dot{I}_B + a\dot{I}_C\right) = \frac{1}{\sqrt{3}K_U}(I_\alpha - I_\beta)e^{j15°} \quad (4)$$
$$\dot{I}_{A0} = \frac{1}{3}\left(\dot{I}_A + \dot{I}_B + \dot{I}_C\right) = 0$$

Where $K_U = \frac{U_{1N}}{U_{2N}} = \sqrt{2}K$, similarly to other two phase of negative sequence component.

Deduced from the known, two types of traction transformers are present negative sequence current. When the two currents power arm are equal, negative sequence current of single phase V/V transformer is half of positive sequence current, and impedance matching balance transformer is 0.

With the comprehensive comparison of the two types of wiring in all aspects, a conclusion is drawn in Table 3[2].

3.2 Electric locomotive

(1) AC-DC electric locomotives

The voltage of 25 kV AC within Transfer contact pantograph web to circuit breaker, then through the vehicle-mounted transformer, 25 kV high voltage is converted to a suitable low-voltage current transformer. After the treatment of thyristor rectifier, it can be driven into a handy DC motor, DC motor to pass through a gear wheel transmission, traction trains.

Table 3. Comparison of the main traction transformer.

Type of connection	Single phase V/V transformer	Impedance balanced wiring
Structure	simple	More complex
Capacity utilization	high	Higher
Phase electric quantity	half	Twice than the single-phase transformer
Adapt conditions	Grid developed regions	Large capacity, high density and weak along the railway grid.
Overall rating	Overload capacity is large but when installing large, difficult to control negative sequence and substation.	Strong overload capacity, and can improve the characteristics of unbalanced and negative sequence, but its complex structure, a larger investment.

The control method of SS4 main circuit is economical four-bridge phase control, SS6B, SS9 uses three sessions ranging phased bridge to control electric locomotives. Above models have been discontinued nowadays

AC-DC electric locomotive semi-controlled bridge rectifier device by controlling the thyristor conduction tube to achieve the regulation of locomotive output regulation. This control method of the cross-straight locomotive power factor is low (about 0.8), and higher harmonic greater. Harmonic produced by different types of AC-DC electric locomotive. Mainly to 3,5,7 times, wherein the third harmonic content of the most, 18% or more.

(2) AC-DC-AC electric locomotive

AC drive electric traction is composed of the pantographs, traction transformers, four-quadrant pulse rectifier, intermediate DC link, traction inverters, traction motors, gear transmission and other components.

Pantograph will deliver 25 kV single-phase frequency alternating current within the catenary to the traction transformers, single-phase AC power supply step-down transformer by the four-quadrant pulse rectifier. Four-quadrant pulse rectifier will replace the single-phase AC electric current. After the DC output of the intermediate DC link to the traction inverter, traction inverter output voltage, current, frequency-controlled three-phase AC power supply three-phase asynchronous traction motors, traction motor shaft torque and speed output gear passing through to the wheel, rotating into the rim traction and line speed.

Locomotive in our country has to pay the rectangular electric locomotive development, is the main force of the future railway.

4 THE ELECTRIFIED RAILWAY LOAD CHARACTERISTIC

Electrified railway load are the three main features of big, negative sequence power and harmonic content is rich, traction load fluctuations makes the problem further complicated.

(1) Harmonic generated by source electrified railway

Electric locomotives produced abundant harmonic, which is a major source of harmonic current. The way of cross-straight driving produces odd and rich harmonic component when it runs, its characteristic is single-phase independence, stochastic volatility. Electric locomotive choose the pay-direct-drive, when it run, high power factor, the harmonic number is small. But the negative sequence and load fluctuation problem still exists, the bullet train basic used in rectangular transmission.

Electrified railway harmonic has the following characteristics: ① single phase independence. Railway power supply system in our country adopt two phase power supply system, relevance is very small, but two phase load is generally thought that load is independent of the two arms; ② stochastic volatility. Traction

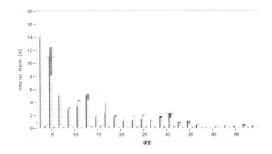

Figure 4. Electrified railway voltage and current waveform.

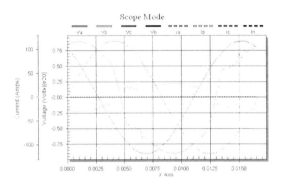

Figure 5. Electrified railway harmonic current spectrum.

power supply system harmonic current amplitude and volatile content along with the fundamental wave load, and the range is large; ③ at runtime only produce odd harmonics, only contain even order harmonic in the flow.

Electrified railway of voltage and current waveform and electrified railway harmonic current spectrum is shown in Figure 4 and Figure 5. The figure shows that the harmonic current range is very large, 3, 5, 7, 9, 11 second-class odd harmonic current is large, and A, B, C each phase unbalance, harmonic current amplitude and content with the fundamental wave load of volatility.

(2) Three-phase imbalance generated by electrified railway

Single-phase traction power supply will inevitably produce negative and unbalanced. Different wiring types of the traction transformer makes power supply voltage drop to 27.5 kV or 55 kV, then through the catenary sends current to electric locomotive. ction transformer have different negative sequence characteristics. Xuzhou Railway Electrification and the Jiangsu section of the Beijing-Shanghai line and the upcoming commissioning of the East Long Hai railway electrification (Xuzhou–Lianyungang) all use impedance matching Balance Traction Transformer. In contrast, the performance of its negative sequence is better. Even so, there are still negative sequence currents. If two power arms load balancing, then there is no negative sequence current. However, the actual situation is not so. The loading of two power arms tend to be independent and unrelated. Although the railway sector by adjusting operational plans to achieve the purposes as much as possible that the load of the two arms are balance, but cannot do the same moment. However, there is no way to make the same all the time. In particular, one arm no load current, the same size as the positive sequence current and the negative sequence current. For example, in July to September 2006, when the Beijing-Shanghai railway electrification was put into use, but the Long hai railway electrification has not been used. Xuzhou North Station an arm pulling operation, fortunately, in two months, the negative sequence current is always the same with the positive sequence current. Since the early commissioning of Beijing-Shanghai railway electrification load is not big, so there is no impact.

(3) Load fluctuations of electrified railway

Trains which go in and out of power intervals quickly and the number is very frequent. It showed the impact properties. Impact load traction station is mainly caused by frequent switching of electric locomotives. Frequent switching of electric locomotives is primarily due to the following two conditions. The first: When the locomotive pit stop, the first locomotive to decelerate to zero, parking, and then start to accelerate to operating speed for two changes. The second: Locomotive transition from one power supply area to another power supply area will go through without electricity. Leaving a power supply area in areas without electricity will cut off power supply, then leaving the power supply area will produce a negative impact on power. Locomotive from area without electricity power supply into a power supply area, system will input power supply, then leaving the power supply area will produce a negative impact on power. Then enter the power of power supply in this area to create a positive impact. According to the locomotive top speed of 350 km/h, traction power supply area up line (or down link) can have up to three locomotives running at the same time. Frequent acceleration and deceleration during the process of driving is not considered. Consider the number of locomotive leaving out of the station and locomotive through area without electricity power supply. Voltage fluctuations generated by traction load is estimated to reach about 5 times/min. So frequent shocks becomes a greater impact on the power system PCC point voltage fluctuations and flicker. Active impact will have an impact on the generator set near PCC.

Table 4: Nanjing-Hangzhou high-speed rail high-pressure side of the reactive phase in the impact of long-PCC point maximum load generated, short-term flicker value calculations.

From the table, the flicker value of PCC point has been more than the allowable values.

5 TREATMENT MEASURE

In recent years, the power quality problems attaches great importance to both at home and abroad in-depth

Table 4. PCC point impact load maximum phase flicker.

Traction station	High side worst hit phase	The PCC point	Short-term flicker value	Long flicker value
Hushu	A	Suzhuang	1.61	1.08
		HuaKe	1.85	1.23
Shangxin	C	Jiuxian	2.77	1.85
		Ganxi	2.23	1.49
Yixing	B	Minzhu	1.01	0.67
		Nanyue	/	/
		Beitang	1.51	1.01

Figure 6. Power quality management both at home and abroad.

exploration. Governance means mainly can be divided into the following two categories[3]:

① traction load characteristics (the operating conditions of the train, locomotive electric property, etc.);
② the electrified railway traction power supply system.

This paper mainly introduces the second scheme of improvement, it mainly includes: optimization of power supply mode, improve the electric locomotive, furnish compensation device, etc.

(1) Optimize the power supply way

The commutation can reduce the asymmetry in the system, in turn, there are two ways of access to the system. Multiple traction station in shifts that are widely used in the following the different system of substation, the other is more than three same railway traction substation station connected to the same system. Taking turns to commutation is simple, but in practical engineering, we can't fully ensure the completely same three traction station load current, therefore can only reduce the load current parts injection.

In addition to phase in turn, can also improve the voltage level, common electrified railway is mainly 110 kV power grids, and high-speed rail, due to the fast speed and high power supply reliability requirements, need to use the power supply voltage of 220 kV and above. After system short circuit capacity, three-phase imbalance of power grid influence greatly reduced, to reduce voltage distortion and voltage deviation.

For one-way connection, the structure is simple, but its negative sequence component is the largest of several traction transformer connection types. Single phase V/V connection corresponding greatly reduces the negative sequence component. Especially, when heavy arm load is twice than light arm load, the negative sequence influence generated by the light arm load is only half of the heavy arm load. The above analysis, when the current of two power supply arms equal, negative sequence current of single phase V/V transformer is half of the positive sequence current, and impedance matching balance transformer is 0. The balance transformer can reduce the negative sequence current of traction station injection system. Actually, it's almost impossible for the two power supply to provide equal current. Sometimes, there will be extreme, with one arm full loading, and another arm too light. In this case, no matter what kind of the traction transformer connection, the influence of negative sequence is the same. So balance transformer is with the higher probability to reduce the influence of negative sequence probability than the other, but cannot solve the extreme situation.

(2) Improve electric locomotive

Electric locomotive is the reason of electrified railway harmonics and low power problem. Installing the electric locomotive compensation device and the filter circuit can improve the quality of traction power supply system of power, and the efficiency is higher than in a power supply side. However, due to the large capacity of compensating device, and the high cost, it wastes more and can be easy interfered to generate bug.

The strategy of improvement of AC drive locomotive control employs the AC-DC-AC electric locomotive, and the power factor can achieve 0.98 and above. Mainly negative sequence generated by the high-speed passenger load component and the impact load, reactive power and harmonic impact is relatively small.

(3) Install compensation device

SVC is most widely used in the application of power quality governance. The TCR + FC, TCR + TSC mixing, etc., usually in busbar side single-phase power supply arm installation. Railway Power Conditioner (Railway Static Power Conditioner, RFC) which can balance the system voltage, eliminate the voltage fluctuation, and reduce the harmonic, is widely used in Japan, and comprehensive compensation device and Unified Power Quality Conditioner (Unified Power Quality Conditioner, UPQC), etc.

6 CONCLUSION

Power quality pollution control needs the providers, suppliers and users make joint efforts. They should keep using the specialized traction line power, rejecting mixing with other production and living power. At present, electrified railway operation management implement the principle of "who pollute, who govern", but still need to strengthen efforts.

REFERENCES

[1] Hou Weiliang & Sun Xulong. 2011. Modern electrified railway on the impact of power quality and its management. Power supply 28 (6): 16–19.
[2] Yu Kunshan, Zhou Shengjun, Wang Tongxun, Qiao Guangyao etc. 2011. Electrified railway power supply and power quality. Beijing: China power press.
[3] Zhang Xin. 2013. Study on the method of electrified railway power quality governance. Zhejiang: zhejiang university.

Experimental research on apparent viscosity behavior of different wormlike micelles

Ni Li & Rui Zhang
Hua Qing College, XI'AN University of Architecture and Technology, Xi'an, Shaanxi, China

ABSTRACT: The wormlike micelles were synthesized by different surfactants and counterion. The first type is anionic surfactant potassium erucic (KEU) and three hydroxyethyl benzyl ammonium chloride (BTHEAC). The second type is cationic surfactant octadecyl trimethyl ammonium chloride (OTAC) and vinyl sodium benzosulfonate (NaSS). After comparing, when the molar ratio of n (OTAC): n (NaSS) = 1:1, the largest viscosity of wormlike micelles is 2530 mPa·s.

Keywords: anionic surfactant; cationic surfactant; wormlike micelles; apparent viscosity

1 INTRODUCTION

Surfactant is a large part of organic compounds. It is important in the field of industry of fine chemicals. The molecular structure of the surfactant with amphiphilic: one is hydrophilic group, the other one is hydrophobic group. Either in theory or in practice, wormlike micelles have a very important position. According to the different polar groups, surfactant can be divided into: anionic surfactant, cationic surfactant, ampholytic surfactant and nonionic surfactant.[1–3] Wang Zhiguo[4] discussed the micro fluid structure and linear viscoelastic of wormlike micelles of cationic surfactant cetyl trimethyl ammonium chloride and sodium salicylate. Liu Jingwei[5] discussed the process and the applications of the surfactant micelles. It described that the surfactant has a good application prospect. Wang Haifeng[6] discussed the synergistic effect of nonionic surfactant, cationic surfactant and oil displacement surfactants. It described that different types of surfactant can compound worm-like micelles. In this paper, we will research apparent viscosity behavior of different wormlike micelles.

2 REAGENTS AND INSTRUMENTS

Erucic acid, Beijing Company of Chemical Reagents, AR; triethanol amine, Beijing Company of Chemical Reagents, AR; benzyl chloride, Tianjin Reagent Factory, CP; octadecyl trimethyl ammonium chloride, Beijing Company of Chemical Reagents, AR; docosyl trimethyl ammonium chloride, Beijing Company of Chemical Reagents, AR; vinyl sodium benzosulfonate (≥98.0%); acryloyloxyethyl trimethyl ammonium chloride, Beijing Company of Chemical Reagents, AR.

Electronic balance JA2003 (0.001 g division value), shanghai laboratory instrument; digital rotational viscometer brookfield LVDV-II+; ultrasonic oscillator SB2200, Shanghai Branson.

3 EXPERIMENT RESULTS AND DISCUSSION

3.1 *Anionic surfactant/counterion*

In aqueous solution, anionic surfactants are ionized to negatively charged hydrophilic radical on the end. There is a strong electrostatic repulsion between the anionic surfactant hydrophilic radical. To promote the surfactant molecules self-assemble into an anionic surfactant wormlike micelles, we need to add the appropriate counterion suppression molecular electrostatic repulsion between the molecules.[7] In the given condition, three hydroxyethyl benzyl ammonium chloride can engage by SN2 type nucleophilic substitution reaction. Potassium erucic and organic counterion three hydroxyethyl benzyl ammonium chloride were mixed at different ratios, the purpose is to get the best ratio to compose wormlike micelles.

The structural formula of potassium erucic (KEU) and three hydroxyethyl benzyl ammonium chloride (BTHEAC) is shown in Figure 1.

We prepared 50 ml molar concentration of 0.1 mol/L KEU solution. Then the solution mixed with different quality of BTHEAC, micelles can be formed. The influence of quality of BTHEAC on apparent viscosity of micelles is shown in Figure 2. As can be seen from Figure 2, when the quality of BTHEAC is less than 0.7 g, because of the strong electrostatic interactions, with the increase of the quality of BTHEAC, apparent viscosity of micelles increases gradually. When the quality of BTHEAC is 0.7 g, we obtain the biggest apparent viscosity, it is 336 mPa·s. Continue

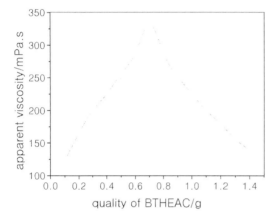

Figure 1. The structural formula of potassium erucic (KEU) and three hydroxyethyl benzyl ammonium chloride (BTHEAC).

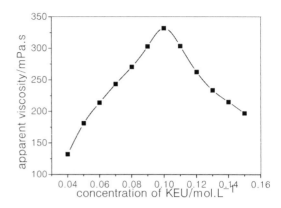

Figure 3. The influence of the concentration of KEU on apparent viscosity of micelles.

Figure 2. The influence of the quality of BTHEAC on apparent viscosity of micelles.

to increase the quality of BTHEAC, because the system is composed of larger asymmetries, the system may transition from a wormlike micelles to spherical micelles, the apparent viscosity of the system gradually reduced.

50 ml different concentration of KEU solution mixed with 0.7 g BTHEAC can form micelles. The influence of different concentration of KEU on apparent viscosity of micelles is shown in Figure 3. As can be seen from Figure 3, when the concentration of KEU is 0.10 mol/L, we obtain the greatest apparent viscosity, it is 336 mPa·s.

Figure 2 and Figure 3 show that, when the mole ratio is n(KEU): n(BTHEAC) = 5:3, the apparent viscosity of wormlike micelles formed by KEU and BTHEAC is the greatest.

3.2 Cationic surfactant/counterion

There are a lot of cationic surfactants, such as hexadecyl trimethyl-ammonium bromide, hexadecyl trimethyl-ammonium chloride, hexadecyl pyridinium bromide hydrate and so on.[8] When the cationic surfactant hydrophobic chain is too short ($n \leq 6$), synergistic effect between the two parts is weak, so aggregation parameters between moleculars changed little. When the cationic surfactant hydrophobic chain is too long ($n \geq 12$), synergistic effect between the two parts is too strong, so it will generate precipitation. Only when the cationic surfactant hydrophobic chain between middle strength, it is suitable intensity to wormlike micelle formation and growth. Cationic surfactant octadecyl trimethyl ammonium chloride (OTAC), docosyl trimethyl ammonium chloride (DTAC) and counterion vinyl sodium benzosulfonate (NaSS) were mixed at different ratios, the wormlike micelles can be formed.

The structural formula of octadecyl trimethyl ammonium chloride (OTAC), docosyl trimethyl ammonium chloride (DTAC) and vinyl sodium benzosulfonate (NaSS) is shown in Figure 4.

When the mole ratio is n(OTAC):n(NaSS) = 1:1, OTAC mixed with NaSS, after sufficient stirring, micelles can be formed. Then, substitute DTAC for OTAC to do the experiment. The influence of concentrations on apparent viscosity of micelles is shown in Figure 5. As can be seen from Figure 5, in all the range, apparent viscosity of OTAC/NaSS system is far greater than the apparent viscosity of DTAC/NaSS system. As far as OTAC/NaSS, when the molar concentration of the solution is in the lower range, as the molar concentration of the solution increases, the apparent viscosity OTAC/NaSS micelles increases rapidly. When the concentration of OTAC is 25 mmol/L, the apparent viscosity of OTAC/NaSS is deviation from the linear relationship. Then, with the increase of the solution concentration, apparent viscosity is basically unchanged. When the cationic surfactant is DTAC, with the increase of the solution concentration, apparent viscosity of DTAC/NaSS is basically unchanged. Summary, selecting OTAC cationic surfactant is better than DTAC cationic surfactant. So, we choose cationic surfactant is OTAC, counterion is NaSS, the mole ratio is n(OTAC):n(NaSS) = 1:1, the greatest apparent viscosity is 2520 mPa·s.

Because of the high hydrophobic interaction of DTAC, making the internal volume is too big. So it

Figure 4. The structural formula of octadecyl trimethyl ammonium chloride (OTAC), docosyl trimethyl ammonium chloride (DTAC) and vinyl sodium benzosulfonate (NaSS).

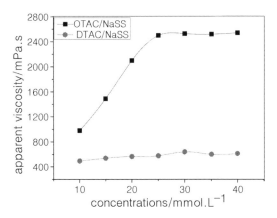

Figure 5. The influence of the concentration of OTAC(DTAC) on apparent viscosity of micelles.

is not easy to form longer wormlike micelles. But the hydrophobic interaction of OTAC is short, so OTAC and NaSS can make long wormlike micelles. When the concentration of OTAC is 26 mol/l, the mole ratio is n(OTAC):n(NaSS) = 1:1, the apparent viscosity of wormlike micelles formed by OTAC and NaSS is the greatest.

4 CONCLUSIONS

In summary, under the right conditions, either anionic surfactant or cationic surfactant can form wormlike micelles with corresponding counterion. When the cationic surfactant is octadecyl trimethyl ammonium chloride(OTAC), the counterion is vinyl sodium benzosulfonate(NaSS), the mole ratio is n(OTAC):n(NaSS) = 1:1, the maximum apparent viscosity of wormlike micelles is 2520 mPa·s. It is greater than the wormlike micelles composed by KEU and BTHEAC. If both are available, we always choose the wormlike micelles of OTAC/NaSS.

REFERENCES

[1] Huang Yinghong, Zheng Cheng. 2014. Progress in research work on synthesis and application of reactive quaternary ammonium salt cationic surfactants. China Surfactant Detergent & Cosmetics. 44(3):155–162
[2] Fan Haiming, Wu Xiaoyan. 2011. Wormlike Micelle with High Salt Resistance Property in Mixed Zwitterionic and Anionic Surfactant System. Acta Chimica sinica. 69(17):1997–2002
[3] Sui Zhihui, Lin Guanfa. 2003. Preparation, Application and Prospect of Surfactant Used in EOR. Chemical industry and engineering progress. 22(4):355–360
[4] Wang Zhiguo, Wang Shuzhong. 2013. The microsturcture and linear viscoelasticity of OTAC/NaSAL wormlike micelles. Chinese Journal of Theoretical and Applied Mechanics. 45(6):854–860
[5] Liu Jingwei, Liu Hongqin. 2014. Performance and applications of surfactants (I)—Surfactant micelles and their applications. China Surfactant Detergent & Cosmetics. 44(1):10–14
[6] Wang Haifeng, Wu Xiaolin. 2004. Research progress on the surfactants for ASP flooding in Daqing oil field. Petroleum geology and recovery efficiency. 11(5): 62–64
[7] Zhao Jianxi, Xie Danhua. 2012. Anionic Wormlike Micelles. Progress in chemistry. 24(4):456–462
[8] Mu Ruihua, Wu Wenhui. 2013. Preparation and properties of wormlike micelles formed by cationic trimer surfactant/NaSS/St. Journal of Chemical Industry and Engineering.

Control of Lithium Battery/Supercapacitor hybrid power sources

Jing Chen & Songrong Wu
School of Electrical Engineering, Southwest Jiaotong University, Chengdu, Sichuan, China

ABSTRACT: Hybrid Light Rail Vehicle (HLRV) with on-board energy storage systems has good characteristic in technical and economical. The control strategy for a voltage-regulated dc hybrid power source employing Lithium Battery (LB) as the main power source and Supercapacitor (SCAP) as the auxiliary power source is proposed. To ensure energy-efficient operation of the system and reduce the current ripple, the use of paralleling Bidirectional Dc-dc Converter (BDC) with interleaved technique is applied. As a high dynamic and high-power density device, SCAP functions to supply energy to regulate the DC bus energy. LB, as a slower dynamic source in the system, functions by supplying energy to keep SCAP charged or discharged with limited current. Simulation results obtained from the control strategy are presented and analyzed.

Keywords: Hybrid Light Rail Vehicle (HLRV); Lithium Battery (LB); Supercapacitor (SCAP); Bidirectional Dc-dc Converter (BDC); interleaved technique

1 INTRODUCTION

As an important part of urban mass transit, the Light Rail Vehicle (LRV) has the advantage of high efficiency and being on time. However, traditional Overhead Contact System (OCS) has such problems as the large quantity and high cost of maintenance. Also, OCS have been considered an eyesore by the general public and hence politicians. The development of LRV is severely limited [1–2]. A lot of research on wireless system is done by urban mass transit enterprises at home and abroad. On-board energy storage system is being paid more attention because it doesn't need wayside equipment and maintenance in theory [3]. Lithium Battery and SCAP energy coupling improves the autonomy and performance of the vehicle. Battery and SCAP have been widely used in the field of Hybrid Electric Vehicle (HEV) [4]. Compared with HEV, LRV requires frequent starting and braking, so the demand for power has greatly changed. There must be good energy management between these devices which enables the reduction of the lithium battery size and improves its life span. To restore the braking energy, on-board energy storage system with SCAP has been used in LRV [5–6]. Taking HLRV as an example, the studied dc power supply is composed of SCAP and lithium battery. SCAP is dimensioned for peak power requirement, and batteries provide the power in steady state.

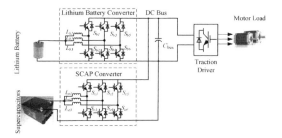

Figure 1. Hybrid Light Rail Vehicle topology.

as LB needs a BDC to adapt the DC bus voltage to the desired dc utility voltage level and smooth the supply current. Nevertheless, a single converter will be limited when the power increases. For the reason of high power, the interleaving technique for paralleling input of the converter is applied [7]. One proposes a power electronics topology for hybrid power-management system, as is shown in Fig 1, in which multiphase BDC are connect at the output of lithium and SCAP, which are then connected in parallel to share the load at the DC bus. The number of parallel converter modules N would depend on the power of load [8], SCAP and LB. In this paper, N is equal to three for the SCAP converter and lithium battery converter.

The topology reduces the number of SCAP as much as possible and improves the system economically. Bi-directional Buck/Boost compensation topology [9] is adopted. The non-isolated structure requires fewer devices, which is not only smaller in size, but also efficient. As is shown in Fig 1, L_{bj} and L_{cj} ($i = 1, 2, 3, 4, 5, 6$) represent the inductors used for energy transfer

2 HYBRID POWER SYSTEM

SCAP and LB need to be interfaced through power electronics converters. Normally, SCAP bank as well and current filtering. The inductor size is classically defined by switching frequency and current ripple.

There are two ways to control the converter, which are independent and complementary pulse width modulation (PWM) [10]. In the independent PWM control, insulated-gate bipolar transistors (IGBTs) at the same bridge (such as S_{b1} and S_{b2}) arm do not act at the same time. The topology can be equivalent to a combination of anti-parallel one-way buck converter and one-way boost converter. In order to ensure the bidirectional power flow, an additional controller is used to achieve a smooth switch between the buck and the boost states. In the complementary PWM control, the IGBTs at the same bridge arm act at the same time, which achieve switching of the two working states without additional controller. As a result, the system gets faster response. Since the short distance between stations, HLRV starts and brakes frequently, results in frequent changes of load power. It's suitable for the use of complementary PWM control.

For safety and high dynamics, the LB and SCAP converter are controlled by inner current regulation loops. A classical current control of parallel converters is illustrated in Fig 2. The current control loops are supplied by two reference signals: LB current reference i_{LBREF} and SCAP current reference $i_{SCAPREF}$, which are generated by the energy-management algorithm presented in section 3.

The interleaving technique consists of phase shifting the control signals of 3 converter cells in parallel, operating at the same switching frequency [8]. A phase shift of $2\pi/3$ from each other is used for the control signals to achieve interleaving algorithm. As a consequence, the interleaved bi-directional converters exhibit both lower current ripple and voltage ripple.

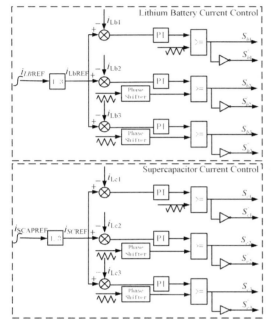

Figure 2. Current control loops of LB/SCAP converters.

3 CONTROL STRATEGY FOR HYBRID POWER SYSTEM

The hybrid control strategy is depicted in Fig 3. It consists of "Lithium Battery Control" block and "Supercapacitor Control" block. SCAP functions to supply energy to regulate the DC bus energy and LB functions by supplying energy to keep SCAP charged or discharged. In the LB control block, v_{SCAP} is the

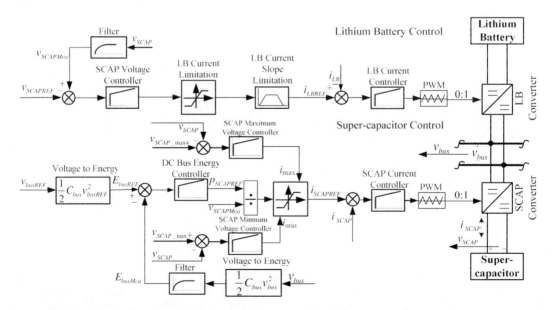

Figure 3. Energy management of the hybrid power source.

SCAP voltage, $v_{SCAPREF}$ the SCAP voltage reference, $v_{SCAPMea}$ the filtered SCAP voltage, i_{LBREF} the LB current reference, i_{LB} the LB current. In the supercapacitor control block, v_{busREF} is the DC bus voltage reference, v_{bus} the DC bus voltage, E_{bus} the DC bus energy, E_{busREF} the DC bus energy reference, E_{busMea} the filtered DC bus energy, $v_{SCAP-max}$ the maximum SCAP voltage, $v_{SCAP-min}$ the minimum SCAP voltage, i_{max} the maximum SCAP current, i_{min} the minimum SCAP current, $p_{SCAPREF}$ the SCAP power reference, $i_{SCAPREF}$ the SCAP current reference, i_{SCAP} the SCAP current. The main objective of the control is to regulate DC bus voltage. The basic principle lies in using SCAP, which is the fastest energy source. The energy is supplied by means of SCAP, as if it's the standard power supply. And LB, although the main energy source of the hybrid system, can be seen as the device which supplies energy to keep SCAP charged or discharged.

Consequently, the SCAP converter is driven to realize a standard DC bus voltage regulation, and the LB converter is driven to maintain the SCAP module at a given state of charge (SOC). As a result, LB is only operating in nearly steady state condition, and SCAP are functioning during transient energy delivery or transient energy recovery. Hybrid power system above has the following features:

1) DC bus voltage is regulated, which is the main objective.
2) LB current must be kept with an interval $[-I_{LBRated}(\text{rated value}), I_{LBRated}(\text{rated value})]$.
3) LB current must be limited to a maximum absolute value in order to guarantee matching the reactant delivery rate and the usage rate.
4) SCAP voltage must be kept with an interval $[v_{SCAP-min}(\text{minimum value}), v_{SCAP-max}(\text{maximum value})]$. Normally, the system attempts to reach the normal voltage $v_{SCAPREF}$.
5) SCAP current must be kept with an interval $[i_{min}(\text{minimum value}), i_{max}(\text{maximum value})]$.

For the DC bus control voltage loop, the principle is analyzed with reference to the block diagram as shown in the "Supercapacitor Control" block. Compared with classical DC link voltage regulation that DC bus voltage was directly considered as control variable, the energy stored in the DC link capacitor $C(E_{bus} = 0.5Cv_{bus}^2)$ has been considered as state variable by means of "Voltage to energy". Also, the SCAP delivered power $p_{SCAPREF}$ which is generated by means of "DC Bus Energy Controller" is used as the command variable. This results in obtaining a natural linear transfer function for the system. If the losses in both the LB converter and the SCAP converter are neglected, the energy E_{bus} stored in the DC bus can be written as the following,

$$\frac{dE_{bus}}{dt} = p_{SCAP}(t) + p_{LB}(t) + p_{load}(t) \quad (1)$$

In the equation above, $p_{SCAP(t)}$ is the SCAP power, $p_{LB(t)}$ the LB power, $p_{load(t)}$ the load power. By power conservative law and no loss assumption, $i_{SCAPREF}$ is calculated by dividing $p_{SCAPREF}$ by the measured supercapacitors voltage. Because the signal v_{SCAP} contains harmonics generated by the converter switching, this measured signal ($v_{SCAPMea}$) must be filtered by low pass filter. In order to keep the SCAP voltage within its limit $[v_{SCAP-min}, v_{SCAP-max}]$, the SCAP current has to be limited. As is shown in Fig 3, $i_{SCAPREF}$ is kept with an interval $[i_{min}, i_{max}]$. i_{min} is generated by means of "SCAP Minimum Voltage Controller" and it's kept with an interval $[-I_{SCAPRated}, 0]$. In the case of SCAP voltage being far from $v_{SCAP-min}$, the discharging current limit i_{min} is equal to its negative rated value $-I_{SCAPRated}$. In the case of SCAP voltage is closed to $v_{SCAP-min}$, the discharging current limit i_{min} becomes greater than $-I_{SCAPRated}$. At the same time, i_{max} is equal to its rated value $I_{SCAPRated}$. With the same principle, the upper boundary of the charging current i_{max} can be analyzed.

For the SCAP voltage control loop, it consists of "SCAP Voltage Controller" limited in value and slope, as is shown in the "Lithium Battery Control" block. For the "LB Current Limitation", i_{LBREF} is limited with an interval $[-i_{LBRated}, i_{LBRated}]$. For the "LB current slope limitation", slope limitation to a maximum absolute value of several amperes per second enables safe operation of the LB during transient power demand especially. In this way, the system can visibly ensure that lithium battery current will gradually increase and decrease, and over rated current will not happen.

Consequently, the main point in hybrid system presented above is to balance the power among the LB main source, the SCAP auxiliary, and the load. SCAP supplies most of load variations through the DC bus voltage regulation; LB power dynamics have been intentionally reduced, results in providing the power in steady state through SCAP voltage regulation.

4 SYSTEM DESIGN AND SIMULATION

In order to verify the feasibility of the control strategy, a HLRV simulation system is built by means of Matlab/Simulink, as is depicted in Fig 4. Load consists of four 150 kw asynchronous motors. The lithium battery device is given a nominal voltage of 400 V and a rated capacity of 1200 Ah to provide the power in steady state. The SCAP bank is 28F, and the maximum voltage is 550 V, the minimum voltage is 300 V, the working (nominal) voltage is 400 V. For the SCAP converter, the inductor depends on voltage boost, average dc current and current ripple. When the BDC acts as a boost converter, the following equations are used to find L_{min}:

$$L_{boost} = \frac{V_{SCAP} \cdot D}{f_S \cdot \Delta i_L} \quad (2)$$

$$D = 1 - \frac{V_{SCAP}}{V_{bus}} \quad (3)$$

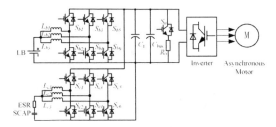

Figure 4. Simulation structure.

When the BDC acts as a buck converter, the following equations are used to find L_{min}:

$$L_{buck} = \frac{(V_{bus} - V_{SCAP}) \cdot D}{f_s \cdot \Delta i_L} \quad (4)$$

$$D = \frac{V_{SCAP}}{V_{bus}} \quad (5)$$

The switching frequency is constant at 20 kHz. The ripple current Δi_L is 10% of average current. Taking the use of paralleling power converters into consideration, the inductors $L_{bj(j=1,2,3)}$ are 0.6 mH. With the same principle, the inductors $L_{cj(j=1,2,3)}$ are 0.24 mH. C_f is the filter capacitor of BDC and $C_f = 4000 \mu F$, according to the following equation,

$$C_{f\,min} = \frac{V_{bus} \cdot D}{2R \cdot f_s \cdot \Delta v_{bus}} \quad (6)$$

R is the equivalent load. C_{bus} is the DC link capacitor and $C_{bus} = 8000 \mu F$. R_0 is the braking resistor and $R_0 = 1.52 \Omega$.

Fig 5 presents the simulation waveforms that are obtained during a load cycle (traction and braking modes). It shows the motor speed, the DC bus voltage, the lithium battery voltage, the load power, the lithium battery power, the supercapacitor power, the lithium battery current, the supercapacitor current and the supercapacitor voltage.

Initially, the SCAP voltage is 400 V. As is illustrated in Fig 5, HLRV accelerates linearly to 36 km/h (corresponding to 1200 rpm) at $t = t_1$. Simultaneously, the load power increases to final value of around +770 KW. Synchronously, the final LB power increases with a limited slope to a maximum power +440 KW. From $t = t_1$ to $t = 28$ s, HLRV runs at constant speed. At $t = t_1$, the load power steps from +770 KW to +310 KW. As a result, the supercapacitor changes its state from discharging to charging. The lithium battery power is still at the limited maximum power of +440 KW to supply the load and charge the SCAP at the moment. At $t = 28$ s, HLRV begins to brake and the load power steps to −700 KW and SCAP is deeply charged. Also, the LB power decreases with a limited slope. At $t = t_2$, HLRV is stopped. After that, the SCAP voltage is controlled at 400 V. One can observe the following.

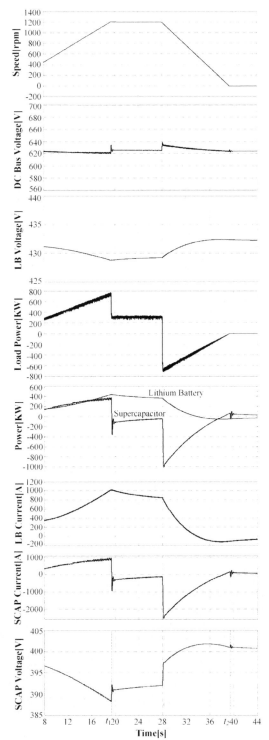

Figure 5. Hybrid source response during a load cycle.

1) The initial supercapacitor voltage (state of charge) is equal to the voltage after a load cycle, as if the energy is provided by the lithium battery only.

2) The auxiliary source (supercapacitor) supplies most of the transient power required.
3) The main source (lithium battery) supplies the average power required indirectly with limited slope by means of supercapacitor voltage regulation.

5 CONCLUSION

The main objective of this work is aimed at the energy management for the HLRV which uses lithium battery as main source and supercapacitor as auxiliary source. The important constraint is to avoid speedy transition of lithium battery current in order to prolong its life by controlling the current slope and absolute value. Instead of controlling the DC bus voltage through the powers, one has chosen an indirect method where the DC bus is regulated by the supercapacitor power, the supercapacitor voltage is regulated by the lithium battery delivered current. As a result, the supercapacitor power provides peak power requirement and the lithium battery power provides the steady power. The simulation results corroborate the excellent performance during a load cycle.

REFERENCES

[1] Margarita Novales. 2011. Overhead wires free light rail systems. In Proc. 90th TRB Annual Meeting:2–8.
[2] Shen Jiqiang. 2012. Modern streetcar vehicle selection and power supply. China Municipal Engineering: 68–75.
[3] Chen yin, Chen Zhongjie 2012. Light rail electric traction energy storage research. Electric Locomotives & Mass Transit Vehicles:5–11.
[4] Gong Junqiang, Den Hao, Xie YingHua. 2012. Energy storage technology classification and domestic large-capacity battery energy storage technology is relatively. China Science and Technology Information:139–140.
[5] Siemens traffic technology group developed a new generation of 100% low floor tram. Siemens Brochure.
[6] Masamichi OGASA. 2005. With a rechargeable lithium ion battery without contact wire of the tram. Foreign Rolling Stock 42(5):19–22.
[7] P. Thounthong, B. Davat, S. Raël, and P. Sethakul.2009. "Fuel cell high power applications," IEEE Ind. Electron. Mag., vol. 3, no. 1:32–46.
[8] D.J. Perreault and J.G. Kassakian. 1997. "Distributed Interleaving of Paralleled Power Converters," IEEE Trans. Circuits Syst. I, Fundam. Theory Appl., vol. 44, no.8:728–734.
[9] Payman, S. Pierfederici, and F. Meibody-Tabar. 2008. "Energy control of supercapacitor/fuel cell hybrid power source,"Energy Convers. Manage., vol. 49: 1637–1644.
[10] Zhang F.H., Zhu C.H., Yan Y.G. 2005. The controlled model of bi-directional DC-DC converter Proceedings of the CSEE 25(11):46–49.

Study on compensation characteristic of ICPT system

Wenyi Tong & Songrong Wu
School of Electrical Engineering, Southwest Jiaotong University, Chengdu, Sichuan, China

ABSTRACT: The basic structure of Inductively Coupled Power Transfer (ICPT) system and its operating principle are analyzed in this paper. Different kinds of compensation topologies are discussed according to the structural characteristics of loosely coupled transformer. In addition, the compensation capacitor parameters are also analyzed according to the structure. The simulation results show that appropriately secondary compensation capacitor can enhance the system power transfer capability, while the appropriately primary compensation capacitor can reduce the demand on system power capacity.

Keywords: Inductively Coupled Power Transfer (ICPT); Primary compensation; Secondary compensation

1 INTRODUCTION

Since direct power transmission still has some problems, such as wires exposed, the contact spark and so on. To overcome these disadvantages, Inductively Coupled Power Transfer (ICPT) system is designed to deliver power from a stationary primary source to one or more movable secondary loads without physical contact. ICPT system avoids contacting with bare conductors and sparks, therefore it can address many traditional power transmission problems. So, it shows a wide application in the military, transportation and other fields.

The efficiency of ICPT system has been attached in recent years. A static capacitance compensation method was proposed [1–2], which series or parallel the specified capacitor on the secondary side to resonant with the inductance in normal frequency. While the specified compensation capacitor was series or parallel connected on primary side can lead the power source current and voltage phase angle to be zero. In this way, the power transmission performance of the ICPT system is greatly improved.

2 SYSTEM STUCTURE

The principle of ICPT system is to achieve energy through space electromagnetic coupling. It contains modern power electronics technology, power compensation technology, resonant inverter technology and switching power supply technology. Figure 1 shows the block diagram of the ICPT system, including the original side DC power supply, the primary component of the inverter, the primary coil, the secondary coil, the secondary converter and the load [3].

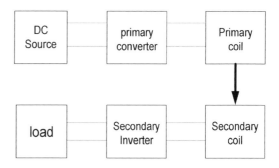

Figure 1. A typical structure of ICPT system.

3 SYSTEM COMPENSATION METHOD

Loosely coupled transformer is a key component to get non-contact power transmission, compared with conventional transformer, whose primary and secondary are separated causing a large air gap [4]. It will generate large leakage inductance when power is transmitted. With improvement of the system operation frequency, the parameters of the inductance of the primary circuit become larger, reducing the load active power absorption, thus decreasing system transmission efficiency.

In order to improve system power supply performance, reactive power compensation on primary and secondary side is necessary. The alternative compensation capacitor in the circuit makes the magnetic field of leakage inductance and the electric energy transform into each other, also makes the inductance capacitance and leakage inductance occur the resonance to maximize the transmission efficiency.

There are two basically kinds of compensation strategy: series compensation and parallel compensation.

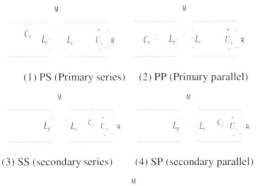

(1) PS (Primary series)   (2) PP (Primary parallel)

(3) SS (secondary series)   (4) SP (secondary parallel)

(5) PSSS (Primary series-secondary series)

(6) PSSP (Primary series-secondary parallel)

(7) PPSS (Primary parallel-secondary series)

(8) PPSP (Primary parallel-secondary parallel)

Figure 2. The diagram of ICPT compensation topology.

If only one on primary or secondary is used, the compensation topology will become four. If the primary and secondary sides are used in series or parallel compensation, the compensation topology is four. So, the total compensation topology is eight [5–6], as shown in Figure 2.

The way of compensation depends on application and actual demands according to the compensation circuit characteristics. When the series compensation is used on secondary side, C_p and L_s are series resonance, then the system load can be equivalent to a voltage source. If parallel compensation is used on secondary side, an equivalent current will be calculated from the load side [7].

4 THE COMPENSATION PARAMETER SELECTION

4.1 Primary compensation on secondary side

Figure 3 shows the equivalent circuit on secondary side of parallel compensation. According to the circuit

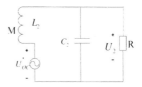

Figure 3. The equivalent circuit of secondary parallel compensation.

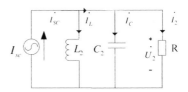

Figure 4. Norton equivalent circuit of secondary parallel compensation.

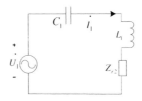

Figure 5. The equivalent circuit of primary series compensation.

equivalent transformation, we can obtain the Norton equivalent circuit of parallel compensation as shown in Figure 4. I_{sc} is a current and can be shown as:

$$I_{SC} = \frac{U_{oc}}{\omega L_2} \quad (1)$$

When secondary side is parallel compensated, the appropriately secondary compensation capacitor can make the secondary winding to complete resonance. Furthermore, the system load can get maximum transmission power, the secondary side compensation capacitor parameters can be determined as:

$$C_2 = \frac{1}{\omega^2 L_2} \quad (2)$$

In this case, the system load can get the available maximum transmission power is:

$$P_{2\max} = I_{SC}^2 R \quad (3)$$

4.2 Series compensation on primary side

The equivalent circuit of ICPT system series compensation on primary is shown in Figure 5, where the secondary impedance Z_{r2} is:

$$Z_{r2} = \frac{\omega^2 M^2}{Z_2} \quad (4)$$

Z_2 is the impedance on secondary side, M is the mutual inductance in loosely coupled transformer.

As shown in Figure 4, when parallel compensation system is used on secondary side, secondary circuit impedance can be expressed as:

$$Z_2 = j\omega L_2 + \frac{1}{\frac{1}{j\omega C_2} + R} \quad (5)$$

The primary impedance is:

$$Z_1 = j\omega L_1 + \frac{\omega^2 M^2}{Z_2} + \frac{1}{j\omega C_1} \quad (6)$$

When the secondary side is compensated according to the primary series compensation and secondary parallel compensation, we can obtain the equivalent impedance of the primary circuit:

$$Z_1 = \frac{M^2 R}{L_2^2} + j\omega L_1 + \frac{1}{j\omega C_1} - \frac{j\omega M^2}{L_2} \quad (7)$$

To reduce the power demand, the inductive reactance of the primary side equivalent circuit should be completely compensated. When the primary side needs to be compensated, the parameters of the compensation capacitor should be designed in condition that the load impedance angle is zero in primary equivalent circuit.

Define:

$$Z_1 = \text{Re}(Z_1) + \text{Im}(Z_1) \quad (8)$$

$$\text{Im}(Z_1) = 0 \quad (9)$$

In this case, since the voltage and current phase to be zero in primary power output, can obtained the primary compensation capacitor parameters is:

$$C_1 = \frac{1}{\omega^2 L_2 - \frac{\omega^2 M^2}{L_1}} \quad (10)$$

Similarly, we can get the compensation parameters in various ways [8].

5 THE RESULTS OF SIMULATION

In this section, we will use the Matlab simulation to verify the above analysis. The system parameter values are shown in Table I.

5.1 Secondary compensation verification

The results on different simulation conditions, including the secondary side without compensation, series compensation, parallel compensation as shown in the following

Table 1. The main parameters value of simulation system.

Variable	Value
Input Voltage V_{in}/V	40
Primary inductance L_1/uH	54
Secondary inductance L_2/uH	85
Load R/Ω	100
Mutual L_m/uH	25
Secondary compensation capacitor C_2/uF	0.119
Primary compensation capacitor C_1/uF	
PSSS	0.046
PSSP	0.052

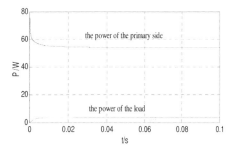

(1) The power of the load and the power of the primary side

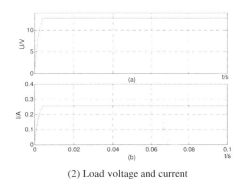

(2) Load voltage and current

Figure 6. System transport power and load voltage when secondary without compensation.

It can be found in Figures 6, 7, 8 that for the same parameters, without the secondary compensation, the load voltage and current have a lower value. When secondary is compensated, the voltage and current values increased substantially, which means the transmission power of the system has also been greatly improved. The power transmission performance by using different secondary compensation is shown in Table II.

5.2 Primary compensation verification

To verify the influence of primary compensation in power transmission performance, respectively for the primary and secondary side uses different compensation topologies simulation. Compensation parameters

(1) The power of the load and the power of the primary side

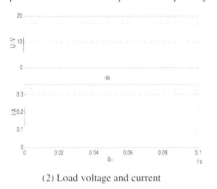

(2) Load voltage and current

Figure 7. System transport power and load voltage when secondary series compensation.

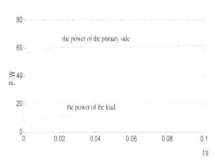

(1) The power of the load and the power of the primary side

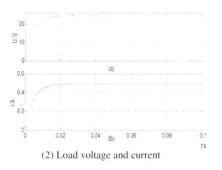

(2) Load voltage and current

Figure 8. System transport power and load voltage when secondary parallel compensation.

as shown in Table 1, when primary and secondary are used in series and parallel compensations, the simulation results of system power transmission performance as shown in the following.

Table 2. Transmission performance simulation results when secondary compensation system in different ways.

The way of secondary compensation	Load voltage V_0/V	Load power P_0/W	The rate of transmission efficiency
No	12.82	3.28	6.51%
Series compensation	16.97	5.75	11.3%
Parallel compensation	24.74	12.2	19.8%

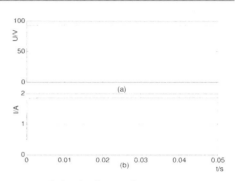

(1) Load voltage and current

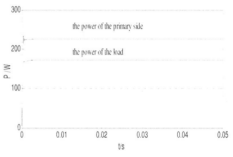

(2) The power of the load and the power of the primary side

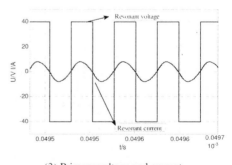

(3) Primary voltage and current

Figure 9. The power transmission performance when series-series compensation system.

Figures 9 and 10 are the power transmission system performance curve when system used series-series and series-parallel compensation topology. Fig 1) is the load voltage and current waveform, Fig 2) is the power the zero phase, which can reduce the need for system power capacity and can greatly improve the systems transmission efficiency.

6 CONCLUSION

Different compensation topologies of ICPT system are discussed in this paper. According to the structural characteristics of loosely coupled transformer, the compensation capacitor is analyzed and calculated. From the simulation results, it is found that the appropriately secondary compensation capacitor can enhance the system power transfer capability, while the appropriately primary compensation capacitor can reduce the demand for power capacity, improving the system power transmission performance.

REFERENCES

[1] Wu Ying, Yan Luguang, Huang Changgang. 2003. Performance analysis of new contactless electrical energy transmission system. Advanced Technology of Electrical Engineering and Energy:11–13.
[2] Zhang Yongxiang, Tian Ye. 2006. The design of loosely coupled inductive power transfer system. Journal of Naval University of Engineering:30–33.
[3] Zhao Zhibin, Sun Yue, Su Yugang. 2012. Primary Side Constant Input Voltage Control and Parameters Optimization of ICPT Systems by Genetic Algorithm. Proceedings of the CSEE: 170–171.
[4] H.L. Li, A.P. Hu, G.A. Covic and C.S. Tang. 2009. Optimal coupling condition of IPT system for achieving maximum power transfer. Electronics Letters 1st January Vol. 45 No. 1.
[5] Zhou Wenqi, Ma Hao. 2007. Design Considerations of Compensation Topologies in ICPT System. Applied Power Electronics Conference: 985–990.
[6] Yang Minsheng. 2009. Dynamic Compensation of Contact-Less Power Transmission System Based on Controlled Reactor. Transctions Of China Electrotechnical Society 24(5): 183–188.
[7] J.T. Boys, G.A. Covic and A.W. Green. 2000. Stability and control of inductively coupled power transfer systems. IEE Proc. Electr. Power Appl: 37–43.
[8] Yong Xiang Xu. 2002. Modeling and Controller Design of ICPT Pick-ups. PowerCon International Conference on (Volume:3):1602–1606.

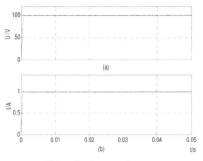

(1) Load voltage and current

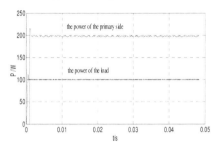

(2) The power of the load and the power of the primary side

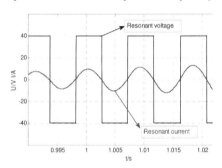

(3) Primary voltage and current

Figure 10. The power transmission performance when series-parallel compensation system.

of the load and primary side; Fig 3) is the primary voltage and current.

From these figures, it is found that in ICPT system, the primary power output voltage and current exists

form
Fiber pullout with stress transfer and fracture propagation

J.Y. Wu, H. Yuan & H.W. Gu
MOE Key Laboratory of Disaster Forecast and Control in Engineering, Institute of Applied Mechanics, Jinan University, Guangzhou, China

ABSTRACT: In this paper, a nonlinear fracture mechanics approach has been introduced to derive theoretical solutions for the shear stress transfer and crack propagation along the interface of the pull-pull and pull-push fiber-matrix pullout model by using the assumed bilinear local bond-slip model. Expressions for the maximum load, shear stress and crack propagation along fiber-matrix interface are derived analytically. Finally, numerical simulations are conducted to discuss the factors influencing the interfacial behavior. Though this paper emphasizes on fiber-matrix pullout test, the analytical solution is equally applicable to similar test between spindly bars of other materials and concrete.

Keywords: crack propagation; FRP; interfacial mechanics; fracture mechanics

1 INTRODUCTION

Fiber Reinforced Polymers (FRPs) have been around for many years, offering us new ways to strengthen or reinforce concrete, masonry and other structures. Making use of FRP, fiber-matrix interface debonding become one of the most common failure patterns. Shear-Lag Model (SLM) or fiber-loading theory was originally developed by Cox (H.L., 1952) for predicting the elastic strength of two-phase composites in which the strong phase (fiber) is embedded in a continuous weak phase (matrix). In order to predict the creep behavior of composites, in the earlier models, Kelly and Street (Kelly, 1972) assumed that the shear strain-rate of matrix increases due to the presence of the fibers and this increase is inversely proportional to the fiber spacing. Lagoudas et al. (Lagoudas, 1989), and Beyerlein and Phoenix (Beyerlein, 1998) extended the shear-lag model to the case of a composite with elastic fibers, visco-elastic matrix and bonded interface. The problem of fiber-matrix pullout test has received considerable attention in recent years and many important results have been achieved. An excellent review of the pullout test analysis has been presented by Shah and Ouyang (P and C, 1991), Bazant and Desmorat (P and R, 1994) study theoretically the size effect in the problem of fiber pullout. In recent years, Yuan and Wu (Yuan, 2001) analyze varieties of nonlinear FRP-concrete interfacial properties, describing the pre- and post-cracking (elastic, softening, debonding) behavior. Quek (Quek, 2004) presents theoretical method to analyze thermal stresses in fibers of finite length embedded in a matrix. Tran et al. (L.Q.N. Tran, 2013) investigate the fiber-matrix interfacial adhesion of natural coir fiber composites by using a physical-chemical-micromechanical approach, and single fiber-matrix pullout tests to measure interfacial shear strength. By using the same method, the problem of pullout of fiber is discussed in this paper.

2 FUNDAMENTAL FORMULAS

Consider pullout test of fiber from the surrounding matrix as illustrated in Figure 1. A cylindrical fiber of radius r_f is assumed to be embedded in an outer cylinder of radius r_m representing the matrix of a composite material. The Young's modulus of fiber and matrix are E_f and E_m, respectively. And the shear modulus of the matrix is G_m. The bond length of the adherends is denoted by L.

Before the derivations, the following assumptions are made in the current study: (a) The adherends (fiber and matrix) are homogeneous and linear elastic; (b) The adhesive (interface) is only exposed to shear stress; (c) Shear deformation doesn't occur in the fiber and matrix, and the normal stress within the fiber and matrix is uniformly distributed over the cross-section.

Based on these assumptions, the equilibrium of the infinitesimal element (Figure 2) can be expressed as

$$\frac{d\sigma_f}{dz} - \frac{2\tau}{r_f} = 0 \qquad (1)$$

For the pull-pull fiber-matrix pullout model

$$\sigma_f \pi r_f^2 + \sigma_m \pi (r_m^2 - r_f^2) = P \qquad (2a)$$

For the pull-push fiber-matrix pullout model

$$\sigma_f \pi r_f^2 + \sigma_m \pi (r_m^2 - r_f^2) = 0 \qquad (2b)$$

(a) Pull-pull fiber-matrix pullout test

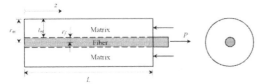

(b) Pull-push fiber-matrix pullout test

Figure 1. Fiber-matrix pullout test.

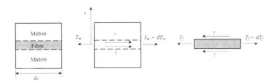

Figure 2. Equilibrium in the infinitesimal body.

where τ is the interfacial shear stress, σ_f and σ_m are the normal stress in the fiber and matrix, respectively.

$$\sigma_f = E_f \frac{du_f}{dz} \quad (3)$$

$$\sigma_m = E_m \frac{du_m}{dz} \quad (4)$$

The constitutive equation for the adhesive layer can be expressed as

$$\tau = f(\delta) \quad (5)$$

The interfacial slip δ is defined as the relative displacement between the two adherends, that is

$$\delta = u_f - u_m \quad (6)$$

Substituting Eqs.(2)–(7) into Eq.(1) and introducing the parameters of local bond strength τ_f and interfacial fracture energy G_f yields

$$\frac{d^2\delta}{dz^2} - \frac{4G_f}{\tau_f^2} \lambda^2 f(\delta) = 0 \quad (7)$$

For the pull-pull fiber-matrix pullout model

$$\sigma_f = \frac{\tau_f^2}{2G_f} \frac{1}{r_f \lambda^2} \left[\frac{d\delta}{dz} + \frac{P}{E_m \pi (r_m^2 - r_f^2)} \right] \quad (8a)$$

For the pull-push fiber-matrix pullout model

$$\sigma_f = \frac{\tau_f^2}{2G_f} \frac{1}{r_f \lambda^2} \frac{d\delta}{dz} \quad (8b)$$

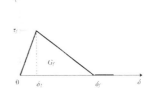

Figure 3. Local bond-slip model.

where

$$\frac{1}{\lambda^2} = \frac{2G_f}{\tau_f^2} \frac{E_f E_m (r_m^2 - r_f^2) r_f}{E_f r_f^2 + E_m (r_m^2 - r_f^2)} \quad (9)$$

3 LOCAL BOND-SLIP MODEL

In this section, the bilinear local bond-slip model is apply to the modeling of interface debonding. A typical bilinear local bond-slip model consists of a linear elastic branch and a linear softening branch, as illustrated in Figure 3. According to this model, the bond shear stress increases linearly with the interfacial slip until it reaches the peak stress τ_f at which the value of the slip is denoted by δ_1. After the occurrence of an interfacial microcrack (softening), the local bond-slip relation is linearly descending with a range from δ_1 to δ_f. The value of shear stress is reduced to zero and an interfacial macrocrack (debonding) occurs when the value of slip exceeds δ_f (Yuan et al., 2004).

$$f(\delta) = \begin{cases} \dfrac{\tau_f}{\delta_1} \delta & \text{when } 0 \leq \delta \leq \delta_1 \\ \dfrac{\tau_f}{\delta_f - \delta_1}(\delta_f - \delta) & \text{when } \delta_1 < \delta \leq \delta_f \\ 0 & \text{when } \delta > \delta_f \end{cases} \quad (10)$$

4 THEORETICAL DERIVATIONS

In the following, two typical kinds of pullout test are introduced to derive the shear stresses along the interface in the bond zone in detail. Before derivation, we assume that $E_m(r_m^2 - r_f^2) \geq E_f r_f^2$.

4.1 The pull-pull fiber-matrix pullout model

4.1.1 Elastic stage
There is no interfacial microcrack or macrocrack along the fiber-matrix interface under small loads, so the entire length of the interface is in an elastic state. Substituting the relation of Eq.(10) for the case of $0 \leq \delta \leq \delta_1$ into Eq.(7), it can be derive

$$\frac{d^2\delta}{dz^2} - \lambda_1^2 \delta = 0 \quad (11)$$

where

$$\lambda_1^2 = \frac{4G_f}{\tau_f^2}\frac{\tau_f}{\delta_1}\lambda^2 = \frac{2\delta_f}{\delta_1}\lambda^2 \qquad (12)$$

The boundary conditions

$$\sigma_f = 0 \quad \text{at} \quad z = 0 \qquad (13)$$

$$\sigma_f = \frac{P}{\pi r_f^2} \quad \text{at} \quad z = L \qquad (14)$$

The solution of Eq.(11) for relative displacement of the adhesive layer can be written as follows:

$$\delta = \frac{P}{\lambda_1 E_m \pi (r_m^2 - r_f^2)}\left\{\left[\frac{1}{\tanh(\lambda_1 L)} + \frac{\beta}{\sinh(\lambda_1 L)}\right]\cosh(\lambda_1 z) - \sinh(\lambda_1 z)\right\} \qquad (15)$$

where $\beta = E_m(r_m^2 - r_f^2)/E_f r_f^2$. Shear stress of the adhesive layer and normal stress of the interface can be obtained from Eq.(15). Setting $z = L$, $\delta = \delta_1$ in Eq.(15) leads to the load at the beginning of softening stage

$$P_s = \lambda_1 \delta_1 E_m \pi (r_m^2 - r_f^2)\left[\frac{1}{\sinh(\lambda_1 L)} + \frac{\beta}{\tanh(\lambda_1 L)}\right]^{-1} \qquad (16)$$

It can be shown that for large values of L Eq.(16) converges to

$$P_s = \lambda_1 \delta_1 E_f \pi r_f^2 \qquad (17)$$

4.1.2 Softening stage

Substituting the relation of Eq.(10) for the cases of $0 \leq \delta \leq \delta_1$ and $\delta_1 \delta \leq \delta_f$ into Eq.(7), it can be derived

$$\frac{d^2\delta}{dz^2} - \lambda_1^2 \delta = 0 \quad \text{for} \quad 0 \leq \delta \leq \delta_1 \qquad (18)$$

$$\frac{d^2\delta}{dz^2} + \lambda_2^2 \delta = \lambda_2^2 \delta_f \quad \text{for} \quad \delta_1 < \delta \leq \delta_f \qquad (19)$$

where

$$\lambda_1^2 = \frac{2\delta_f}{\delta_1}\lambda^2, \quad \lambda_2^2 = \frac{2\delta_f}{\delta_f - \delta_1}\lambda^2 \qquad (20)$$

By solving Eqs.(18)–(19), we can obtain

$$P = \frac{\frac{\lambda_2}{\lambda_1}\tanh[\lambda_1(L-a)]\cos(\lambda_2 a) + \sin(\lambda_2 a)}{\frac{\lambda_2}{\lambda_2^2 \delta_1 E_f \pi r_f^2}\left[1 + \frac{\cos(\lambda_2 a)}{\beta\cosh(\lambda_1 L - \lambda_1 a)}\right]} \qquad (21)$$

P reaches its maximum value when $dP/da = 0$. For large values of L, the expression for a at the maximum load can be simplified to

$$\tan(\lambda_2 a) = \frac{\lambda_1}{\lambda_2} \qquad (22)$$

Substituting Eq.(22) into Eq.(21), we obtain for large values of L

$$P_{\max} = \sqrt{2}\lambda\delta_f E_f \pi r_f^2 \qquad (23)$$

For simplicity we discuss only the case of large values of L in the following. The validity of Eq.(23) requires that shear stress at the left end be still less than τ_f at the maximum load.

$$P = \lambda_1 \delta_1 E_m \pi (r_m^2 - r_f^2) \qquad (24)$$

Comparing Eq.(24) with Eq.(24), we conclude that Eq.(23) is valid and softening does not appear at the left end if $\delta_f \leq \beta^2 \delta_1$; softening appears at left end before the maximum load is $\delta_f > \beta^2 \delta_1$.

4.2 The pull-push fiber-matrix pullout model

According to the pull-pull fiber-matrix pullout model discussion before, we can solve ordinary differential equations of the pull-push fiber-matrix pullout model.

4.2.1 Elastic stage

The shear stress of the adhesive layer:

$$\tau = \frac{P\lambda_1 \cosh(\lambda_1 z)}{2\pi r_f \sinh(\lambda_1 L)} \qquad (25)$$

Setting $z = L$, $\tau = \tau_f$ in Eq.(25) leads to the load at the beginning of softening stage

$$P_s = \frac{2\pi r_f \tau_f}{\lambda_1}\tanh(\lambda_1 L) \qquad (26)$$

4.2.2 Softening stage

The shear stress of the adhesive layer:

For $0 \leq \delta \leq \delta_1$ i.e. $0 \leq z \leq L-a$

$$\tau = \tau_f \frac{\cosh(\lambda_1 z)}{\cosh[\lambda_1(L-a)]} \qquad (27)$$

For $\delta_1 \leq \delta \leq \delta_f$ i.e. $L-a \leq z \leq L$

$$\tau = -\tau_f\left\{\frac{\lambda_2}{\lambda_1}\tanh[\lambda_1(L-a)]\sin[\lambda_2(z-L+a)] - \cos[\lambda_2(z-L+a)]\right\} \qquad (28)$$

Before debonding, the load P continues to increase as the length of the softening zone a increases.

$$P = \frac{2\pi r_f \tau_f}{\lambda_2}\left\{\frac{\lambda_2}{\lambda_1}\tanh[\lambda_1(L-a)]\cos(\lambda_2 a) + \sin(\lambda_2 a)\right\} \qquad (29)$$

P reaches its maximum value when $dP/da = 0$. For large values of L, the expression for a at the maximum load can be simplified to

$$\tanh[\lambda_1(L-a)] = \frac{\lambda_2}{\lambda_1}\tan(\lambda_2 a) \qquad (30)$$

Substituting Eq.(30) into Eq.(29), we can obtain

$$P_{max} = \frac{2\pi r_f \tau_f}{\lambda_2} \frac{\delta_f}{\delta_f - \delta_1} \sin(\lambda_2 a) \quad (31)$$

For large values of L, Eq.(31) can be simplified to

$$P_{max} = \frac{\sqrt{2}\pi r_f \tau_f}{\lambda} \quad (32)$$

If $\beta = 1$ (i.e. $E_m(r_m^2 - r_f^2) = E_f r_f^2$), Eq.(32) can be simplified to

$$P_{max} = \pi r_f \sqrt{2G_f E_f r_f} \quad (33)$$

If $\beta \gg 1$ (i.e. $E_m(r_m^2 - r_f^2) \gg E_f r_f^2$), Eq.(32) can be simplified to

$$P_{max} = 2\pi r_f \sqrt{G_f E_f r_f} \quad (34)$$

4.3 Load-carrying capacity

For the pull-pull fiber-matrix pullout model, the cases of $\beta = 1$ and $\beta > 1$ give the same formula $P_{max} = \sqrt{2}\lambda \delta_f E_f \pi r_f^2$ provided that the bond length L is sufficiently large. It can also be expressed as

$$P_{max} = 2\pi r_f \sqrt{G_f E_f r_f \left(1 + \frac{1}{\beta}\right)} \quad (35)$$

It can be concluded that P_{max} is independent τ_f and δ_f.

A similar conclusion can be obtained for the pull-push joint. According to Eq.(32), we get

$$P_{max} = 2\pi r_f \sqrt{G_f E_f r_f / \left(1 + \frac{1}{\beta}\right)} \quad (36)$$

5 NUMERICAL SIMULATIONS

To gain a clear understanding of the interfacial shear stress distribution and crack propagation along the fiber-matrix interface, numerical simulations are conducted to simulate the behavior of the pull-pull fiber-matrix pullout model, based on the ANSYS finite-element program and theoretical derivations. In the numerical simulations, fiber and matrix are modeled with linear bar element LINK8 and solid element SOLID65, respectively. Both are connected by nonlinear combine element COMBIN39, and the simulations are performed by the displacement control method. The parameters of the local bond-slip model $\tau_f = 40$ MPa, $\delta_1 = 0.006$ mm, $\delta_f = 0.06$ mm (Wang, 2003), so interfacial fracture energy $G_f = 1200$ N/m. The detailed dimensions and material properties are given in Table 1.

Table 1. Dimensions and material properties in numerical simulations.

E_f (GPa)	r_f (μm)	E_m (GPa)	r_m (μm)	β	
72	5	0.7273	50	1.0	–
72	5	1.4545	50	2.0	$\delta_f > \beta^2 \delta_1$
72	5	2.2998	50	3.1623	$\delta_f = \beta^2 \delta_1$
72	5	2.9091	50	4.0	$\delta_f < \beta^2 \delta_1$

5.1 Shear stress distribution and crack propagation

Different loading stages of shear stress distributions are demonstrated in Figure 4. At small loads the shear stress at both ends is less than τ_f, and there is no interfacial debonding or softening along the interface. Once the shear stress reaches τ_f, the softening zone appears. The length of the softening zone a will increase as the load P increases. The maximum load P_{max} appears at this stage. Crack propagation happens, and the shear stress peak τ_f moves toward the other end.

In Figure 4(a), softening and debonding appear at both ends at the same time for $\beta = 1.0$. After debonding appears at the right end, softening will appear at the left end for $\beta = 2.0$ (i.e. $\delta_f > \beta^2 \delta_1$), as seen from Figure 4(b). After the crack propagation happens at the right end, the shear stress at the left end will reach τ_f but softening doesn't appear at the left end for $\beta = 3.1623$, as shown in Figure 4(c). For $\beta = 4.0$, the shear stress at the left end won't reach τ_f, and softening doesn't appear at the left end, as demonstrated in Figure 4(d).

5.2 Load-carrying capacity

According to Eq.(35) and Eq.(36), we conduct a parametric study on load-carrying capacity for both pull-pull and pull-push fiber-matrix pullout model, as shown in Figure 1.

The bond length $L = 3$ mm, material properties and geometry parameters are $E_f = 72$ GPa, $E_m = 2.5$ GPa, $r_m = 50$ μm, $r_f = 5$ μm. For large value of bond length L, the relation between the maximum load P_{max} and interfacial fracture energy G_f, the relation between the maximum load P_{max} and the Young's modulus of fiber E_f, and the relation between the maximum load P_{max} and β are shown in Figures 5–7. It can be concluded that P_{max} increases with increase of interfacial fracture energy G_f and the Young's modulus of fiber. The values of P_{max} in the pull-pull fiber-matrix pullout model are twice that the pull-push fiber-matrix pullout model joint when $\beta = 1$, and the values of P_{max} in the pull-pull fiber-matrix pullout model and in the pull-pull fiber-matrix pullout model tend to approach each other with increase of β.

By comparing Eq.(35) with the finite element simulations, we concluded that theoretical derivation is correct, as illustrated in Table 2.

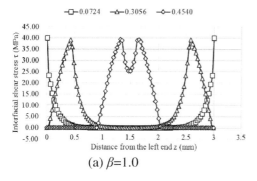

(a) $\beta=1.0$

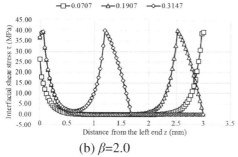

(b) $\beta=2.0$

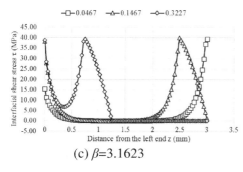

(c) $\beta=3.1623$

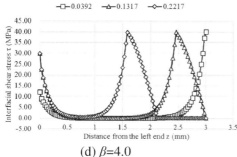

(d) $\beta=4.0$

Figure 4. Different shear stress distribution modes the pull-pull pullout model.

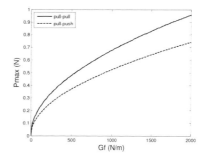

Figure 5. The relationship between P_{max} and G_f.

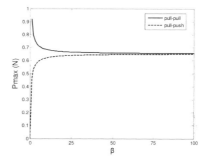

Figure 6. The relationship between P_{max} and E_f.

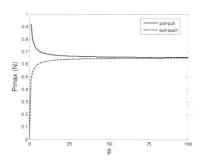

Figure 7. The relationship between P_{max} and β.

Table 2. Comparison of theoretical P_{max} and FEA P_{max}.

β	FEA P_{max} (N)	Theoretical P_{max} (N)	Error (%)
1.0	0.9272	0.9234	0.412
2.0	0.8022	0.7997	0.313
3.1623	0.7514	0.7491	0.307
4.0	0.7314	0.7300	0.192

6 CONCLUSIONS

In this paper, a nonlinear fracture mechanics approach has been introduced to derive theoretical solutions for the shear stress transfer and crack propagation along the interface of the pull-pull fiber-matrix pullout model by comparing with the case of the pull-pull fiber-matrix pullout model by using the assumed bilinear local bond-slip model. The following conclusions may be drawn:

(a) Provided that the bond length is large enough, the maximum load P_{max} is only dependent on the values of interfacial fracture energy, the Young's modulus of fiber and the radius of fiber and β;

(b) The maximum load P_{max} increases with increase of interfacial fracture energy G_f and the Young's modulus of fiber E_f;

(c) The shear stress and crack propagation is different in the pull-pull and pull-push fiber-matrix pullout model. In general, load-carrying capacity of pull-pull pullout model is larger than that of pull-push pullout model.

REFERENCES

[1] Beyerlein, I.J., Phoenix, S.L. 1998. Time evolution of stress redistribution around multiple fiber breaks in a composite with viscous and viscoelastic matrices. *International Journal of Solids and Structures*, 3177–3211.

[2] H.L., C. 1952. The elasticity and strength of paper and other fibrous materials. *Br. J. Appl. Phys.*, 72–79.

[3] Kelly, A., Street, K.N. 1972. Creep of discontinuous fibre composites:II. Theory for the steady-state. *Proc. R. Soc. London, ser. A328*, 283–293.

[4] L.Q.N. Tran, C.A.F., C. Dupont-Gillain, A.W. Van Vuure, I. Verpoest Ashton. 2013. Understanding the interfacial compatibility and adhesion of natural coir fibre thermoplastic composites *Composites Science and Technology* 80, 23–30.

[5] Lagoudas, D.C., Phoenix, S.L., Hui, C.-Y. 1989. Time evolution of over-stress profiles near broken fibers in a composite with a viscoelastic matrix. *International Journal of Solids and Structures*, 45–66.

[6] P, B. Z. & R, D. 1994. Size effect in fiber or bar pull-out with interfacial crack in slip. *J of Engineering Mechanics, ASCE*, 120, 1945–1962.

[7] P, S. S. & C, O. 1991. Mechanical behavior of fiber-reinforced cement-based composites. *J. Am. Ceram. Soc.*, 74, 2727–2738, 2947–1953.

[8] Quek, M.Y. 2004. Analysis of residual stresses in a single fibre-matrix composite. *International Journal of Adhesion & Adhesives*, 24, 379–388.

[9] Wang, H.-W. 2003. *Experimental and Theoretical Study on the Interfacial Bonding Strength of the Glass-Fiber/Polymer Matrix Composites*. master, Wuhan University of Technology (in Chinese).

[10] Yuan, H., Teng, J.G., Seracino, R., Wu, Z.S. & Yao, J. 2004. Full-range behavior of FRP-to-concrete bonded joints. *Engineering Structures*, 26, 553–565.

[11] Yuan, H., Wu, Z.S., Yoshizawa, H. 2001. Theoretical solutions on interfacial stress transfer of externally bonded steel/composite laminates. *Journal of Structural Mechanics and Earthquake Engineering*, 18, 27–39.

Problems and suggestions on Physical Education and Training methods in the new era

Peifang Cao
Martial Arts Academy, Wuhan Sports University, Wuhan, Hubei

ABSTRACT: This work advanced thinking on improving Physical Education and Training methods of China's universities in the new period. On the basis of problem analysis, it put forward thinking on Physical Education and Training methods in the new era, especially for problems in reality. Then some thinking and suggestions on its development were made on this basis. Physical Education and Training methods in the new era should pay attention to absorbing theoretical knowledge of various related disciplines without losing their own characteristics. The innovation in this work used Subject Penetration Theory to analyze and clarify Physical Education and Training methods in the new era.

Keywords: Physical education training; New era; Problems; Universities; Suggestions

1 INTRODUCTION

Only constant innovation can enable a discipline to develop continuously as well as meet requirements of the next era and overall environment. Physical Education and Training is actually a branch of the Sports Science category. The advancement of disciplines such as Physical Education and Training actually play an important role to trigger the whole Sports disciplines constantly evolving. And of course that includes physical education and sports training [1]. After their establishment in China's universities, disciplines of the physical education category have played an irreplaceable role in the continuously sustainable development of China's sports and nurturing large numbers of sports talents. In the development of the times, only through constantly eliminating things that do not meet the requirements of the times and slowly adjusting subject knowledge structure can a discipline really promote the advancement of China's academic knowledge. It is also a strategic essence in the sustainable development of China. This work analyzed problems of Physical Education and Training methods in the new era, and proposed thinking on solutions.

2 PROBLEMS IN PHYSICAL EDUCATION AND TRAINING AND ITS METHODS IN THE NEW ERA

2.1 *Too saturated settings of physical education and training category currently*

China is actually a sports country as well as sports power country of diverse development. Since the founding of China's new regime, China has made a lot of achievements in domestic and international competitive sports. In addition, the settings of sports disciplines in universities witness the importance China attaches to them [2]. Therefore, Sports Science currently has longer time of establishment in universities and relatively complete development. However, in recent years, China's universities have continued to expand their enrollment. Meanwhile demand of the whole society for talents in sports has enlarged compared to the era before. As a result, it provides a better job market environment and, more than that, proposes a number of new requirements and challenges to cultivating physical education talents in universities [3].

Hence many universities in China have started to reconstruct disciplines of physical education according to current trends. In particular, the Fourteen Key Universities in China have also begun to upgrade the cultivating level of sports talents. Besides, some Key Normal Universities have also enhanced their degree-granting ability. But these aspects show that universities only consider their professional setting issues from their own development status with rare consideration of the subsequent impact on the entire market. In fact, the current settings of physical education and training disciplines in many universities conform the positioning of sports colleges themselves. Nevertheless, some universities of Science and Technology also follow the trend of establishing such disciplines. As a matter of fact, it is detrimental to and thus negatively impacts the development of Physical Education and Training. Such rapid development is easy to cause a serious bottleneck in the development of Physical Education and Training.

2.2 Uneven construction standards of physical education and training

Nationally, the construction level of Physical Education and Training in many universities differs unevenly. China has continually adjusted its higher education policy to fit its demand for sports talents under times development. Additionally, the scale of cultivating undergraduate and graduate students in physical education in China is also increasing. However, some universities seem to be too hasty in setting Physical Education and Training. Subsequently, it occurs that these universities cannot find more scientific orientation in daily training methods, despite their establishment of Physical Education and Training. As a consequence, some of the training process and the actual condition of the universities appear disjointed. Such a situation is actually a Great Leap Forward movement in China's physical education. Several universities possess deficient instructors as well as substandard hardware and software in Physical Education and Training. Instead, they quickly build the disciplinary structure of Physical Education and Training through various virtual-high tactics. Such kinds of construction method only present a framework but do not substantially improve the construction level of Physical Education and Training. Thus, such a waste of teaching resources as well as its certain distance with the overall development of Sports Science cannot adapt to the future development of education.

2.3 Current methods of physical education and training generally lacking characteristics

The current teaching contents of Physical Education and Training in universities of China generally have no real features. And basically they are common sports subjects such as track and field, volleyball or basketball, as well as football, as the main types. Especially, the training process of these common sports subjects presents no significant features. Meanwhile, basically following established rules, the training ways and means cannot reflect the innovation in Physical Education and Training methods of universities. From the aspects of reasons, the past sports researchers and practitioners in China, as well as expertise of sports in the world, have already got relatively mature researches and ideas in various sports skills especially training tips. In particular, the motor skills of many common sports have gradually formed a complete system. Therefore, many universities have inertia in exploration of Physical Education and Training methods. Such physical education and training process without innovation will inevitably lead to decline of physical education advancement and consequent elimination by constant evolution of the times. This is a huge potential crisis of disciplines.

3 SUGGESTIONS ON WHAT PHYSICAL EDUCATION AND TRAINING IN CHINA CURRENTLY SHOULD PAY ATTENTION TO

3.1 Offering more services for athletic sports in universities

In the early twenty-first century, China hosted the Olympic Games and Asian Games successively. It made the masses of people and the country experience the development of sports in China. Besides, it also motivated the builders of sports to rethink economic sports athletes training mechanism of the entire country, as well as more considerations about the development goal of China transforming from a sports country to a sports power country. Such a construction approach mainly promotes national development from a sports angle. From the aspects of sports power countries such as the United States, most of its athletes are students accepting education in universities. China currently still adopts the professional closed training approach to cultivate athletes, unfavorable for the development of athletes. Therefore, the combination and joint development of universities with athletic sports should actually become an important trend, focused on by China in the process of expanding its sports career. In the past, Physical Education and Training in China mainly paid attention to cultivating high-level athletes. But in the future, it will gradually turn to cultivate university students under their interests. Thereby a batch of talents with sports practical skills and research capabilities will be trained from students learning in universities.

3.2 Enhancing the comprehensive degree of physical education and training

Physical Education and Training and its related aspects in China should further enhance their overall level of education. In particular, Physical Education and Training itself should comprehensively combine more with other disciplines. Currently, this discipline in China is at a stage of comprehensive building and development. And its future innovation in methods should combine with human and social sciences of a wide range, such as Education, Sociology, Psychology, Medical Science and other aspects. Therefore, the development of future Physical Education and Training will be in a system state, multi-level and interdisciplinary. Meanwhile, Physical Education and Training should not only be an accumulation of experiences in training methods, but also a combination of knowledge in mathematics and science. It is across multi-disciplinary knowledge. Thus, faced with various developing knowledge fields in China, Physical Education and Training and its related aspects should absorb more disciplinary knowledge to enhance their disciplinary competitiveness, also integrating such disciplinary knowledge to strengthen their own comprehensiveness.

3.3 Integrating the content of each system within physical education and training

Physical Education and Training is with relatively high degree of interdisciplinary knowledge. But it is also an integration of other internal disciplinary knowledge within Sports Science, according to the specific development trend and the internal development requirements of the times, as well as the developing trend of future Sports Science. Currently, leisure sports are popular in China's sports development. With the evolution of Chinese economy and improvement of people's living standards, the life values of people have changed from simple enjoyment of materiality to currently healthy development direction of life. In the course of conversion of people's ideology, Sports Science should not only focus on enhancing athletic sports skills, but also pay attention to its combination with daily life. It should enhance the life delight of people and adjust the mental health of individuals. Daily physical exercise should enhance both physical and psychological health of people. Leisure sports should be an important integrated item in the development of Physical Education and Training. Related personnel should pay more attention to this direction and add this element to Physical Education and Training.

The national and traditional sports elements are one important aspect formed in the development of Chinese national culture, and also one significant wonderful work of Chinese traditional culture. Chinese government has put more emphasis on the development of cultural soft power compared to the previous stage. Thereby the heritage of sports traditional culture should be focused on. National and traditional sports culture should be included in the sports culture of the new era. Therefore, the integration of national sports culture should be emphasized, whether from the perspective of development of athletic sports or development of universities' sports culture. Adding national and traditional sports culture into physical education and training can be full enrichment and support for sports culture, as well as important theoretical sources of China's sports development. It will continually enrich the innovations of both methods and theory in Physical Education and Training.

3.4 Maintaining independence of physical education and training in the new period

Times are continually changing and society constantly developing in China. Physical Education and Training and its methods should turn into a comprehensive development process. Its external and internal disciplines are constantly penetrating into physical education and training processes in the new era. However, it should be realized that whatever types of sports culture must maintain its relative independence. Physical Education and Training should emphasize more on showing its own characteristics, presenting its independence in the future development process. In the course of development, only through expanding more of its own characteristics can this discipline better promote its potential in the new period. Thus, it will be more stamina in future development and better enrich its own form.

It should be noted that Education, Psychology, Physiology and Medical Science mean a lot in the content development of Physical Education and Training. Nevertheless, it cannot be allowed that Physical Education and Training becomes a vessel of other disciplinary knowledge. It should also be noted that knowledge out of Sports Education is still just value of a reference and an auxiliary equipment to Physical Education and Training. Physical Education and Training in the new era should focus on its own reform, development and promotion. Ultimately, a Physical Education and Training mode with Chinese characteristics will come into being.

4 CONCLUSIONS

Physical Education and Training and its methods should be an open system, with a longer developing process. Therefore, in the development of Physical Education and Training, its optimization and integration with disciplinary knowledge structure, as well as maintaining disciplinary independence, should be promoted. Thus, it will provide strong support for the development of China's sports, and promote the sustainable development of Sports Science.

ACKNOWLEDGEMENTS

This work was funded by *Hubei Social Science Sports Humanities Project* (2009b360). The project name is *Countermeasure Research on Current Situation and Development of Wushu Routine Teaching in Hubei Universities*.

REFERENCES

[1] Wang Xiaodong. On the Logical Starting Point of Physical Education and Training Research. Shenyang Institute of Physical Education, 2006 (4):73–75
[2] Tan Guang, Ma Weiping. Analysis on Study Characteristics of Postgraduates of Physical Education and Training. Sports Culture Guide, 2009 (1):88–89+92
[3] Sun Dechao. Humanities Thinking on the Current Situation of Postgraduates of Physical Education and Training. Zhejiang University of Technology, 2005 (1):24–25

Impact of Roman Mythology on English literature

Huiqing Wang
Nanjing City Vocational College, Nanjing, Jiangsu, China

ABSTRACT: As an important basis of English literature, Roman Mythology plays an important role in the development of English literature. There has been an important historical and cultural relationship between ancient England and Roman. The impact of Roman Mythology on English literature embodies many aspects, such as creative material and creative inspiration. The process of influence on English literature affected by Roman Mythology is the process of language communication, as well as the process of shaping ways of thinking.

Keywords: Roman Mythology; English literature; Cultural Edification

1 INTRODUCTION

English literature is a form of literature under the influence of Greek and Roman cultures. Roman culture, especially the works of Roman Mythology, has had a great impact on English literature during long periods and it has played a role in the development of English literature since ancient times [1]. In this work, it will analyze conditions, connotation and influence process that Roman Mythology influences on English literature.

2 PREREQUISITE OF ENGLISH LITERATURE INFLUENCED BY ROMAN MYTHOLOGY

As an important source of modern European civilization, various achievements of Roman civilization are inherited by modern westerners through language, literature and philosophy. The development process of English literature has fully reflected the influence of Roman Mythology. The prerequisite of English literature influenced by Roman Mythology can be summarized as political and cultural impact. The Romans conquered Europe and spread their own culture and philosophy, which was the best comment on English literature influenced by Roman Mythology [2].

Firstly, Great Britain is the cradle of English literature development and also a space for the British survival and development. English literature depends on the forming of English and the English people. Having long been occupied by Romans and being the most Northern provinces in Europe controlled by them, Great Britain, as well as other Roman footprint places, was influenced by Romans who were in the leading position among early European nationalities. The 300-year-old rule by Romans not only brought dawn of civilization into the Anglo-Saxons, but also Roman Mythology. As part of British primitive belief, the early Roman Mythology was inherited and developed by British ancestors through word-of-mouth. In fact, the inheritance stemmed from the entire nation's huge change in production and lifestyle. The Romans' occupation and invasion played a great role in the development of these peoples. The advanced production and lifestyle brought by Romans was a historical progress for Nordic nations who were still in a state of ignorance. The worship for Greek and Roman civilizations formed during this period became the social psychological foundation of the wide absorption of Roman Mythology [3].

Secondly, in the Middle Ages, European nations had gradually formed through collision and fusion among them and the language used by the Romans became a carrier of culture dissemination in the process of national association. The dissemination of Latin among educated class from various countries made Roman Mythology nourishment for various nationalities, including the British people. Many intellectuals started their knowledge journey in the process of getting Roman Mythology exposure. Massive of active Latin etymology in modern English reflected that Roman Mythology was imperceptibly inherited and studied by later English writers. The elegant wording and troublesome diction techniques of Roman Mythology were gradually absorbed and grasped by English writers, and finally transferred into their creative inspiration and material, which played an important role in English literature development. Learning and absorbing from Latin had a very important significance on the forming of English. Currently, numbers of Latin root in English not only enrich English expression, but also establish vocabulary basis of English further development. Taking Roman Mythology as the carrier, Roman culture leads to dramatically leap in the British form of language.

Finally, the Renaissance time, an era of prosperous literature and art of all forms in Europe, provided a free space for dissemination of Roman Mythology. At the same time, as an important carrier and tool, literary forms, such as poetry, comedy and fiction, were eventually produced by writers from all European countries. During the same period, increasingly weaker durance by feudal forces, support and expansion from the feudal monarchy could also provide market and soil for the development of various literary forms. The Renaissance had become inexhaustible nourishment for the development of European literature and English literature was also flourishing in this context. The European intelligentsia tended to pay special attention to Southern Europe—Rome and used various spiritual wealth left by their predecessors as a direct power for creation. In Renaissance times, faced with relatively large and strong European feudal forces, writers expressed their ideas by ancient Greek and Roman Mythology, which became an important means for writers with new ideas and concepts. British writers like Shakespeare showed their attention and praise for human values and dignity by this way so as to avoid social exclusion because of content and themes.

In a word, being prerequisite of important force for English literature, Roman Mythology should include three aspects. British long history as a province of the Roman Empire became the historical premise of English literature influenced by Roman Mythology. Roman Mythology, which was brought by wide use of Latin, was popular with elites of English-speaking countries. The Renaissance relied on extensively absorbing the contents of ancient Roman mythology, and human spirit contained in literature became the mainstream of European literary and artistic development, which was the historical motivation of English literature influenced by Roman Mythology.

3 EMBODIMENTS OF ENGLISH LITERATURE INFLUENCED BY ROMAN MYTHOLOGY

Along with the development of English literature, the impact of Roman Mythology upon English Literature lay in creating materials and imagination. Rich characters and storylines that appeared in Roman Mythology can provide much inspiration for English writers. The social customs outlined in Roman Mythology seem to be distant and kind of mysterious when compared with modern European social life. People can meet their needs for imagination and gain more creative freedom by reading these works. It means a lot to the highly realistic English people.

The impact of Roman Mythology on English literature firstly embodied creating materials for numerous English writers. In fact, it was also the common feature that appeared in the process of overall European literature and art affected by Roman culture. English writers such as Shakespeare widely applied characters and plots of Roman Mythology to their works. To a certain extent, characters in Shakespeare's works, such as Hamlet and Macbeth, were similar to heroic figures in Roman Mythology. At the same time, the feature that Romans paid attention to real life, which appeared in Roman Mythology, was also learned by later English writers.

Preceded by Greek Mythology, and clearly influenced by the imagination of the Greek, Roman Mythology was gradually learned and absorbed by English writers. As the romantic leading figures in the field of English literature, Shelley and Byron obviously transferred style full of unconstrained imagination in Roman Mythology to their own creative styles. Heroic figures in Roman Mythology such as Prometheus were often praised by British romanticism literary masters. The powerful imagination needed for romantic literature can be obtained through Roman Mythology. As a result, Roman Mythology played an important role in the development of English literature after the Renaissance.

The difference between Roman and Greek Mythology was that the latter put greater emphasis on deepening its own interesting plot. In other words, similar to human beings, god itself also had rich emotion and could create a friendly atmosphere for people on this basis. The former reflects more of god groups with social relations, among which were clearer classification. In fact, it provided an important premise for reflecting real life through literature. English literature paid special attention to showing writers' thoughts by depicting story plot, and aimed to influence the public.

In short, Roman Mythology could provide richer creating materials and much more imagination for English writers. The more realistic social relations among Roman Mythology could offer a comparable literary language system for English writers.

4 HISTORICAL PROCESS OF ENGLISH LITERATURE AFFECTED BY ROMAN MYTHOLOGY

The impact of Roman Mythology upon English Literature traversed a long road. For English writers, the acceptance of Roman culture and language was an important premise. The allusion referred by them directly reflected the start of impact. And the overall shaping of English Literature by Roman Mythology had already finished when the plot and structure of Roman Mythology could be freely controlled by English writers.

Firstly, the impact of Roman Mythology upon English Literature began when the Roman culture and language was widely accepted by English writers. Classical Latin and Vulgar Latin used by Romans were compulsory courses for British intellectuals during long periods. After accepting Roman language, their works of literature and mythology became readable material. The whole society's worship and respect for Latin caused a trend toward reading Roman literature works among the educated class, which enabled the spread of Roman Mythology among the British upper class in this period. In modern English literature, some traditional writers still hoped to bring the strict syntax specification of Latin into English. From them, we can see the great impact of Roman literary language on English writers.

Secondly, in order to increase meaning for their works, the English writers who were familiar with story plots that appeared in Roman Mythology began to use Roman Mythology as allusion. It was another important stage of English literature affected by Roman Mythology. As the descendant and branch of Nordic Germanic peoples, the English people and Latin civilization formed and developed in Southern Europe were different from each other in terms of cultural and psychological features. Under this premise, bringing in various allusions in Roman Mythology can provide an opportunity to demonstrate knowledge for the educated class in England. Therefore, citing content related to Roman Mythology became the common interest of English writers during the Late Middle Ages and the Renaissance.

Finally, it was also the most important historical stage that English writers referenced and cited Roman Mythology in terms of plot and structure, and imitated its creative ideas. It could be said that Roman Mythology had reshaped English literature after this stage. Roman Mythology features combining with real life, which reflected Romans' emphasis on real life and rational analysis. Actually, it was quite similar to characteristics of English people, who took business as their important life style. As a result, English writers could directly accept the people activities and ideology in Roman Mythology so as to provide inspiration and material for their creative activities. In fact, the feature shown in the national literary tendency lay in the evolution of creative ideas. After entering into the era of critical realism, succession of Roman Mythology had become a national literature feature.

In a word, the impact of Roman Mythology upon English Literature traversed a long road. Taking rich expression forms in Latin as their national literary language marked the beginning of the impact. Greatly bring in allusions of Roman Mythology, promote the romanization of English literature, absorb and reference narrative style and creative architecture of Roman Mythology, and finally help English writers in reengineering their own creative methods, all of which led to the high development and prosperity of English literature today.

5 CONCLUSIONS

The impact of Roman Mythology upon English Literature was produced in the process of mutual blending between two nations during long periods. Reference and bring in Roman Culture, widely use Latin, help English writers in transferring Roman Culture into national literature style, translate narrative style of Roman Mythology into their own creative model, use social class relations among gods and realistic thinking in Roman Mythology as tool for showing spirit and feature of the times, all of which led to English literature with a long history, unique appearance and rich connotation. Roman Mythology plays an important role in English Literature and finally helps boost its prosperity because of numerous information and various thinking on real life within Roman Mythology during various periods of English literature. As a popular saying by westerns that great belonged to the Roman.

REFERENCES

[1] Liu Huming, Impact of Greek and Roman Mythology upon English Literature. Journal of Shanxi College for Youth Administrators, 2009(3):85–87
[2] Peng Fang, Li Jiaxin, Influence of Greek and Roman Mythology on English Culture. Age of Literature, 2012(3):145–146
[3] Li Chuanjing, Ren Rui, the Protagonist's "Narcissism" in Doyle Gray's Portrait from the Perspective of Mythological Archetype Theory. Reading and Writing (Last Third of the Month), 2010(7):53–53

Practical research on mode of combining learning with working in higher vocational English education

Zhirong Zhang
Ningxia Polytechnic, Yinchuan, Ningxia, China

ABSTRACT: Optimization of a practical way of higher vocational English education should be proposed according to the mode of combining learning with working. On the basis of problem analysis, the problem in English teaching process in higher vocational college is analyzed, and the optimization of practice is proposed based on the mode of combining learning with working. It is demonstrated that the combination of the English course and professional knowledge is extruded, and the knowledge structure is optimized based on the educational mode of combining learning with working. In the work, it is innovated that the mode of combining learning with working in higher vocational English education is variously analyzed in the way of multilayer.

Keywords: Combining Learning with Working; Higher Vocational College; English Education; Practice

1 INTRODUCTION

Higher vocational education has becoming a viable teaching mode in the current Chinese national education system. And the inside building and the innovation of the course have been reinforced for decades. Many students, having passed through higher vocational education, are popular because of better vocational abilities. Higher vocational education still stresses on employment. The comprehensive adaptive ability to the market is made and developed by higher vocational education, and then the employment competitive power is exerted in the situation of rigorous employment. So, the market competitive power should be the main direction of higher vocational education. English, as a very important course, can strongly advance the employment competitive power of students with the development of higher vocational education [1]. The development of English courses in the market and the improvement of the comprehensive quality of students are important for innovation in the higher vocational college. So, combining learning with working is a new developing mode for English teaching and an important method aiming at the employment market.

2 PROBLEM OF HIGHER VOCATIONAL ENGLISH EDUCATION AT PRESENT

2.1 *Students with poor basics in English and teachers of single structure in higher vocational college*

The English level of the students in higher vocational colleges is lower than those in other types of higher education, and the levels are not the same. With the development of teaching scale of higher education, the number of students enrolled in higher vocational colleges is increasing and the competition between higher vocational colleges is becoming more and more intense. However, a lot of higher vocational colleges constantly expand enrolment, but ignore the teaching. Specially, the admission requirement is reduced after the expansion of enrollment [2]. So, the enrollment quality of the whole higher vocational education is decreasing constantly. Many students have passed through a series of English education in higher vocational colleges, but the English level of the students in higher vocational colleges still cannot be entirely improved [3]. And it is difficult for many students to listen, read and write in English.

The lower enrollment quality affects the teacher with more pressure on teaching. The English education in higher vocational colleges aims at making the students adapt the development of the market. However, the method of holding the degree is becoming a problem of teaching process in higher vocational colleges, proposing a rigorous challenge for the English teachers in higher vocational colleges. Most of the teachers in higher vocational colleges are majors in English, with a better ability of application of the English language. Actually, they are uncertain to have rich experience in English teaching, especially as they can only use the traditional way in the teaching of the professional English course. This will cause disjunction of the application and professional knowledge and the bottleneck of improving application of professional English when the students are learning related professional English in higher vocational colleges.

2.2 Influence of traditional teaching mode to that of higher vocational colleges

Although many higher vocational colleges are paying attention to innovation of English teaching of higher vocational colleges, it is still affected by traditional teaching mode. The actual condition that the teachers are the teaching center should be changed to the condition that the students are the subject of the English course in higher vocational colleges. In fact, English examination in higher vocational colleges is not really changed, but the students are required to have a certain level of English knowledge and the certificate to prove it. So the English course in higher vocational colleges cannot be actually changed. It still focuses on English level evaluations and college English tests. The basic knowledge of the course is mainly taught by the teacher, and ESP teaching cannot be efficiently implemented. It takes much time and energy for the students to learn grammar and recite the texts. The indiscriminate English teaching cannot promote the language level of the students in the professional region, making the students not adaptive to the market.

2.3 Impractical English teaching materials of higher vocational colleges

Higher vocational colleges mainly develop some higher technical talents for construction and production, requiring the course system of higher vocational colleges to be practical in the direction of the development of market demand. To strengthen the practicability of English courses in higher vocational colleges, the teaching process and the teaching material should be professional. Nowadays the English teaching materials in higher vocational colleges are not professional and cannot be applied to actual problems. At present the students learn through the English teaching materials, consolidating English grammar. However, it cannot form the professional English knowledge face to the market, reflect the practicability and application of the teaching material and aim at the developing direction of the career of the students.

3 ENGLISH EDUCATION IN HIGHER VOCATIONAL COLLEGES IN THE MODE OF COMBINING LEARNING WITH WORKING

The thought and suggestion of improving the teaching quality of higher vocational colleges are proposed by the ministry of education. It is considered that combining learning with working and cooperation of colleges and companies should be advanced. Especially, combining learning with working, as the personnel training mode, is also the point of innovation of talent education in higher vocational colleges. Actually, the innovation begins with the classroom teaching and the college teachers, and the center of teaching is changed from teachers and classroom to students, advancing the teaching efficiency of higher vocational colleges constantly. In this context, the English teaching should be connected with the idea of combining learning with working, especially in the condition of conversation about foreign trade business or foreign merchandiser negotiation. English can be connected with not only the basic conversation of language, but also the actual application, making a better professional service. So, the teaching way of the teachers, the mode of the teaching system and the teaching materials are studied to realize the mode of combining learning with working and improve the English level and learning direction of higher vocational colleges for the requirement of society and the market.

3.1 Combining learning with working while the teachers are teaching and the students are learning

From the above paragraphs, it is analyzed that the English levels of student sources are not ideal in higher vocational colleges, but the plasticity and creative awareness of the students are stronger. And the students are young and good at showing themselves off. So, the English teachers in higher vocational colleges will teach according to the characteristics and interests of the students, making the students join in the daily teaching process more often and providing them a richer space for learning and developing. Based on this, the vocabulary and grammar, including the knowledge of all kinds of languages and works, should be added to the learning content in higher vocational colleges. Then the students will be able to deal with problems about English knowledge when they are engaged in professional jobs.

After developing certain teaching activities, the teacher should find opportunities to make the students join in the daily practical process and learn for the purpose of application. So the interests and passions of the students can be increased, and the students can get something useful at last. However, the point considered by the English teachers of higher vocational colleges is the organized forms of classroom and the way of developing teaching for the purpose of making the students join in the English classroom. The English teachers should constantly increase comprehensive quality themselves, learn the related higher vocational knowledge and the enterprise practice in the direction of higher vocational colleges, understand the knowledge in the usual courses and optimize the method for English classroom teaching.

3.2 Advancing combining learning with working of higher vocational English teaching mode

Combining learning with working, as a new English teaching mode of higher vocational colleges, is optimized on the basis of the traditional English teaching mode, laying stress on the professional characteristic of higher vocational colleges, contents of different professions and the teaching contents in accordance with professions. On basis of this, the English teaching system of higher vocational colleges is strengthened and the degree of combination is promoted. The teaching idea of combing learning with working is used to improve the professional skills of higher vocational students. So, in the daily teaching process of the English course, the skill of English knowledge is trained and strengthened, and the main parts are taught and trained repeatedly. The teacher in higher vocational colleges should pay attention to the employment trend of the students and combine the professional knowledge with the language knowledge. Then the teaching task in the classroom can be arranged based on the current professional work to realize a seamless join of practice and theory.

In addition, according to the cognitive level and the rule of learning knowledge, the students schematically participate in the related enterprise and master more comprehensive knowledge during the actual operation, achieving transferring knowledge to practice and improving comprehensive professional ability and employment competitive power of the students at last.

3.3 Combining learning with working of the English teaching materials of higher vocational colleges

The English teaching materials of higher vocational colleges should be different to those of common higher colleges, especially in practicability, so that they can be applied to the actual work. So, on the one hand the English teaching materials should contain the introduction of English basic knowledge, on the other hand it should stress the interesting teaching. The professional teaching materials are selected according to subject characteristics of different higher vocational colleges. The refining teaching of English grammar and theory is reduced and the practicability of the teaching materials is strengthened. Especially, some professional teaching materials including English terminology, style and text are selected for professional work. The language texts selected in the English teaching materials cannot be too difficult and the actual operation contents should be added to in-class exercises behind the books in order to encourage the students in the freedom or team discuss and mock drilling.

The profession and temporal spirit of the content should be extruded and the texts out of time should be removed when the English books are selected in higher vocational colleges. Then the students can feel the time of the teaching materials and be more interested in learning English knowledge. The teachers should reduce or add some content of classroom teaching according to the characteristics of professional learning and post requirements. Besides, they should consider the English teaching materials as the basic of teaching without limiting the content in the range of text and introduce some daily English materials of companies as the teaching samples, optimizing the teaching materials and advancing combining learning with working.

4 CONCLUSIONS

Innovation of the English courses in higher vocational colleges should be oriented to the world and the future in the direction of thought of combining learning with working. With the development of the English course in higher vocational colleges, including the related teaching method and teaching materials, it will become an important driving force for training talents.

REFERENCES

[1] Kang Chunyan. Innovation of English Teaching in Higher Vocational Colleges in Cooperation Mode of College and Enterprise, Science and Technology Innovation Herald, 2013(04): 194–195.
[2] Cui Chunping. Exploration and Practice of Talents Training Mode "Combining Learning with Working" in English Teaching in Higher Vocational College, Higher Education Exploration, 2012(05): 150–152.
[3] Zhou Yongxiang. Research on Effectiveness of English Classroom Teaching in Higher Vocational Education in Context of Combining Learning with Working, Vocational and Technical Education, 2012(20): 31–33.

Improvement of computer multimedia technology of English education

Nan Shi & Nan Li
Xingtai University, Xingtai, Hebei, China

ABSTRACT: The rapid development of scientific technology continuously accelerates and updates computer multimedia technology, bringing great changes to various fields of human life. Meanwhile, education also experienced many changes with the help of computer multimedia technology. Particularly, students' English skills such as listening, speaking, reading and writing become the focuses of English teaching. Therefore, the development of multimedia technology has greatly accelerated English teaching. On this basis, the work analyzed the promoting effect of computer multimedia technology on English teaching.

Keywords: computer technology; multimedia technology; English teaching

1 INTRODUCTION

The development and improvement of multimedia information technology has brought revolutionary changes to English teaching. Compared to traditional classroom teaching, the initiative and enthusiasm of students has been greatly improved in multimedia teaching. In traditional English teaching, students usually accept knowledge first, and then digest what they've learned and finally put the knowledge into practice. In this process, the learning methods and contents are always the same, easily causing a decrease of learning interests and efficiency. As a result, it is difficult for students to achieve substantial improvement of their English abilities and skills [1]. However, things have been changed ever since the introduction of multimedia technology into English teaching. Nowadays, students' initiative and interest in English learning can be inspired and stimulated through videos and other vivid visual methods. Therefore, the learning efficiency and teaching effect are greatly improved. It is of great help for teachers to make full use of computer multimedia to assist their teaching and stimulating students' learning enthusiasm [2]. Besides, this new teaching mode enables students to understand and control learning progress and efficiency, thus enhancing their abilities to digest and apply knowledge in the process of learning. Therefore, the English teaching level can be fundamentally improved, and students can apply the knowledge they've learned from class in real life.

2 ADVANTAGES OF MULTIMEDIA COMPUTER TECHNOLOGY IN MODERN ENGLISH TEACHING

2.1 *Enhancing the enthusiasm of students to learn English*

With the continuous development of multimedia computer technology, teachers pay more attention to train the listening and speaking skills of students in English classroom teaching. They make use of multimedia to play some listening materials or animation videos in English class. On one hand, students can have an immersive feeling in the dialogue scenes with the stimulation of listening to audios or watching videos [3]. Thus, students don't feel bored in the classroom, greatly improving their learning enthusiasm and interest. On the other hand, teachers can make the content of listening exercises into situational dialogues, animations or other video materials. Therefore, when practicing English listening skills, students also generate interest to do imitation practice. In traditional English teaching, teachers usually spend a lot time and energy preparing teaching content. However, students have little interest and enthusiasm in English learning due to the single teaching mode. With the aid of multimedia technology, teachers can show students the teaching content of that class through some videos or audiovisual materials. Therefore, through the training of watching, listening and speaking, students' enthusiasm and initiative of English learning are greatly improved, and their English abilities and skills also have substantial improvement at the same time. Moreover, multimedia computer helps to activate the classroom atmosphere, prompting students to focus on class and develop active thinking.

2.2 Creating an authentic English learning environment

English teachers usually adopt the one-way-output mode in traditional English teaching, passing on the teaching content prepared before class to students without much interaction. For example, teachers usually introduce basic grammar knowledge to students and then play some listening materials for exercising. This kind of single-way teaching mode is likely to limit students' thinking. On one hand, students always accept knowledge from teachers in a passive state, and they gradually feel bored in this process. As a result, students will ultimately lose their interest in English learning. On the other hand, students' mind activity is limited due to passive acceptance of knowledge. Moreover, humans have a divergent thinking mode. Therefore, this kind of single teaching mode causes certain restrictions for developing students' open mind and thinking mode.

The ultimate goal of English teaching is to help students be able to apply English in their actual life through the training of English skills: listening, speaking, reading and writing. However, current English teaching is still relatively weak in the training of English listening and speaking. So the emergence of multimedia computer technology enables students to feel an authentic environment of English through audio or video materials. Moreover, students can imitate the dialogue scenes that appear in video materials such as movies or other related situational dialogues. For example, students can simulate social dialogues of restaurants or hospitals so as to practice their communicative language. In addition, students' thinking mode can be developed through the visual stimulation of videos, enabling them to think about how to express themselves in English in similar scenes and exercising their ability of divergent thinking. Meanwhile, students' memory can be strengthened through this kind of visual stimulation. Thus, students will unconsciously remember those English expressions in the same or similar scenes in real life. At the same time, students can also practice their listening when watching videos.

Scene simulation of multimedia audios and videos can achieve multidimensional construct so as to enable students to actively join English learning and always stay relaxed with a happy mood in the process of learning. Quickly learning and mastering knowledge in a relaxed environment thus improves the efficiency of learning and promotes the progress of English learning for students.

2.3 Strengthening the cultural awareness of students

Any kind of language is the output carrier for a particular culture. Therefore, as for language learning, it is very necessary to understand the corresponding cultural background of a foreign language, including the history, geography, customs and habits, and places of interest of that country. English learning is no exception. Understanding the corresponding cultural background of English does not only help the English learning of students, but also helps the application of English in real life. Multimedia technology is applied to display students with related data about cultural background information. Combined with English explanation, those audio and video data not only expands students' vision of multinational culture, but also practices their English listening ability. At the same time, it is of great help for students to achieve good English learning and application and to understand the cultural background.

3 DEFICIENCIES OF THE TRADITIONAL ENGLISH TEACHING MODE AND THE PROMOTING EFFECT OF MULTIMEDIA TEACHING ON IT

The traditional English teaching mode is based on teachers' lecture teaching, and students are in a passive state of accepting knowledge. Besides, the traditional English teaching mode shows the characteristics of singleness in both teaching content and method. On one hand, teachers mainly explain the content of textbooks to students and then assign some exercises to them in order to strengthen students' learning after class. Therefore, the knowledge students learn is mainly from textbooks. However, this kind of teaching mode is relatively boring, resulting in little interest of students in English learning. On the other hand, English learning resources for students are relatively deficient due to classroom lecture teaching, unable to satisfy the requirements of students to enhance and apply their English knowledge. As a result, students usually get half the result with twice the work. Traditional English teaching is difficult to create an authentic English learning environment for students, and thus students are usually weak in the application ability of practical English skills. However, with computer multimedia to assist English teaching, the deficiencies of traditional English teaching have been greatly improved.

3.1 Changing the mode of students to accept and learn things

Traditional English teaching is to pass on the basic knowledge of textbooks to students first and then let students digest, understand and grasp what they've been taught in class. After they understand and grasp the knowledge, they need to deepen and consolidate the knowledge, and finally apply what they've learned into actual life. For example, in traditional English learning, students firstly read the model essay in textbooks, then do a lot of practice after class, and finally use scenario expressions and daily English that commonly appear in textbooks in their actual life. This kind of learning process is relatively single and inflexible. However, ever since the emergence of multimedia computer technology, students' English learning does not necessarily follow this process. Instead, they can flexibly master and apply English knowledge while watching videos and situational dialogues. Therefore, this kind of new learning mode is relatively relaxing and easy, greatly improving the efficiency of English learning at the same time.

In addition, multimedia technology teaching also makes it more flexible for students to consolidate and apply knowledge. For example, students can remember words and articles by audio materials rather than by textbooks and vocabulary books. Besides, audio materials can also deepen their impression of English words and articles. Moreover, the ways for teachers to conduct class exams or tests become diverse with the assistance of multimedia computer technology, no longer limited to the traditional examination mode of paper test.

3.2 Changing the relationship among teachers, students and textbooks of traditional teaching

With multimedia technology, the traditional teaching mode, based on lecture teaching of textbooks, has experienced great changes. In multimedia English teaching, teachers can make full use of computer and network technology to download the latest teaching materials to assist their teaching in classroom, helpful to timely update their knowledge structure. At the same time, students can also take the initiative to download some learning materials from the rich resources through different multimedia means according to their actual situation and practical English levels. In traditional lecture teaching on the basis of textbooks: on one hand, students with good basic English knowledge find it is not of great help, soon losing the interest in class; on the other hand, students with poor English basic knowledge find it difficult to understand the textbook content, showing little interest in English learning. However, those problems are largely solved since the emergence of multimedia teaching. The learning initiative and enthusiasm of students in the classroom is fully stimulated, greatly improving the efficiency of English learning. In addition, the rich and colorful content of multimedia teaching mode includes images, audios and animation videos, optimizing the output of teaching information and stimulating students' thinking and interest.

3.3 Accelerating students to feedback and master knowledge in class

Multimedia computer technology enables students and teachers to have timely interactive communication in class, which is conducive to the feedback of knowledge. In a multimedia class, teachers can make use of computer network resources to quickly find abundant teaching resources, and then display to students with the help of multimedia. Moreover, if students or the teacher meet difficulties in the process of learning or teaching, they can search for and find corresponding solutions through computer network resources, largely improving the efficiency of teaching and learning.

In addition, combining other information technologies such as audio, images and animation videos, multimedia teaching can stimulate students' learning enthusiasm and initiative through visual and auditory feelings. Therefore, multimedia teaching enables students to study in a relaxing and happy learning environment, helping them to grasp corresponding knowledge quickly and also greatly enhancing their training of listening and speaking skills.

4 INEVITABILITY AND CURRENT PROBLEMS OF MULTIMEDIA COMPUTER ENGLISH TEACHING

The combinations of multimedia computer technologies with classroom English teaching are the inevitable development trend of current English teaching. The first is the combination of English teaching textbook with the computer. Computer multimedia technology can make use of storage means such as hard drives to display audio and video materials such as situational dialogue, listening and vocabulary training. The second is the effective combination of the learning process of students with computer multimedia technology. In the traditional teaching mode, teachers only explain the content of textbooks to students, and students' study enthusiasm and interest cannot be fully stimulated, thus resulting in inefficiency of English learning. With the help of multimedia computer technology, teachers can stimulate students' study enthusiasm and create a relaxing and lively atmosphere through vivid situational dialogues and video materials, helping students to easily grasp the learning content in the process.

However, the major problems generated from multimedia computer technology in English teaching should also be taken into consideration when emphasizing its advantages. Firstly, most English teachers at present are English graduates with few computer skills and little computer knowledge. As a result, once multimedia equipment fails to work or has some problems, English teachers cannot solve the problems timely due to the limitation of their computer knowledge, thus affecting students' learning progress. The main solution to this problem is to comprehensively improve the multimedia computer technology level of teachers, so they can check relevant equipment before class and find potential problems in advance. Secondly, some teachers depend on multimedia technology too much, because some of them play videos to students the whole class. Although multimedia computer has many advantages, it is just an auxiliary teaching means. Teachers still play the main role in class to guide students to learn, making use of video materials as well as explanation to assist teaching in class. Moreover, the communication between teachers and students can be enhanced, and students can feel humanistic emotions as well, thus fully mobilizing the study enthusiasm of students.

5 CONCLUSIONS

Multimedia computer technology has brought a revolutionary change for current English teaching, possessing absolute advantages compared with traditional English teaching mode. Teaching resources and content are richer in the multimedia English teaching mode, greatly improving the English teaching environment. This kind of teaching mode can greatly improve students' learning initiative and enthusiasm as well as the efficiency of students' English learning. In a word, multimedia computer technology has greatly improved modern English teaching.

REFERENCES

[1] Li Qing. Functions and Characteristics of Metaphor in Computer English. Journal of Qinghai Normal University (Philosophy and Social Sciences Edition), 2003(03): 93–104

[2] Qu Hui, Cheng Peng. New Terms of Computer English. Exam Week, 2009(29): 120–120.

[3] Wang Xiaosi. Cognitive Semantics of INTO Sentence Patterns. Journal of Xinxiang Teachers College, 2001(02): 27–29.

Industrialization management of sports economy under market economy

Aining Qu
Department of Physical Education, Shandong University (Weihai), Weihai, Shangdong, China

ABSTRACT: For a long time, the sports cause has been government-leading public products in China. Along with the development of market economy, it has become a common issue of how to achieve the industrialization of the sports industry. Based on economic theory and general rules of market economy and according to the objective requirements of the transfer from sports cause to sports industry, this work proposed principles and strategies on the development and management of sports industry. The innovation of this work lies on the analysis of sports products, classifying it into private products and public products.

Keywords: market economy; market building; industrialization; sports market

1 INTRODUCTION

Sports industrialization is the process of gradual marketing of sports products, with the automatic adjustment of price mechanism and allocation of sports resources. For a long time, China has classified sports into the sort of public utilities and developed it in a concentrated and rapid way. Tremendous as the achievements are, this pattern of development has also led to the ossification of sports management mechanism and imbalance on the allocation of sports resources [1]. It is suggested that sports should be market-oriented so that the masses can genuinely enjoy sports as public events, thus promoting the industrialization of sports.

2 OVERALL OBJECTIVE AND PRINCIPLES OF SPORTS INDUSTRIALIZATION

The core objective of sports industrialization lies in the marketing of main sports products, distributing sports resources under market economy mechanism and reducing direct state intervention. The direct meaning of this objective is to make numerous sports organizations transform into manufacturers of sports products; and make its scale can be acknowledged by the market. Over years, various sports organizations, which make sports events as their main tasks, exist as state public institutions and accept the state budget appropriation. Although, to some extent, these sports events deserve to be watched, they excessively waste the national resources. In order to optimize the allocation of sports resources and achieve the goal of rational use of sports resources, it is imperative to transfer numerous sports institutions into sports industry [2].

The general principles of sports industry include two elements: to establish a relatively perfect market mechanism and to reduce sports organizations' dependence on the national financial resources. Establishing a market mechanism raises requirements for those products with no public properties. Namely, the prices of those products should be market-oriented, and trading quality, shape and function of products should be determined by the law of supply and demand [3]. To gradually get rid of the excessive reliance on national finances, sports organizations need to be transformed into sports enterprises, thus building an independent and self-financing management and operation mechanism.

3 MEASURES TO ESTABLISH PRICE MECHANISM OF SPORTS PRODUCTS

Under a market economy, the first step to realize sports industrialization is to establish price mechanism on the allocation of sports products; the primary part of sports industry management is the commoditization of sports products, providing sports products in accordance with market price for the public.

According to Marxist Political Economics, goods should be the unity of value and use value. Use value is the property of goods that can meet the market needs and the value that comes from undifferentiated human labor injected by producers of goods. Therefore, no matter whether they are tangible sports goods or intangible events, sports products should have the objective attributes to meet the market needs. To achieve the goal of establishing price mechanism of sports products, it is important to provide market-oriented products. The construction of price mechanism of sports products causes from the commercialization of sports products.

In the marketing process, price mechanism expands gradually with the products entering the market. Thus, due to the chronicity of market formation, price mechanism should be established in a certain phase. In the moment, broadcast industry of sports events has gradually changed from the industrialization of mass media to commercialization. Hence, this part of sports products can be more successful in commercial operation. While some products with less use and lower market acceptance should gradually increase the degree of market acceptance and wait for proper opportunity to enter, thus ensuring ordered marketing of related industries. In fact, the marketization of professional contingents, the major providers of sports supplies, can be helpful to the marketing of these industries. Thus, the establishment and extension of price mechanism should be advanced gradually, along with the propelling of industry chain and supply chain.

4 ACCELERATING ENTERPRISE REFORM OF SPORTS INSTITUTIONS

In a modern market economy, the essential measure of building property rights system is restructuring sports institutions. After determining the market price mechanism, suppliers of sports products are bound to face the problem of cost-benefit. Only when the operation of suppliers is in accordance with the principle of profit maximization, can the market of sports products really function well.

Sports enterprises transformed from sports institutions and should be independent. However, for a long time, sports organizations have been regarded as sports institutions and directly managed by the government. As a result, lack of autonomy has also been obvious throughout the development of sports institutions. To establish independent enterprises, sports institutions should gradually separate from government and be transferred into independent legal entities. Establishing a modern enterprise system is the core of the reformation of entire sports institutions; and the core of modern enterprise system, the enterprise management and operation mode. The first point of sports institutions corporatization is the clarification of the identity, and then the property boundaries and relationship between ownership and management right.

Sports enterprises should become a self-financing economic entity. Formerly, in the process of operation and management of state-owned enterprises and institutions, there still exists soft budget constraint problem in the transfer of sports institutions to enterprises. The original sports institutions, as the organizations supported by state finance, are trend to have a close link with higher authorities. This phenomenon lasts for a long period of time, thus directly impacting the survival and development of enterprises. Soft budget constraint not only squeezes the financial resources of government, but also impulses severe expansions of sports enterprises. The consequence of soft budget constraint is that their operation target is more budget funds from superior governments rather than profits. The main method to get budget funds is the blind expansion of production scale and staffing scale. Due to this phenomenon, it has brought out a serious problem of overcapacity in the field of industrial production. So it will definitely give rise to the blind expansion of sports industry, if sports enterprises cannot straighten this contradiction in the early stage.

To solve soft budget constraint, more rational property relations should be established. It is suggested that modern joint-stock enterprises should separate management right from ownership. Property has resource allocation function and incentive function. In the industrialization of sports industry, the first step is to clarity property relations and management right of related enterprises, ensuring the rights of enterprises with a strict legal system; then clarity resource-dominating rights of enterprises and resource-allocating rights of ownership, preventing higher authorities from abusing the resources of enterprises. More importantly, residual claims of enterprises should be clearly shared among particular subjects to make operators get more effective incentive from enterprise management, thus mobilizing the enthusiasm and creativity of administrators. In terms of separating ownership and management right, modern professional managers should be introduced, thus forming the management mode of separating government functions from enterprise management.

5 BREAKING ENTRY BARRIERS OF SPORTS INDUSTRY

Sports cause, especially sports events, has long been the national public activities, carried out by the governments. Although some events have gradually been taken over by enterprises, the private capital intensity is rather low in sports industry. While, due to the social influence of these events, they cannot avoid the official interventions, let alone promoting the industrial development of sports industry.

The core phase of completing sports industrialization is to promote the free flow of resources. Price mechanisms the most effective way to ensure the free flow of resources. The function of supply-demand law in sports industry cannot be truly used to allocate resources effectively, if capital and labors fail to timely enter the related industries when sports industry trends to prosper. Actually, though not obviously, sporting goods industry and sports events industry belong to scale economy. They can accommodate many manufactures. Therefore, market power cannot explain the barriers and internal requirements of sports industry.

The first factor of free flow is the free entry of capital. Free capital investment is the main phrase of forming fixed capital in fixed costs and variable costs in production process. The free entry of capital not only exists in direct investment, but also in indirect investment—the currency of credit and various ownerships and debt certification. Only when both areas liberate to the external capital can the later be truly able to enter sports industry and have a stable development in a long period of time, thus realizing rational allocation of resources and marketization of sports industry.

The free flow of labor should also be included in this process. In the previous management mode of sports industry, practitioners are staff of public institutions; hence they have no initiative to participate in market competition. But it is difficult for external labors to join sports industry. This narrows sports industry and restricts the development of sports talents. All these go against the law of free flow in market economy and require to be improved in the marketization process.

6 DIFFERENTIATING PUBLIC SPORTS PRODUCTS AND PRIVATE ONES STRICTLY

In the industrialization of sports industry, it needs to differentiate private products and public ones. In fact, it clarifies the responsibilities and relations between government and enterprises, returning the part that can be allocated by the market and the part need governments' support.

There are distinct differences between public products and private ones, for public products are non-exclusive and non-competitive. Non-exclusiveness refers to products with no feasible technologies for not having been paid; non-competitiveness, to products that cannot determine the unit costs. For the inability to determine the costs and exchanging with price mechanism, it is impossible to let public products be completely allocated by the market. Meanwhile, when it comes to public products, for market has failed to allocate them, it is important to make use of government. While the private ones are exclusive and competitive, thus it is appropriate to product and allocate through market mechanism and cost-benefit mechanism. It is important to differentiate public sports products and private ones in the industrialization management of sports industry. Only by distinguishing them rationally can it be possible to have the final determination. That is to say, determining those enterprises that can enter the competitive market, and those should continue to exist as a national funding institutions and reformation objections.

Semi-public products, which are between public products and private ones, are the focus and difficulty of reformation. These items bear characters of both public products and private ones; therefore they cannot be fully provided in either types. These items can be produced by private enterprises and provided by governments, thus realizing the efficiency and fairness on the supply of public products.

The products that could be thoroughly provided by private sectors should never be sold completely. Neither should the share rights. If the corresponding industry belongs to natural monopoly industry, it tends to give rise to imbalance on the efficiency of enterprises. Meanwhile, it can be provided by state-owned enterprises, if the products do not have proper production and delivery method, such as market competitive structure. In developed countries, it is ever pervasive that the pure public products are open to the market and reduce financial costs by government procurement. Frankly, it is an instrumental problem-solving approach when it comes to the development and management of the sports industry in China.

7 ESTABLISHING SPORTS INDUSTRIAL STANDARDS

Base on the executive law, industrial standards are a kind of government-issued regulation of management model for products within a certain field. To sports industry, industrial standards are very important. Sports industry is closely linked with the bioscience and human medicine. The use of these products can directly benefit people's health. Therefore, it needs to regulate products through standard regulations, thus reducing insensible products and protecting lives and health of the masses. In the process of formation of industrial standards of sports industry in China, national physical fitness and conditions should be taken into consideration, avoiding fully copy of the administrative rules and provisions of Western countries.

In short, to meet the requirements of sports cause under market economy, it is important to realize sports industrialization. To finally establish a sports industry model does not only need to rely on market mechanism and promote sports products which deserve to be allocated in the market, but also needs to make sure the suitable system of capital and labor currency. Only in this way can China eventually realize it.

REFERENCES

[1] Zhang Fengli. Current Situation and Development Trends of China's Sports Industry andIts Intrinsic Relationship with Harmonious Society. Value Engineering, 2013, (21):300–301
[2] Li Ying. Analysis on Regional Sports Industrialization in New Economic Era. Education Science and Culture Magazine, 2013, (15):93–96
[3] Huang Ruiyuan, Zhu Jinghua, Cong Ling. Study on Branding and Industrialization Development Model of College Sports. Journal of Shandong Institute of Physical Education, 2011, 27 (12):35–39

Analysis of sports news transmission and its significance

Wenteng Zhan
Department of Physical Education, Shandong University (Weihai), Weihai, Shandong, China

ABSTRACT: As an important news category, sports news has attracted extensive attention from the public. This work aims to explore the significance of sports news and its route of transmission. Nowadays, with the rapid development of social media, more and more people are exposed to various news through diverse approaches. Besides, the emergence of new media also provides a good platform for the development of sports news. The innovative point of this work lies in the analysis of sport news transmission through new media to prove its significance and make a further exploration on its future development.

Keywords: sports news; the route of transmission; new media

1 INTRODUCTION

With the rapid development of society and economy, people's living standard has improved a lot. Our lives are no longer limited in the work and routines; instead various leisure-time activities bring us a lot of vitalities and color. Sports activities become quite prevalent in daily life. Since sports activities are very interesting and loved by many people, we tend to be eager to hear about news reports of our favorite sports [1]. In order to spread news to these sports fans timely, a kind of new industry arises at the historic moment, which is sports journalism. This industry has been playing an important role so far. However, only news is not enough in journalism. We are still in need of proper means and approaches to transmit the news and share them with the public, which is exactly the meaning of sports news transmission. This work illustrates some problems about the transmission of sports news and its significances, and I hope readers could acquire a broader sense of this issue through reading.

2 ROUTES OF SPORTS NEWS TRANSMISSION

2.1 *TV media*

TV media, including TVs, video recorders, DVDs and cameras, takes TV sets as the carrier to exchange and spread information. Recently, an increasing number of media appear in our society, and TV media is just one of them, the oldest and the most basic one. For a long time, TV media has developed its unique features and advantages, such as timely dissemination of information, clear picture, visual intuition and understandability. Since almost every family is able to afford a TV set, the audiences of TV cover all ages. In addition, with the development of technology, an audience now does not only watch TV at one time, but can also participate in the interactive activities while watching TV programs and enjoy more benefits brought by TV media [2].

2.2 *Network media*

The function of network media is the same as that of TV media and newspaper, all aiming to spread news timely and extensively. However, it is different from the latter two in the mode of transmission. With high popularity of the Internet, people's way of life has changed a lot. Gradually, we are adapted to the present life under the influence of the Internet and learn to make use of its characteristics to serve ourselves better [3]. The network media today has fostered its own advantages for development: it is open to the world, easy and flexible to operate; its cost is relatively low; it is highly efficient and economical; it has a strong sensuality and so on, among which the high interactivity is the biggest advantage. It breaks through the situation where audiences receive information passively and creates a new interactive mode. Audiences can get information they need on the Internet, and meanwhile, the network media can also get suggestions and scores from the audiences by taking information feedback. In such a way, websites are allowed to upgrade according to the audiences' advice, and thus contribute to improve the economic profits; at the same time, benefit the audience by providing them with better programs.

2.3 *Mobile transmission*

With the development of technology, mobile technology is continuously upgrading as well. Mobile phone has become a necessity in everyone's daily life. After the popularization of smart phones, the mobile phone is equipped with more functions and has become the main source of information. Mobile media, serving as the transmission carrier, takes the mobile phone as an audio-visual terminal and mobile Internet as the platform. It is a mass media, targeting in mass decentralization, regarding orientation as communication effect and interaction as the carrier. Digital application, portability and interactivity are all its main features. Nowadays, there are hundreds of millions of users of mobile phones. Such an enormous number is far beyond that of the network media. So it's obvious to see that the domestic mobile market still has a great potential. Through smart phones, people could learn more about sports events. Moreover, owing to the recent prevalence of wireless networks, the cost people spend on Internet recreation is much less than before, which can much facilitate the development of mobile media. People can also browse sports news on their mobile phones and learn about the relative sports games anytime and anywhere.

2.4 New media—Microblog/Weibo

Recently, the emergence of Weibo raises a big wave in the media circle and is expanding its influence on people's everyday life. The purpose of Weibo is to promote people to share news around them anytime and anywhere. People can make comments on sports news freely and interact with others through the media of Weibo. As a new tool for communication, Weibo turns to be a best place for entertainment. At the same time, it becomes the top platform for many organizations and companies to release news. Weibo, with the standard of "140 words", is influencing the whole world.

3 ANALYSIS OF THE SIGNIFICANCE OF DIFFERENT TRANSMISSION ROUTES

3.1 Significance of TV media

With a history of nearly one hundred years, TV media, as the most influential media in the world, goes on developing rapidly and starts a great reform in people's way of life. In modern society, various media are emerging constantly. Some people partially point out that TV media will be taken place by network media or some other media in the future. Though new media takes a large proportion in news media, TV media won't be replaced totally.

Firstly, for audience, the price of a normal TV set is much cheaper than that of a mobile phone or computer. Many people in undeveloped areas still couldn't afford a computer. So purchasing a TV remains to be the best choice for them. As for many sports fans, watching games live is a kind of entertainment, while watching live on the Internet is much limited by the Internet speed. Secondly, for the whole society, TV news represents a kind of social responsibility to some degree. It can pass positive energy to the whole country in the form of public service advertising and advocate the society to develop better, which is just what the other media is not able to do. Finally, with the development of economy and technology, TV media can easily broadcast important sports events, which can fulfill the audiences' desire to experience a strong sense of presence in the scene. Therefore, for many audiences who are keen on sports activities, choosing TV media to watch sports programs is their best choice.

3.2 Significance of network media

Network media is a medium form where news is spread based on the Internet. Since sports news is an important part of news, it cannot be broadcast on TV only, and network media is such a complement of TV media. It has its own advantages. For example, firstly, people can search information on the Internet without any limitation, so people like to watch live or replay shows on the Internet. Secondly, there is no advertisement inserted when people are watching programs, which can save them a lot of time. Last but not least, people can search any information both at home and abroad on the Internet without going outside, which is the biggest advantage that TV media can hardly achieve.

However, network media also has some disadvantages. For instance, some accidents, like unclear broadcasting picture or fractured programs, often occur if the Internet speed is not so fluent. As for an audience, this will be the biggest challenge for them to watch real-time sports games, especially those with a great expectation to watch the show live. Similarly, as far as the sports program itself, this kind of broadcasting may affect its ratings and expected targets.

As network media is developing rapidly, various new media have sprung up, exerting an influence on people's lives. Weibo, as one of these new media, gives rise to a huge change towards people's lifestyles. In the recent two years, there is an increasing number of Weibo users, and progressively Weibo takes up an important position to exchange information. Furthermore, companies, enterprises, individuals, and organizations are paying more attention on Weibo, because Weibo can spread information and news timely and easily. Most audiences of sports news are youngsters who are the main users to share and spread sports information. In the mean time, they are also enthusiastic to have interaction with others. Consequently, it can attract people of different ages and areas to gather in Weibo, and play an important role in sports news transmission.

3.3 Significance of mobile media development

Under the highly developed Information Age, a variety of new media is constantly evolving. The mobile phone, as one of the important tools for communication, has generated a great impact on people's life. It is portable, highly interactive, relatively cheap and easy to be accepted by the public, etc. With the development of technology, mobile phones are attached with more functions, especially smart phones. People can browse websites and watch videos on their phones. Sports events are so attractive that they are fascinating to a great many sports fans. In order to make profits in such a broad market, mobile phone manufactures strive to seize this opportunity to transform their products and provide sports fans with first-hand information through mobiles. In such a way, it can also improve the transmission of sports news, inspire people's passion and attention on sports, and enhance their awareness of health.

4 ANALYSIS OF SPORTS NEWS TRANSMISSION THROUGH NEW MEDIA

Sports news, as an important part of news, has greatly influenced society and people's everyday life. Since the founding of new China till now, the government has put arduous efforts to promote its development at all levels so as to raise people's awareness of health. It is essential for harmonious society construction by advocating the public to take more physical exercises and comprehensively enhance their physical fitness. To better develop the sport undertakings, the government has invested a big sum of money on the construction of sports infrastructure and the rebroadcast of sports news. Additionally a batch of high-quality athletes is also cultivated to make a contribution. People begin to transmit sports news in other routes, apart from TV media, in order to transmit sports news to more people. The emergence of these new media has thoroughly changed the way of broadcasting, which is a great reform on the sports news transmission.

Firstly, new media provides a bigger data space for sports fans, which allows them to search the match result they are interested in as well as the latest news of sports stars. Secondly, people transform to watch sports matches on new media instead of TV little by little. In the past, people used to watch sports events on TV sets, but now they can watch them on their mobile phones or PC at any time they like. Finally, it provides a broader platform for governments to propagandize and develop sports undertakings and an efficient approach to know about people's attitudes towards some related policies. These measures can not only facilitate the interaction between government and the public, but also create a better condition for development of the undertakings of physical cultures and sports.

As the rapid development of economy in the 21st century brings about huge transformation to people's lifestyle, the advancement of technology also accelerates an unprecedented revolution to the sports news transmission. Economy never fails to progress ahead, and neither does the innovation of new media. We have reasons to believe that a series of superior communication means would be invented in the near future, which can broadcast sports events and news more easily and efficiently.

5 CONCLUSION

Since sports are very crucial to our life, the transmission of sports news means quite a lot to the development of sports undertakings. To better broadcast sports news, different new technologies are being unveiled to address alteration on people's lives. New media has received great attention and recognition from the public for its convenience, efficiency and globalization. However, it also faces fierce competition that will help invigorate the spread of news through high-tech means of new media. The impact generated by new media on traditional media is unprecedented. But the right track is that the new media and traditional media should coexist with each other harmoniously and build a better environment for sports news transmission.

REFERENCES

[1] Ma Teng, Deng Jianwei. *Research on the Characteristics of Weibo in Sports News Transmission. Shandong Sports Science & Technology.* 2013(6):36–40
[2] Tan Xiuhu. *The Influence of Mobile Media on Sports News Transmission. Editors Monthly.* 2014(1):83–86
[3] Xiao Rong. *On the Communication Strategies of Sports News by Paper Media in the Era of Weibo. Sports Science Studies.* 2013(3):11–14

Analysis on the role of ideological and political education in the cultivation of sense of worth

Hongjuan Ma
Xi'an Physical Education University, Xi'an, Shaanxi, China

ABSTRACT: The ideological and political education is significant in the cultivation of sense of worth. By analyzing the optimal measures of ideological and political education, this study is aimed at improving the role of ideological and political education to help people form a correct sense of value and outlook on life. With the development of market economy, the social competition is increasingly fierce and more and more factors are influencing people's sense of worth. In order to better demonstrate the role of ideological and political education in the cultivation of sense of worth, it is needed to briefly analyze the essence of ideological and political education and its impact on the cultivation of sense of worth.

Keywords: Ideological and Political Education; Sense of Worth; Cultivation

1 INTRODUCTION

The ideological and political education serves for the formation of a person's sense of worth, outlook on life and world view, so strengthening the ideological and political education is significant for each person. Considering the current thoughts of the Chinese, enhancing the ideological and political education is important since this education can not only facilitate the formation of correct sense of worth, outlook on life and world view, but also help people make good choice in their lives. As time changes, there are different things and matters in the whole society, which cannot be excluded and influences people' growth and progress to some extend [1]. Therefore, face with the complex social environment, it is vital for people to grow in peace and make correct choices in their lives. The ideological and political education can create a good growing environment for people to help them form enterprising mentality. Including the ideological and political education in people's daily life and work can help people relate the ideological and political education to the real problems, strengthening their understanding with correct judgment abilities and facilitating the formation of correct sense of worth.

2 NECESSITIES OF FACILITATING THE FORMATION OF CORRECT SENSE OF WORTH

People's development in modern society needs more respect to social environment and requirements of social development, pushing the social progress by individual development. Correct sense of worth is significant for both society and individuals [2].

Firstly, the development of society needs the support and assistance of correct sense of worth. Considering the current Chinese society, the social culture is the core value system of social development since the better development of society is based on the demonstration of social culture. In political science, the essential character of people is social attribute, which decides the life conditions and development conditions. People grow with the social development, which means human beings develop when the society progresses and the relation between human beings and the society is mutual promotion and mutual development [3]. Therefore, only when people's ideas and senses of value mature can they develop better in a complex social environment, facilitating the social reform and social progress.

Secondly, individuals should establish correct sense of worth and have self improvement and self innovation on the basis of reasonable ideas and senses of values, which is more beneficial to personal growth and progress. Each person has some kind of value assessment, whose construction is closely related to the ideological education that can transmit correct values to people and teach them to make judgments. As the society develops, the individuation of a person is gradually appearing. In a complex environment, if people want to adapt to the society environment better and resist harmful factors, they have to receive good ideological and political education. Whenever in primary school, junior high school, senior high school and even university, people cannot have the ideological and political education separated from their lives.

Under such a condition, people can better demonstrate their values and personalities, promoting their comprehensive development.

3 ANALYSIS ON CURRENT IDEOLOGICAL AND POLITICAL EDUCATION

As the society develops and progresses, the society needs more powerful ideological and political education to support its development and demonstrate the relation between individuals and the whole society. However, current ideological and political education in China is not optimistic. Therefore, to better conduct the ideological and political education and magnify its role in facilitating the formation of people's sense of worth, we have to analyze the present situation of the ideological and political education in China and find a scientific way to solve some problems, thus promoting the maturity of the ideological and political education.

3.1 Family ideological and political education

Family ideological and political education is the most fundamental and the most immature ideological and political education. From the very beginning of a person's birth, he or she has become a member of one family, in which the adult members will influence him or her in mode of thinking and sense of worth. On the other hand, when adult members are saying "This is wrong" or "That's not impossible", they are telling their children what is wrong or what is right. In the current situation, it is not perfect for the mode of ideological and political education in China.

3.2 School ideological and political education

The ideological and political education appears in all periods of curriculum design in Chinese schools, so the ideological and political education has been a compulsory curriculum for different levels of students. Compared with family ideological and political education, school ideological and political education is more systematic and greatly satisfies people's requirements in different growing stages. Therefore, it is the main stage for people to establish their sense of worth.

3.3 Social ideological and political education

In essence, the society is a grand classroom where family ideological and political education and school ideological and political education will be tested. Through the test of the society and further transmission, people's sense of worth can be more consolidated. Meanwhile, harmful factors can also lead to the distortion of people's sense of worth.

4 ANALYSIS ON THE ROLE OF IDEOLOGICAL AND POLITICAL EDUCATION IN THE CULTIVATION OF SENSE OF WORTH

The ideological and political education has played a unique role in modern society. Through the platform of ideological and political education, a communication channel will be more effectively formed between the educators and those being educated. With the help of this communication channel, on the one hand, people can learn the ideological and political knowledge and conduct corresponding activities; on the other hand, they can have their own understanding of ideology and politics and broaden their thoughts, helping them establish a corresponding outlook on life, sense of worth and world view.

4.1 Improvement of moral cultivation of the ideological and political education

As a practical activity, the ideological and political education is conducted under certain goals and organizations to help people form some kind of concept through particular educational form. This concept is beneficial to the development and progress of people's thoughts. The original and final goal of the ideological and political education is to cultivate talents with high moral sentiment. By cultivating talents, the ideological basis of receiving education can be laid. It is especially important to have the ideological and political education directed against people's thinking activities in modern society. People at all ages are curious about strange or new things, which lead them to discover and explore new matters. However, not all new things and matters are good. When people's thoughts and behaviors are influenced by harmful things, there can be some changes to their behaviors. If people have a strong curiosity to a certain thing, they will tend to discover it. However, if people in modern society act without the guidance of correct sense of worth, their behaviors will be totally free from any constraint, which can lead to some problems or even crimes. In that way, having people in modern society receive ideological and political education can facilitate the formation of good moral sentiment, possessing a correct judgment rule.

4.2 Formation of correct value orientation by strengthening the ideological and political education

Nowadays, many people are accustomed to a cozy life, uninterested in their life pursuit. They focus more on games and dressing styles and do not show any interests in their study and work, let alone sense of worth or value orientation. Due to the circumstance, strengthening the ideological and political education for modern people will help them form correct value orientation and make correct choices in their lives so as to adapt to the society and conduct reasonable social reform. People in modern society are faced with multidimensional things and matters, so comprehensive ideological and political education can help people in modern society have rational recognition to things and think rationally, thus making correct judgment. It is important to help people make rational choices and correct judgments in modern social environment by strengthening the ideological and political education since it can better facilitate people to form correct value sentiment.

4.3 Enhancement of discipline awareness by strengthening the ideological and political education

Lack of sense of organization and discipline is reflected in many people in modern society. They tend to be more self-oriented and emphasize more about their self realization. These people often neglect collective benefits when they engage in collective activities, lacking certain sensitive of organization and discipline. The first duty of a soldier is to obey orders while the obedience to orders should have some reasons. Not every one who joined the army could command the true essence of the obedience to orders. Likewise, not every soldier can absolutely obey orders. Comprehensive ideological and political education lies behind the obedience, which is the demonstration of the soldiers' sense of organization and discipline. As members of modern society, common people need a certain sense of organization and discipline, which can enable them to take the collective benefits into consideration and seriously accomplish the tasks given by the collective. To conduct reasonable ideological and political education is to help people actively accept the existing social principles and use correct sense of worth to face social reform and social rules.

4.4 Formation of legal awareness by strengthening the ideological and political education is beneficial to the

In a society with rule of law like China, citizens have the duty to have the knowledge about law and learn to use law to protect their legitimate interests when constraining their behaviors according to legal norms. The moral codes are ahead of law and are the foundation of the rule of law. The ideological and political education in modern society is to help people form correct moral ideas, thus upgrading moral ideas and legal concepts. Specific ideological and political education should be given to those who have ideological problems, helping them solve concrete problems and make rational judgment on illegal activities. Therefore, they can use correct sense of worth and ideas of value to observe law and discipline, thus further improving their value concepts.

4.5 People's moving closer to the party through ideological and political education

As a focal point of China's development, the ideological and political education is powerful to encourage educatees to actively care for the affairs of the country and devote themselves in China's development and progress. For the Communist Party of China (CPC), the ideological and political education is also the power to increase national cohesion and can improve China's international status. For people in modern society, strengthening the ideological and political education is good for them to trust the CPC and understand its policies, avoiding deviation of their thoughts. In addition, the ideological and political education transmits modern people a correct recognition to the CPC, which can make them have more confidence in CPC and closely follow the policies and pace of the CPC, moving closer to the party.

4.6 Laying theoretical basis of sense of worth through ideological and political education

In ideological and political education, some correct and positive ideas are always conveyed to the public. The ideological and political education is an exchange or communication of correct value attitude and value concept. In modern society, people are faced with complex things; in order to distinguish these things, it is needed to have a strong theoretical basis, which is also indispensable in forming correct sense of worth. The ideological and political education is a long-term education, which runs through the whole stage of a person's growth. From family, school to the society, one has to receive different kinds of ideological and political education. Although these educations may take on different forms, the contents they want to express are same. It can be said that the ideological and political education is an education throughout the whole process of a person's growth. This kind of education demonstrates the essence of education and lays a good moral and theoretical basis for other kinds of education. The sense of worth can only be correct, appropriate and scientific on the basis of good ideological and political education.

5 CONCLUSIONS

To better play the role of ideological and political education in the cultivation of people's sense of worth and facilitate the formation of it, the system of ideological and political education has to be further optimized, demonstrating the inevitable connection between the ideological and political education and the cultivation of sense of worth. Based on this inevitable connection, comprehensive constructing the connection among social development, the ideological and political education and sense of worth can better demonstrate the progressiveness of the ideological and political education.

REFERENCES

[1] Mo Chunju. *To the Public: the Inevitable Choice of Modern Ideological and Political Education of College Students*. Jiangsu Higher Education, 2014(2): 131–132

[2] Wang Chen. Strengthen Ideological and Political Education to Promote the Sense of Worth of College Students. The Guide of Science & Education, 2011(22): 158–159

[3] Cheng Fan. Opportunities, Challenges and Countermeasures of Ideological and Political Education for Vocational Students under the New Media Environment. Education and Vocation, 2014(9): 63–55

Development strategy of internet platform based on bilateral market theory

Shangyou Ju
Ningbo City College of Vocational Technology, Ningbo, Zhejiang, China

ABSTRACT: With the development of an e-commercial platform, the Internet and mobile information platform play an increasingly important role in competitive activities of manufacturers. It has become realistic problems for enterprises to change the former operation mode of the market and enhance the inherent competitiveness of enterprises through e-commercial systems by using the Internet and mobile device platforms. This work put forward relative strategies to develop the mobile Internet information platform, in line with theories of bilateral market and industrial economics.

Keywords: Bilateral Market, Industrial Organization, Competitive Model, Mobile Internet

1 INTRODUCTION

In the world where e-commerce is widely used, the prototype of a new business model, relying on the mobile Internet platform, has come into view with a gradually increasing developing speed. In fact, the E-commercial platform has established a bilateral model of market business, and raised new requirements for competitive behaviors of manufacturers. Analyzing competitive behaviors between manufacturers, new environment generated from the Internet information platform should be regarded as a fundamental concept of relevant theoretical researches [1]. This work will combine with the traditional industrial economics and propose related strategies to develop a mobile Internet platform from the perspective of applicability of the bilateral market theory in the mobile communications market, according to the theory of bilateral market.

2 APPLICABILITY OF BILATERAL MARKET THEORY IN MOBILE INTERNET INFORMATION PLATFORM

Bilateral market theory describes one business form in which a manufacturer or a kind of manufacturer integrate suppliers with demanders by building a kind of information platform, with costs of using platforms being the main source to earn money. This form is characterized by that external effect of cross-network will occur through trading platforms of two markets. Manufacturers' revenue in the market will be directly affected by the number of consumers in the corresponding market on the information platform, thus forming the corporate scale-up revenue model different from the traditional Internet external effect and scale economies effect. Contacts between the two markets not only include tangible products and reverse movements of funds, but also mutual influences between them mainly achieved through bilateral information platforms [2].

The mobile Internet platform is the technical foundation and prerequisite for operation of e-commerce. Trades through electronic information platforms will connect suppliers to demanders effectively by electronic network. In order to use these networks, the supply side and the demand side have to pay fees, and there are external effects of cross-network between the two. When online buyers increase, corporate average costs of business transactions through e-commercial information system will reduce, achieving economy of scale [3]. However, with more manufacturers to release supply information through e-commercial systems and electronic information systems, customers will spend less time searching products, indicating that both of them are dependent on the other's size of the market to a certain degree.

Thus, the mobile Internet platform will establish a bilateral market structure through the popularity of e-commercial systems. Suppliers and demanders will be gathered into industries with cross-network externalities, thereby reducing the operating cost of enterprises and time cost of consumers. Nevertheless, certain fees need to be paid to use the mobile internet platform, primarily in the form of membership fees and charges. Bilateral market theory can be applied to mobile Internet platform.

3 DIVERSIFIED INTERACTIVE MODES OF MOBILE INTERNET PLATFORM

The mobile Internet platform, as an exchange platform of business information, will integrate suppliers and demanders. Then this integration will create an operation mode of bilateral market only by achieving bilateral network externalities. The premise of cross-network externality is that the seller and the demander can interact through the information platform. This interactive relationship is not a simple scale economy, but a process of mutual attraction and feedback. Demanders should be able to learn information about manufacturers to pass this information to more buyers through the network platform. In the purchase process, the demonstration effect of other consumers' behaviors will lead the consumers to dependency and obedience.

To achieve this goal, the mobile Internet platform should have more extensive functions to provide and integrate information. The function of providing information is a one-way transfer of information supplied by sellers and information demanded by buyers, and this unidirectional transfer depends on visual display of transaction outcomes, thus a more detailed reference for the development of enterprises and the cognition of consumers are provided. The realization of information integration, in fact, relies on a monitoring effect of the mobile Internet platform. Mobile Internet, a carrier of trading activities, can reduce the difficulty of the transaction, and carry the basic information of trends of market and society. This information is crucial for both the supplier and the demander.

Diversified interactive mode should not only include the process of integration and transmission, but also the function of resource allocation. The information of supply and demand will be classified, and certain subdivision, according to the characteristics of the market demand, is an important prerequisite for this diversified interactive mode. The mobile Internet platform helps the seller to complete a part of market segmentation, and consumers can select suppliers of their products with more purposes.

4 REQUIREMENTS FOR MOBILE INTERNET PLATFORM TO ENRICH SERVICES

For the purpose to practically build up an effective bilateral market, the mobile Internet platform should also enrich services. Its function as an information platform will obviously be revealed by helping manufacturers to judge the price more reasonably. Actually, this is the prominent embodiment and illustration of the integration of mobile Internet platforms and modern information consulting business.

Manufacturers set prices in accordance to cost-benefit, and it can be explained by cost-plus pricing including cost and flexibility under the traditional industrial organization perspective. But under the bilateral market structure, companies can carefully consider the structure of cost and benefits, allowing a certain negative balance between their prices and costs. Because the decline in product prices will bring certain increase of sales, especially for companies with high price elasticity of demand. To adjust their prices below the marginal cost is still the balanced decision to maximize profits in the presence of cross-network externality, since increased demands will enlarge corporate profits and produce networked externality. This externality totally generated from the other corresponding market on the Internet platform.

In order to turn cross-network externality into the competitive edge, what mobile Internet platform companies should do is to transfer their services from provision and transmission of information to multi-services including consulting, management and planning. These manufacturers should endeavor to take the range of price fluctuation as an information resource and provide it for the enterprise. With the cross-network externality, it is possible for manufacturers to adopt a lower residual demand curve in restricting prices, allowing manufacturers to exclude external manufacturers. In fact, a Gordian knot of restricting prices is that manufacturers are unable to accurately judge the cost type of potential manufacturers, and time length of implementation of restrictive pricing is not easy to determine. As for potential manufacturers, chained backward inference could be used to ensure that they will be accepted by incumbent firms. The cross-network externality will help companies to expand the range of deflating prices. For incumbents, they are able to further the implementation of predatory pricing in the process of implementing the restrictive pricing strategy. For potential manufacturers, they need to take a greater decision-making risk in deciding whether they should enter this industry.

The mobile Internet platform will estimate the level of overall consumers' willingness to pay through assessment of the number of consumers and judgment of market demands. It will enable companies within relevant fields to implement a more accurate pricing. The multipart pricing is essentially that it bases profits of an enterprise on the usurpation of surplus consumers; the price per unit of product is not the source of profit while fixed costs will transform all consumers' willingness to the corporate revenue. The multipart pricing depends on the full knowledge of market demands, and it is difficult for companies to work out a more reasonable price measure of fixed fee if trends and the status quo of market demands cannot be understood.

5 WEAKER MARKET POWER OF MOBILE INTERNET PLATFORM UNDER THE SAME CONDITION

For providers of mobile Internet platform, the users are highly complementary in terms of demands. Companies can truly establish a bilateral market structure when two markets are interdependent, and both manufacturers in two markets use the mobile Internet platform. Therefore, the mobile Internet platform provider should be aware that their platform licensing is highly elastic, and reactions of the market to price changes will convert to changes in demand with a positive correlation between the two markets.

Manufacturers' market power is the ability to set prices above marginal cost, which has an inverse relationship with the price elasticity of demand. From the above analysis, the mobile Internet platform has a high price elasticity of demand, which will weaken the manufacturers' market power. Naturally, the competitive pressures will inevitably increase. To decrease excessive competition, it is necessary to count on product differentiation to integrate industrial organizations within the industry.

In the modern market economy, the ability to reduce technical risks should be a prominent feature of platforms. Reducing technical risks of companies is a process in which possible economic losses owing to the improvement of techniques will be resolved by market operations through a variety of ways. There are two ways to resolve technical risks on the mobile Internet Platform, including calibration of information and external scale. Bilateral market has the function of transferring funds, as well as transmitting information. Through analyzing the data from transaction platforms, rules of consumption data alternation can be figured out. They reveal the tendency of consumers' preferences, revising the development of product research.

When lowering the corporate technical risk, the effect of external economy of scale formed by bilateral market should also be valued. The reason is that external economy of scale is a united supply channel and technology spillover through interactions of manufacturers. As a matter of fact, consumers in the seller's market can reap the benefits from increased manufacturers in the market where the supplier is in. Consequently, there is a more extensive but poor network externality between the two markets. The external economy of scale, generated from the division of labor and collaboration on the supply chain, will expand consumers' interests of the other party in the bilateral market, so as to provide a broader space for the development of the industry.

6 ENCOURAGEMENT OF CO-OPERATION

In previous relationships among manufacturers, non-cooperative game on price and product is what economists are mainly concerned about, but cooperative behaviors are inclined to be affected by the cartel operating model. These theories usually just focus on the framework of a single market, without concerning about the complementary of needs and interaction between markets, leading to a simple analysis.

Under the current bilateral market structure, manufacturers have to not only consider the cost and demand within the single market into consideration, but also understand that the bilateral market is a specific market structure depending on growing manufacturers to form competitive advantages. As a result, external manufacturers will be the incentive to increase corporate competitive advantages. Then the traditional competition mode of the zero-sum game tends to be reduced in a secondary and replaced position.

Bilateral market structure, an e-commercial information platform depending on market structure, is a method of market operation strongly relying on the market. The characteristic of bilateral market lies in the complementary relationship between the cross-market externality and industry. The mobile Internet system will reflect specific effects of bilateral market structure after the completion of an e-commercial platform, thus relative theories will be able to promote the development of electronic information and the mobile Internet platform.

7 CONCLUSIONS

To develop the mobile Internet platform, manufacturers should focus on the complementary of two markets or multiple markets linked with the platform. Relying on cross-network externality, various companies will be encouraged to cooperate with each other to change the original pricing mode. The advantages of electronic information of the mobile Internet platform will be used to reduce business risk and improve technology. It is needed to increase external revenues with mutual cooperation between companies. Interaction of platform providers can be understood as a broader partnership. Coexistence of manufacturers can increase external revenues, while the division of supply chain can provide a vector of technology and production for the mutual coexistence of different manufacturers in the same market. In this process, the competition of price between companies can be transformed into the competition for customer base, more customers standing for that more depreciate rooms will be can enjoyed. Bilateral market theory has changed the way of competition and survival, the relationship between manufacturers changing into symbiotic relationship from the relationship of zero-sum game. The premise of symbiotic relationship is the relationship of forward kinematics between market returns and the number of manufacturers.

REFERENCES

[1] Wang Jiang, *Using Mobile Internet Platform to Better Explain Credit Card Offers—Analysis of Channels to Display Credit Card Offers*. China Credit Card, 2013(12):50–50.

[2] Cheng Guisun & Cheng Hongming & Sun Wujun, *Analysis of Welfare Effects of Television Media Platform Merger in Two-sided Market*. Journal of Management Sciences in China, 2009(2):9–18.

[3] Cao hui & Bian Yijie & Sun Wjun, *Economic Explanation for Operation Mechanism of Innovation Rely Center Based on Two-sided Markets Theory*. Study in Science of Science, 2010(11):1731–1736.

… in the case of a planned economy. Therefore, it is necessary to further promote the reform of art and design education.

Concept of art and design education

Hongquan Sun
Hunan City University, Yiyang, Hunan, China

ABSTRACT: It is with important practical significance to strengthen the study on the concept of art and design education since the education not only is linked with art, but also belongs to education industry. Through the practice of educators, this work mainly analyzes the confidence-building education of students majoring in art and design, combining with the new requirements for art and design at this stage of social life. In addition, based on the Bauhaus idea, there is further study—it needs three levels for art and design education closer to reality, aiming at exploration of its educational concept.

Keywords: Art and design education; confidence; Bauhaus; life

1 INTRODUCTION

Art and design differs from traditional disciplines in many ways—learning methods, education approaches, application fields and academic goals. In addition, this discipline applied to modern life, which has gradually been in close connection with the economy, technology and daily life of ordinary people, can be described as a beauty-oriented learning with aesthetic pleasure integrated into mindless and regular life reasonably. The art and design education as entry and guide of this discipline bears enormous responsibilities. It is a key point for students majoring in art and design to gain systematical and accurate theory and skills with appropriate education methods [1]. Meanwhile, education details concern art thinking extension and improvement of aesthetic attainment as well as results of relevant education activities. The concept of art and design plays a leading role in the whole of art teaching. The artistic design, as an act of creation aiming at pursuit of inspiration, needs higher-level guidance and support of the art and design concept provided by educators.

2 PRESENT SITUATION OF ART AND DESIGN EDUCATION

With the development of artistic design connotation, the transformation of relevant educational concept is inevitable. Only in this way can sustainable development of artistic design and targets of art education be achieved. Up till now the teaching concept in the field of artistic design has gradually changed from merely seeking for further artistic attainments to serving daily life and reality. This trend is also true of multiple subjects towards science and theory. Some learning activities are held aiming at improvement of the deficiencies and finding a solution to those difficult problems in reality [2]. For scholars and students, they spare no effort to achieve the goal of making knowledge more adaptable in serving daily life. It is gratifying that these subjects free from traditional restrictions play a positive role in all aspects of daily life, including plenty of theory knowledge as well as stereotyped arts and crafts. Thus there is emergence of pleasant changes aroused by these factors in the world.

In the development of art and design, education of this field remains a strong sense of China's planned economy, such as school names, major names and course syllabus subject to strict rules. For art education, despite the mode characterized by planned economy making a great contribution to maintenance of stability and fundamental functions in a certain period of time, it is obvious the mode has hindered the development of art and design education by analyzing the current situation. China has a large population, vast territory and abundant resources but the labor force lacks optimal allocation, thus arousing a serious imbalance on the speed of economic development [3]. Every year, plenty of talents majoring in art and design complete their studies, such as graduates, masters and doctors who are skilled in relevant professional theory. In spite of this, these talents cannot apply the knowledge to practice, which is incapable of solving the problem of manufacturing in China—lacking the capability of independently intellectual innovation. The pressure will be largely relieved in China, including intensive labor, low productivity and lack of innovation, if there are some designers of car, dress and landscape with practical ability. In fact, however, these designers cannot transform their knowledge into productivity of national economy because they have never been exposed to the new educational concept of art and design in terms of ideology, without gaining reasonable and practical skills at the stage of professional learning. In addition, they are unable to devote themselves to art following the willing. The significance of art and design education cannot be taken on in society, whose education effect is disappointing.

3 DEVELOPMENT AND REFORM ON CONCEPT OF ART AND DESIGN EDUCATION

The artistic design, an inspiration-oriented learning, will become meaningless once it is divorced from real life. Educational concept as a thinking mode and value system plays an important role in this inspiration-filled learning. With the change of times, people have continuously changed their life and living environment, resulting in the emergence of art and design education characterized by people-oriented thoughts and being more responsive to social demands. The sublimation in the ideology lays a foundation for reform and practice of art and design education, with vivid demonstration of development in human's thoughts and wisdom. The new concept of art and design education has gradually experienced improvement by educators, thus bringing life into a new look under subtle influence of its significance.

3.1 Concept of confidence-building education

It is generally believed that art and design has been regarded as a major of being flashy without substance. Some students are compelled to choose art and design (enrolling students according to artistic score plus professional score), subject to low professional score. However, multiple students of this major and parents have mixed feelings about the employment situation when they graduated from university. Even "it is in vain to study art very hard for several years" some students helplessly and disappointedly said. In fact, there are great chances of success for these students according to requirements for interdisciplinary talents in the social economy. Artistic accomplishment can make people in a position of more advantages and potential value like a lot of Nobel Prize laureates who are also painters or musicians.

For students of this major, a confidence-building concept lays a foundation for realization of their self-recognition. Only in this way can reasonable self-evaluation and direction for hunting a job be gained at the present stage, contributing to the implementation and success of systematic education methods among these students. In fact, the purpose of the confidence-building concept is to cultivate their positive values and view of life. They can rationally set the goals of short and long term only when gaining the correct self-evaluation. In addition, teaching will go well if they regain the learning motivation automatically. Their positive attitude and passion for learning make these students in good condition where they can design some more creative and much better artistic works following the inspiration. Actually confidence-building education as a spur to students' enthusiasm for learning bridges a virtuous cycle between artistic design studying and education activities while both students and teachers benefit from the new teaching concept. The concept characterized by wide application and obvious functionality is also true of other teaching fields, with the accurate demonstration of requirements in popular ideas keeping pace with times. Even if modesty as some traditional virtue has been inherited in China for several thousand years, it is found that self-confidence and modesty are not contradictory in modern society. With regard to individual development, appropriate control of the two aspects provides a prerequisite for realization of their dreams. In addition to this, they usually tend to maximize their marginal utility in the social division of labor, thus playing a central role in solving problems and accomplishing tasks. It is necessary for the talents majoring in art and design to be equipped with excellent psychological quality and iron will because of characteristics of this major, great changeability and subjective judgment. Therefore this point for these students is with great practical significance when they hunt a job experiencing social test. In addition, they can gain great psychological support influenced by the confidence-building concept while the teaching mode of artistic design field can take on a vigor-filled new look.

3.2 Development of Bauhaus idea

The Bauhaus idea put forward by Walter Gropius has an important impact on thought and behavior of designers in various design-related fields, with the demonstration of Germany's modern design. The idea has experienced three main stages, including Gropius's idealism, Meyer's communism and Mis's pragmatism, thus intertwining with the idealistic Utopian spirit of intellectuals and political pursuit of communists as well as preciseness and practicality in architectural design. Bauhaus as an important microcosm in the development of design concept is rich in connotation.

With the Bauhaus idea applied to practice, a new annotation with times characteristics to its significance has been given in modern education fields. The idea requires a perfect unification of art and design, especially learning and education in the major of art design. Given its characteristics of this major, it is necessary for educators to add the idea to their teaching where they should not only adopt the existing knowledge of artistic design, but also achieve design innovation with technology. The continuous emergence of various science and technology is conducive to the development of artistic design. In addition to this, all kinds of seminars and exchange meetings can provide a platform for an art-design new combination. For example, the *2014 Contemporary Art and College Education of Fine Arts Forum* was held in the library, Nanjing Arts Institute. In this forum, the method of driving the innovation of artistic design was discussed while some problems in the practical teaching of artistic education were analyzed. What's more, there was a summary and assessment about their practice of educators of artistic design major, thus enriching content of teaching in the technological level. New concepts were proposed in order to make art and design keep pace with the times here. Consequently it is a good way to explore teaching and education by such topic-targeted exchange participated universally. The core of Bauhaus idea has been a transformation from product-oriented design to that of people-oriented with the good demonstration of its development. The transformation is some reflection of its practical significance—art should meet the needs of society, thus creating economic and social value. The aesthetic essence of art originates from life while serving life, in other words, returning to the source. It is the only way to realize the sustainable development of art.

3.3 *Requirements in art and design, closer to life*

The key of sustainable development of art is closer to life according to the further extension of the Bauhaus idea, which is true of art and design education. At the beginning of art and design education, the needs of real life should be paid much more attention to, thus laying a foundation for improving sensitivity of demand and art as well as drawing inspiration from life. At present, the art and design education is confined to some specific subjects—interior design, environmental design, landscape design, architecture design, and city planning design. These subjects set up are very professional and typically targeted so as to largely meet the needs of life and production. However, whether knowledge and skills can be used for improvement of life depends on the mode of concept conveying and teaching by educators.

For example, some seniors from the faculty of art and design in some university, Guangzhou, showed their graduation works in the exhibition with the theme—design, harmony, personality and life. Their life-oriented works was consistent with the needs of daily life, with the demonstration of their education achievements for four years. In addition, these design works attracted outside enterprises, other students and teachers in this university, thus showing the public these works closer to life, while offering some convenient product models with a sense of beauty and design. Students are supposed to draw from life in order to design some art works closer to life, which is also true of educators. As teachers of art and design, it is necessary to possess keen observation ability, higher-level application ability and innovation ability for conveying of correct teaching concept in class. In the practical teaching, educators not only can explain some design works at hand to students, but also ask them questions by observing some necessities and details of daily life, while guiding and inspiring them. In fact, it is essential for the scientifically-macro system of educational concept to use this convenient and easy method of drawing from life. Only in this way can the embarrassing situation of artistic design be ameliorated, making connection with other fields, seeming to be no relation with artistic design, such as manufacturing and cultural industry. Thus visible economic benefits or strong cultural repercussions can be gained. This can provide a great counterforce for art and design, in order to achieve the sustainable development of art as an independent subject in the traditional view.

4 CONCLUSIONS

It is a great challenge for educators' teaching of art and design because this education is faced with double pressures and restrictions from development needs of art and education. Thus the enrichment and innovation of educational concept (art and design) plays a leading role in overcoming those difficulties. With the development and practical exploration of the educational concept, it needs people to keep students' psychological construction, distillation of traditional concept and social reality balanced, according to requirements of modern times-characterized social life. Only when the educational concept forms abstract instruction keeping pace with times correctly and effectively can the substantive leap in practice of art and design education be achieved.

REFERENCES

[1] Deng Han, *Sensory Elements of the Products Design*. Hunan University. 2012 (12):23–34
[2] Ha Liyang, *Thoughts on Higher Education*. Proceedings of the seventeenth academic conference held by China Hospital Association of Medical Record Management Professional Committee. 2010:34–38
[3] Zhang Zhijun, *Designers Demanded to Read, Learn and Study More*. China Construction News. 2011 (2):78–80

Points of the Party's mass line education and purity construction

Bin Du
Sichuan Agricultural University, Ya'an, Sichuan

ABSTRACT: Mainly taking the Party's mass line education and purity construction as research emphasis, the work analyzes their features by combining theory with practice. It can be concluded that points of the Party's mass line education and purity construction should be expanded as much as possible in the new period, so as to better highlight the positive connection between the above two and fundamentally reflect the emphasis and difficulty of them in the modern social environment.

Keywords: mass line; purity; construction; point

1 INTRODUCTION

Essentially, there is a positive connection between the Party's mass line education and purity construction. The two are of interdependence and inseparability. On the one hand, mass line education can effectively improve purity construction. On the other hand, purity construction can lay a foundation for mass line education. In the modern social environment, the Party has fully reflected its purity and carried out mass line education at the same time, which fundamentally led to its development, as well as better fulfillment of subjective initiative, improvement of sustainable development and the reflection of great power from masses. Focusing on the Party's mass line education and purity construction, the work briefly analyzes specific issues.

2 CONNOTATION OF THE PARTY'S MASS LINE EDUCATION AND ITS PURITY

2.1 *Purity of the Party*

The Party's purity refers to that all party members should be consistent with its nature in terms of thought and behavior. That is, the unity and liberation of all Party members' thought and behavior are guaranteed by various ways. With the continuous social development, the Party's purity should embody many aspects, such as life, work, as well as economic construction [1]. From the perspective of social development, relying on the joint effort of the Party members, the Party's purity construction adapts to social and economic development, and fulfills each member's subjective initiative, which ensures its leading role in social development. Meanwhile, in China's increasingly integrating into international economic environment, the Party member should absorb the advantages of international factors and keep clear of factors that are not conducive to the purity construction in order to conduct a better and comprehensive purity construction.

2.2 *Mass line of the Party*

The Party's mass line refers to "doing everything for the masses", carrying out the principle of "from the masses, to the masses". With continuous development of the Party, its mass line becomes more obvious [2]. As the fundamental working line for the Communist Party of China (CPC), mass line can reflect the Party's thought. Founded by the Chinese Communists in the long course of revolution and practice, the Party's mass line, reflecting the Party's fundamental purpose and basic nature, demands that the Party should trust the masses, rely on the masses and solve their problems by themselves in all work. Fundamentally, the mass line, a positive and powerful mean supported by the Party, plays an important role in the CPC since its foundation, better coordinates the relationship between the Party and the masses, improves party-masses relationship and builds a platform for the Party's development and people power. With continuous development of market economy, it is necessary that the Party's mass line and its education are further implemented, thus essentially emancipating the mind and serving the masses, and fulfilling the masses' subjective initiative in order to pull our weight on the world stage [3].

3 RELATIONSHIP BETWEEN THE PARTY'S MASS LINE EDUCATION AND PURITY CONSTRUCTION

The Party's mass line features keeping pace with the times. Essentially, as an important carrier of the Party's purity construction, mass line plays a proper role in a larger range. In the market economy environment, mass line education can effectively improve purity construction. Purity construction can lay a foundation for mass line education. The relationship between mass line education and purity construction is of inseparability in the modern social environment.

3.1 Mass line as basis and demand

Dated back to 1929, mass line, first appearing in a letter from the Central Committee of the CPC to the Fourth Frontline Committee of Red Army, was proposed according to Marxism Materialism at that time. The important connotation reflected in mass line emphasized service and dependency on the masses, which laid a theoretical foundation for the Party's purity construction and becomes the theoretical requirements and guidelines for purity construction. From the perspective of materialism, existence of any political party must rely on the masses because they are the creators of history and of unlimited power. As a result, the Party's purity construction relies on supports from the masses, as well as the theoretical basis and practical requirement of mass line. In mass line education, essential requirements and nature of mass line firmly demand each CPC member that only if the Party relies on the masses and always adheres to the attitude and mission of serving the people can the Party development be truly improved. As the most fundamental requirement of purity construction, mass line education makes the Party members aware of the power of the people and regulates their specific behavior and work.

3.2 Performance of purity construction

Purity construction is divided into ideology purity, organization purity, working style purity and probity construction. In purity construction, the CPC with timeless self-warning ensures its pioneering role and reflects the fundamental aim of serving the people according to the situation and social economic development in China. It can clearly be seen that the CPC always put the people's interests in the first place in purity construction. Purity construction is undoubtedly a practice of the mass line and reflects the importance of mass line by all means of purity construction. Meanwhile, in order to solve some inevitable problems occurred during the purity construction, mass line is needed to provide the theoretical basis for purity construction. The final result of purity construction is to ensure each party member can truly serve the people. Therefore, the Party's purity construction is the reflection and result of mass line.

3.3 Common development of the two

The party's mass line education and purity construction are of inseparability. The latter relies on the theoretical support of the former, while theoretical support of the former also depends on asylum of the latter. Only if the Party's purity construction combines mass line can the Party's integrity and pioneering role be guaranteed, as well as the people's principal position in the country and fundamental idea of "everything from the masses, to the masses, serving the people whole-heartedly".

4 SPECIFIC MEASURES ON THE PARTY'S MASS LINE EDUCATION AND PURITY CONSTRUCTION

The key factor—the people—is inevitable whether to carry out mass line education or purity construction. As a result, subjective position of the masses, as well as the fundamental idea of serving the people and basic working thought of "everything from the masses, to the masses", is required to carry out a better mass line education and purity construction. A rational construction based on the actual requirement of modern social development will eventually lead to modernization of mass line education and purity construction.

4.1 Optimization of mass line education

The Party's mass line is the basic line for its development. Mass line education lays a theoretical foundation for its development in the Socialist social environment. To enrich mass line and fulfill its leading role, the essence and practical significance of mass line construction should be reflected in mass line education.

Firstly, to enhance the cognitive level of mass line, the party's mass line education should be based on ideological education. In mass line education, the Party's ideological education should be firmly carried out to fully help party members in understanding the new style and development demand of times. As a result, mass line education can be implemented along with times and social development.

Secondly, comprehensiveness should be embodied in the Party's mass line education. In the first place, the Party's mass line education features service. In modern society, as an important standard of the Party's mass line, service-oriented society demands the full reflection of party members' service and establishment of service awareness. Then, the Party's purity construction should be fully reflected and combination between mass line education and purity construction should be further improved. Finally, the vital interests of the masses should be fully respected in mass line education.

Finally, the Party's mass line education should fully respect combination of theory and practice. It is not only a kind of theoretical education, but also a practice for party members. Theory combines practice to fulfill theorization and practicality of mass line education. On the one hand, practice is used for testing the theory of mass line education. On the other hand, theory is used to strengthen the practice of mass line. In order to better fulfill the role of mass line education, theory and practice should be fully implemented.

4.2 Strengthening of the Party's purity construction

For the CPC, strengthening of the Party's purity construction is of importance. Purity construction demands timeless self-warning and serving the people wholeheartedly. In modern society, characteristics of the times and requirement of the masses are needed to be a true purity construction.

Firstly, ideological emancipation should be reflected in purity construction. As an important precondition of the Party's ideological line, ideological emancipation needs to continuously integrate into modern society so as to conduct true ideological emancipation. Only if erroneous ideas and concepts have been emancipated can the innovation and development of ideas and serving the people be achieved. With social development, in order to improve purity construction, the Party should follow the idea of ideological emancipation, and take it as the precondition.

Secondly, the democracy in the CPC should be consolidated. As an emancipation and development for a party member, the importance of inner-party democracy should be strengthened. In modern society, the Party's purity construction needs the involvement of each party member. Only if inner-party democracy is achieved in the CPC can each party member's enthusiastic participation in the Party's purity construction be mobilized. In other words, only if the individual behavior of each party member is fully respected and the accuracy and importance of individual behavior are guaranteed can the Party's purity construction be truly implemented.

Thirdly, the Party's purity construction should positively bring responsibility of mutual monitor between party cadres and ordinary party members into play. The Party members, as well as other ordinary people with their own way of thinking, are also vulnerable to the erosion of sugar-coated bullets. In addition, the party members being in an unstable environment, which can easily lead to deviation from their thoughts, go against the Party's purity construction. Therefore, in the party's purity construction, supervising and being supervised should be implemented so as to fulfill the obligation of each party member and broaden supervision. What's more, close ties with the masses, emphasis on active and important role of the Party members and deep implementation of the Party's pioneering role are also required.

Finally, the Party's purity construction includes ideology construction, politics construction, organization construction and working style construction, which are of inseparability and should be followed in purity construction. From the perspective of ideology, only if ideology construction is continuously persisted can the Party's purity construction be guaranteed. At the same time, practical implement of politics, organization and working style construction provides a good environment for purity construction and positively reflects the advantage of the Party, thus making the Party's purity construction be achieved.

5 CONCLUSIONS

In a word, according to specific measures on the Party's purity construction and mass line education, the Party's mass line education should be further implemented and improved so as to strengthen concept, relationship between the Party and the masses, and commonly maintain the Party's purity construction. Moreover, better reflection of superiority and development of the Party, development of the masses driven by the Party development and the Party development maintained by the masses will finally lead to comprehensive progress in China's politics, economy and culture.

REFERENCES

[1] Zhao Qun, Sun Haitao, *Mass Line Education and Purity Construction of the CPC*. Journal for Party and Administrative Cadres. 2013(12):36–38
[2] Ju Jiqing, Shi Gongpeng, *Multidimensional Analysis of the Party's Mass Line Education Practices*. Hubei Social Sciences. 2013(11):17–19
[3] Han Qiang, *Maintenance of the Party's Purity Needed to Deepening Understanding of the Five Major Relationships*. Theoretical Study. 2013(1):121–125

Dynamical and biomechanical model of side kick in free combat

Shugen Zhang
Chinese Martial Arts School, Wuhan Institute of Physical Education, Wuhan, Hubei, China

ABSTRACT: As an important event of sports, free combat is one of the fortes in China. Side kick is an important way to attack and score in free combat, largely affecting the athletes' achievement. The work mainly studied the dynamical and biomechanical model of side kick in free combat. Based on the importance of side kick, an in-depth research has been done on its dynamical and biomechanical principle and performance. It was found that the application of these theories to practice can highly improve the training effect.

Keywords: free combat; side kick; dynamics; biomechanical; model

1 INTRODUCTION

In free combat, attacks are mainly done by hands, legs and body. Among them, the hand is the most agile part but weak in power, while the body is just on the contrary. Only the legs are balanced in power and agility. Thus utilizing legs is very important in actual combat. As the most powerful move for legs, side kick is usually used as a nivose. Being hit by side kick, the opponent's body will be highly influenced, which lowers his chance to win the game. With the internationalization of China, sports events with other countries are increasing [1]. The centuries-old history of China gains the popularity of Chinese Kungfu around the world. Benefiting from this, Chinese players did well in free combat. After studying the match of free combat, it is found that side kick plays an important role in it.

2 IMPORTANCE OF SIDE KICK IN FREE COMBAT

2.1 *Influencing factors of side kick*

As a traditional strong sport, Chinese athletes perform well in free combat of all kinds of sporting events. According to a survey, such achievements highly depend on utilizing side kick. As side kick is bending and extending legs sideways, huge power can be released by this move. With its power and agility, side kick is favored by many free combat players. Intensive training aimed at side kick are done during daily workouts to improve the speed and strength. In order to optimize the training effect, some targeted moves are introduced by the coach to improve players' speed of bending and extending legs [2]. However, according to the survey, limited by China's economy, sports training like free combat are in lack of modernized equipment. Only some traditional methods are used to strengthen players' physical quality, hindering China's development of sports to a certain extent. As for free combat, which requires less for equipment, a satisfying training effect can still be achieved by some basic moves without any help from the equipment. Therefore, sports like free combat are very suitable for Chinese players. For example, influencing factors of side kick such as leg strength and turning speed can be improved by repeating practicing one single move.

2.2 *Application of side kick in free combat*

Because of its strength and speed, side kick is popularly used in free combat. Compared with other attack modes, side kick has a longer attack range. Because the leg is the longest part in the human body, attacking by extending the leg can reach very far. Moreover, side kick has various attack points. The attack point is unpredictable even if the opponent predicts the side kick. As a frontal attack, side kick combines both offense and defense, making it a method to attack and defend at the same time. Different from other defensive methods, side kick can beat back the opponent's attack. It is common to see two players using side kick at the same time in a free combat match. That's because they both choose side kick for offense and defense. Side kick is a simple move with a lot of variations that can be used in various situations [3]. In free combat, in order to beat down the opponent, the move has to be fast, accurate and hard, which can be well achieved by side kick. Specific analysis shows that side kick is widely used not only in free combat, but also some other fighting matches because of its power and speed.

3 DYNAMICAL ANALYSIS ON SIDE KICK

3.1 Dynamical principle of side kick

In dynamics, the power and speed of side kick come from the strength of legs. The whole body leans forward with crotch as the pivot. Side kick has a lot in common with axe kick to a certain extent. But focusing all the strength in one point makes side kick more powerful than axe kick. And side kick is also very agile because the pointed attack can aim at any point on the opponent. Therefore, side kick is widely used in free combat. Players on both sides will use side kick frequently because one hit on frontal area will influence the opponent a lot, which indicates the importance of side kick in free combat. In order to improve the effect of side kick, all kinds of method will be taken to strengthen players' strength and agility in daily training. With the development of technology in recent years, dynamics and biomechanics are introduced into practical training, with significant achievement on the effect. Influenced by special historical factors, China's economy and technology started late. The application of modern science falls far behind some developed western countries. As a country with the highest population, the Chinese team performed poorly in early sports competition. In order to change this situation, China referenced some advanced training experience accordingly from the western countries. Some scientific training methods are introduced, achieving satisfying effects.

3.2 Dynamical manifestation of side kick

As a frontal attack, the force of side kick can be calculated by Newton's second law F=MA, where F is the force, M is the application of force, usually the person's weight, which is unchangeable, and A is the acceleration. According to this equation, in order to maximize the force of side kick, it is better for the player to be heavier. However, there is a trade-off between player's weight and agility, and many fighting games are classified according to weight. Thus improving the power of side kick can only be achieved by increasing acceleration, which is related to leg muscles. And the strength of leg muscles mainly depends on training. Therefore, some targeted moves can be introduced in daily training to increase the explosive force of the player's muscle, which will highly improve the power of side kick. In dynamics, as the player stands on the ground, obviously the M of side kick cannot simply be counted by the player's whole weight. Instead, the gravity point is considered. If the player leans his body, his gravity point moves. In order to increase the force of side kick, the gravity point should be controlled by leaning body and swinging arms and legs to increase M in the equation. Therefore, the effect of side kick is related to multiple parts of the body.

4 BIOMECHANICAL ANALYSIS ON SIDE KICK

4.1 Biomechanical principle of side kick

With the development of biology in recent years, people's understanding of life comes to a new level, bringing out biomechanics. Limited by the developing time, in biomechanics, only specific movement can be studied. No general rules can be proposed to suit everything because, influenced by gene and environment, every creature is unique with its own characteristics. In sports training such as free combat, players are the subjects. Regularities and characteristics can be found by studying their moves in biomechanical methods. Targeted training plans can be proposed according to this study to improve the training effect, which is of significant importance to the players. As an important move in free combat, side kick is the key to scoring or even knocking down the opponent. Chances of winning can be highly increased if the player can master the skill of side kick. The biomechanical principle of side kick is the combination and match-up of all parts of the body. For humans, a specific movement requires coordination of the whole body. Any changes of a single part will influence a lot on the movement. Therefore the movement of legs is very important for side kick. The power of it can only be maximized by coordinating the body and legs.

4.2 Biomechanical manifestation of side kick

Side kick seems to be simple, but it is actually very complicated considering all parts of the body. In order to accomplish this move, the player has to lean his upper body backward to hold back his gravity point, bending one of his legs up to the height of his crotch, and then he only needs to extend the leg and kick forward with his tiptoe valgus to finish this move. This is the full biomechanical manifestation of side kick. During attacking, the side kick has to be adjusted according to the opponent's situation, because different forces and angles of any part will result in different kicking effects, although they all follow the same biomechanical principle. Therefore in free combat, variation should be made in every side kick. If the move of side kick is fixed, the opponent will easily predict it and give an effective counterattack. Moreover, as the player has to stand on one leg during side kick, staying balanced is very important. Some targeted moves should be introduced to improve players' balancing ability in daily training. In biology, human's balance mainly depends on vision, sense and the central nervous in cerebral cortex.

5 APPLICATIONS OF BIOMECHANICS TO SIDE KICK TRAINING

5.1 Optimization of side kick based on biomechanics

Side kick has to follow biomechanical principles to maximize its power, despite individual differences and different abilities of every part. In recent years, large-scale sports events are very frequent, which is a good chance for every country in peacetime to show their comprehensive strength in the competition. Especially in the worldwide Olympic Games, ranking gold medals in the first place indicates a strong national strength. As an important competition event in many sports matches, players need a lot of training if they want to perform well in free combat. Scientific training can be very helpful, and training with the application of biomechanical theories is the best one. In traditional training, only some basic moves are introduced to improve the reaction of players' muscle, which is effective but harmful to the players' bodies in the long term. However, this problem can be avoided in training based on biomechanics. The coach will make a training plan according to the players' situation. Decomposition training, auxiliary training and full training are introduced to help the players to understand every detail of side kick, and coordinate all parts of the body to optimize the effect of this attack.

5.2 Targeted training

Considering the individual differences, the biomechanical manifestation, for example the strength and speed, of acting the same move by players will change accordingly. Therefore instead of general training, a targeted training plan has to be made according to the players' situation. The biomechanical analysis on side kick is mainly for the general principle, while the actual move of side kick differs from player to player. Therefore targeted training should be introduced based on the individual differences. For example, an intensive training is suggested for the player weak in leg muscle, and the player slow in reaction requires training on balancing ability and the cerebral cortex. This kind of training can strengthen physical functions to increase the power of side kick and the chance of winning the game. For an experienced coach, if he has trained a certain player for a long enough time, he will be able to make training plans according to the physical and mental condition of the player to improve training effect. Thus a stable relationship between coach and player is very important.

6 CONCLUSIONS

According to the discussion in this work, because of the power and agility of side kick, it is very popular in free combat. In every free combat match, side kick will be frequently used by both sides of the players, indicating its importance. During daily training, intensive training on side kick is introduced according to the players' situation. Through massive researches and practices, it is found that applying dynamics and biomechanics to side kick training can highly improve the training effect.

ACKNOWLEDGEMENT

The work was supported by the project "Characteristics and Training Countermeasures of the Application of Free Combat Strategy under the New Rules" of Martial Art Research Institution of General Administration of Sport under Grant No. WSH2012D12.

REFERENCES

[1] Li Xielong, Liu Fushun. Biomechanical Analysis on the Side Kick Strategy of Chinese Free Combat Player. Zhejiang Sport Science, 2009(03):104–107.
[2] Liu Nan. Discussion on the Strategy of Side Kick in Free Combat. Science & Technology Information (Academic Research), 2006(12):431–432.
[3] Zhou Tianfen, Liu Fushun. Biomechanical Principles and Training Guidance of Outstanding Free Combat Players' Side Kick Strategy. Wushu Science, 2007(02):45–46.

Application of engineering technology in chemical production

Xingrong Jiang
Sichuan Technology & Business College, Dujiangyan, Sichuan, China

ABSTRACT: Chemical engineering technology has been widely used in practical production. On basis of concept and feature of chemical engineering technology, and its application in chemical production was studied in this work, combined with the influence factors and problems that occur in China's chemical engineering technology application. Results show that only by using high level technicians can advanced devices and reasonable catalysts improve application of chemical engineering technology.

Keywords: chemical engineering, chemical production, science and technology

1 INTRODUCTION

With the development of modern natural science and technology, industry production efficiency has been improved greatly. Meantime, the appearance of new science and technologies such as chemical engineering has enriched the industry production. On the above basis, western countries first entered into the industry era after two industry revolutions. Lots of new materials that occurred during chemical engineering application have promoted industry to a great extent. Affected by special historical factors, China's science and economy started late, so there is still some distance comparing with some developed western countries [1]. China has become the second largest economic unit after more than thirty years of reform and opening. To keep the sustainability of economic growth, great attention has been paid to the development of new technologies such as chemical engineering, a great deal of human and material resources are given to research such new technology, and good effect has been obtained.

2 BRIEF ANALYSIS ON CHEMICAL ENGINEERING

2.1 *Concept of chemical engineering*

Chemical engineering is a new technology, which gradually formed with the development of modern natural scientific technology. Chemical engineering has its important application during practical industry production. Corresponding courses are set in many universities. A large number of chemical engineering talents are educated to meet the requirement of industry production. Because wide areas are involved in chemical production, chemical engineering can be used in many product manufacturing. Chemical engineering is to produce by using chemical reaction or chemical features of material. Besides better physic features, chemical parameters are also important [2]. Product quality can be greatly improved if material chemical features are used fully. In practical production, applying level of chemical engineering is already high through many years' practice. In contrast, China's chemical engineering is lower, some advanced devices and technologies are usually imported from overseas in practical chemical production, thus affecting China's development of chemical engineering to a large extent.

2.2 *Features of chemical engineering*

Comparing with other technology, chemical engineering has its own features. Firstly on the aspect of new material R&D, new material can be produced using chemical features and reaction of materials. In the beginning of industry development, R&D of materials is very important. Especially the development of industry areas requires higher feature level to traditional materials. Under this situation, better materials need to be developed. Fewer new materials are developed after many years' R&D, nowadays, the application of chemical engineering is focusing on material purification and improving material feature by changing material composition. For example, flexibility and strength of steel can be changed by adding different microelements to meet different requirements [3]. Coal and oil, as a traditional energy sources, their utilization efficiency are important to modern industry. Coal and oil are known as non-renewable resources. With the development of industry level, more energy sources are consumed, thus causing energy crisis if using them without thinking of results. One of the biggest effects of chemical engineering is the purification technology of oil and coal. In recent years, with many large petrol chemical centers built, it can be seen that chemical engineering plays an important role in energy usage.

2.3 *Progress of chemical engineering*

Chemical is a long historical subject. Human has made study on chemical reaction phenomenon and used some simple chemical technology to produce ever since ancient time. Nevertheless, there was no such concept of element and combination limited by technical level at that time, only rough study on reaction phenomenon. With the development of modern natural science, especial the occurrence of element periodic table has promoted progress of chemical engineering to a large extent. Simple substance can be purified to a percentage of 99.99% from compound and mixture. All objects in nature can be analyzed by chemical engineering. Chemical features can be obtained through analyzing detailed composition in practical industry production. According to the requirements of practical use, material performance can be improved by using some chemical reactions, and this is very meaningful to industry production. The present chemical engineering is already very high through many years' progress and upgrading and chemical engineering can be divided into many subjects according to practical requirements. Corresponding courses are set in many universities, and a large number of chemical engineering personnel are educated. However, limited by China's education level, educated personnel have lower professional quality, and it is hard for them to complete their job after graduated.

3 PRESENT SITUATION OF APPLICATION OF CHEMICAL ENGINEERING

3.1 *Factors that affecting the application of chemical engineering*

Considering the importance of chemical engineering, every country pays more attention to progress of chemical industry. Because of the late start in economy and technology in China, there is a quite a distance of technical level compared with some developed western countries. For shortening such distance, a large number of human and material resources are invested to the study of chemical engineering. Through lots of practical researches and studies, it is found that the biggest affecting factor in the practical application is the professional quality of technical personnel. The technical level mainly depends on it. They are all controlled by personnel whether operation of devices or design of chemical reaction. Knowledge of chemical engineering cannot be mastered if personnel quality is low, if they are lacking understanding of corresponding devices, then chemical engineering cannot plays a role. For example, if the purification concentration requires to reach 95%, it cannot be satisfied if personnel quality is low. Besides the factor of personnel, device advantage and material can also affect the application of chemical engineering to a certain extent. Especially with the development of electronic information technology, intellectual chips are embodied into traditional mechanical devices. Certain automation can be fulfilled for devices. At the present, automation situation can be set after corresponding materials are added into chemical engineering device, thus greatly improving the application of chemical engineering.

3.2 *Problems in application of chemical engineering*

At present, it is found that chemical engineering has wide application in China, especially for improving utilization efficiency of an energy source like petroleum. Many petro-chemical centers are set in China. By-products like asphalt can be acquired in the process of petroleum purification, thus showing the importance of chemical engineering. China has accumulated lots of experience. While compared with the mature application mode of western developed countries, there are still some problems in the application of chemical engineering. The first problem is the low personnel quality, and this issue cannot be solved in a short time limited by the education level. To solve this problem well, some enterprises hired high quality personnel from foreign countries. Nevertheless, foreign technical personnel usually need some time to get used to because of difference of device advantage and ambient, and many of them also cannot complete corresponding jobs after quite a while. So, the issue of personnel quality is a severe problem in the application of China's chemical engineering. The advantage of used device is low in China limited by industry level, especially limited by the cost, many companies choose lower devices. These devices can meet the practical application requirement, but certain time need to wait, causing low efficiency of chemical engineering application. This is also a common phenomenon in China's chemical engineering application.

4 MEASURES FOR APPLICATION OF CHEMICAL ENGINEERING IN CHEMICAL PRODUCTION

4.1 *Improving technical quality of personnel*

Considering the importance of technical personnel to chemical engineering, professional quality of personnel should be improved for better utilization of chemical engineering in the practical production. But it is hard to get changed in a very short time limited by China's education level and professional quality of students. In recent years, China is under education revolution, hoping to replace traditional education with quality education to improve the practical ability of students. Anyway, it is found that current quality education does not go well, and a better examination method has not been found yet. It can be seen that the assurance of personnel quality in application of chemical engineering mainly relies on enterprise itself to a certain extent. For instance, attention is paid to personnel training in practical work. Try to hire higher quality personnel with abundance working experiences if cost permitted, and corresponding devices can be better operated and current work mode can be upgraded. Because of the rapid developed modern science and technology, continuous study is needed to assure the quality of personnel themselves. However, many people thought they did not need to learn anymore after they graduated from school, and this thought affects China's application of chemical engineering to a large extent. To solve such problems well, daily training is necessary for technical personnel to acquire the latest knowledge of chemical engineering.

4.2 *Using advanced devices*

Because chemical engineering is a subject with a strong practicality, the using of devices is very important; only simple devices like measuring cylinder can be used in the initial developing period of chemical engineering. With the development of modern nature science, especially the emergence of electronic information technology, devices have certain automation like microscope which can observe micro reaction phenomena. Devices can automatically control reaction velocity after corresponding material are added into device, thus greatly improving chemical engineering level. Upgrading the speed of chemical devices is rapidly affected by developing speed of electronic information technology, and advanced devices are mostly controlled by foreign companies. China has to import from overseas if needed, and this causes a high using cost to a large extent. Some advanced devices are explained by English, which requires high personnel quality. Considering of cost, some Chinese companies usually bought lower devices. With the lower performance, these devices can also meet the practical requirement. However, the reaction speed of device is low and cannot control more details, which affect efficiency of chemical engineering. Only adopting advanced devices to solve above issues well, the case will be saved from other aspects.

4.3 *Using rationally of catalysts*

Application of chemical reaction is the core issue for the chemical industry. Using catalysts plays an important role in application of chemical reaction, not only improving reaction velocity, but also making sufficient reaction. Kinds of types of catalysts can be selected for chemical reactions, but each catalyst has great differences. The most reasonable catalysts have to be selected to maximise the efficiency of chemical reaction. There are many studies on catalysts through many years' development of chemical engineering, and there are a large number of corresponding data about the using situation of each catalyst. Considering that different environment can affect chemical reactions, corresponding experiences can be done before the practical reaction to assure the effect of catalysts. This is to check if the effect can be the same as expectation; choose this catalyst if so, otherwise, replace the catalyst. It can be seen that catalysts play an important role in chemical reaction; reasonable catalysts have to be selected to improve the application of chemical engineering in chemical production. Especially in some important chemical reaction, expected effect can be reached after adding catalysts, even if there is no harsh environment. Certain costs can be saved, that is why catalysts are used widely in the application of modern chemical engineering.

5 CONCLUSIONS

It is known that chemical engineering plays an important role in practical chemical production through analysis in this work. Chemical features can be changed by using chemical engineering, and also energy sources like petroleum can be refined. The use of chemical engineering gets involved in almost modern industry production. The level of chemical engineering can be the important symbol of judging the industry. There is still great distance compared with western developed countries; to solve this problem well, advanced performance of devices have to be confirmed, while improving professional quality of personnel, also, reasonable catalysts have to be selected.

REFERENCES

[1] Shi Qingbin, Zhao Qiwen. *the Present Situation and Developing Trend of Qinghai Fertilizer Industry,* Journey of Qinghai University(Nature Science) 2006(02): 34–37
[2] Sun Hongwei, *Developing Trend of Chemical Engineering——Focus on Chemical Engineering—— Understanding Space – time Multi – scale Structure,* Chemical Industry and Engineering Progress, 2003(03): 224–227
[3] Liu Zheng, Jin Yong, Wei Fei, Li Yourun, Luo Guangsheng, Yuan Naiju. *Recent Development of Chemical Engineering Science,* Chemical Industry and Engineering Progress, 2002(02): 97–91

Study on ideological and political education evaluation system of universities

Yuping Qian
Jiangsu University, Zhenjiang, Jiangsu, China

ABSTRACT: In recent years, university students are facing unprecedented employment pressure, and the attention-degree of the whole community on ideological and political quality of students is rising. Therefore, the ideological and political education evaluation system in each university is facing enormous challenges, and it is imperative to strengthen the evaluation system. This work analyzed the current situation of evaluation system in colleges, as well as the shortage and causes of it. And the work also made some corresponding suggestions for the improvement. It is known that university plays a very important role in the ideological and political education of students. Therefore, each university should seriously treat its shortage, thus improving the whole system appropriately according to its own characteristics.

Keywords: University; Ideological and political education evaluation system; Current situation; Improvement program

1 INTRODUCTION

In the 21st century, the rapid development of global economy has driven social progress and improved people's living standards and quality. Meantime, it also brought unprecedented pressure to contemporary university students. In recent years, in order to adapt to the rapid development of the economy, the enrollment expansion of college students has never ceased. With the shortage of employment opportunities, more and more college students will undoubtedly lead to a competitive job market. However, with the development of society, the requirements of modern enterprises for university graduates also increase [1]. The requirements not only limit to the professional knowledge and ability, but also include the personal character and thinking characteristics of college students. With the rising attention-degree on ideological, great importance has been attached to the ideological and political education in universities to help graduates find satisfactory jobs.

2 IMPORTANCE OF IDEOLOGICAL AND POLITICAL EDUCATION EVALUATION SYSTEM CONSTRUCTION

2.1 *Beneficial to Chinese characteristic socialist construction*

From a macro perspective, university's focus on the development of ideological and political education evaluation system is beneficial to Chinese characteristic socialist construction. Firstly, the main content of Chinese characteristic socialist construction includes two points: on the political guarantee of socialist construction, sticking to the socialist road, the People's democratic dictatorship, the leadership of the CPC, Marxism-Leninism and Mao Zedong Though are emphasized. These four principles, the foundation of China, can ensure the healthy development of the reform and modernization, which possess new era contents. On the road of socialist development, taking the Chinese own road is emphasized, without taking books as dogma and copying foreign models [2]. Taking Marxism as a guide and practice as the sole criterion for testing truth, emancipating the mind, seeking truth from facts, then the Chinese characteristic socialist will be built. Both of the two points emphasize the importance of guiding ideology in Chinese characteristic socialist construction. Contemporary college students are an important human resource for the future development of China. The ideological and political education they accepted affects their behaviors, which affect the development of the whole society, so strengthening their ideological and political education is necessary.

2.2 *Beneficial to the development of socialist market economy with Chinese characteristics*

The socialist market economy with Chinese characteristics, guided by the important thought of three representatives, includes economic development and reform. With the development of socialist market economy, the requirements of enterprise for employees also increase. Instead of the professional knowledge and ability, the ideological and political quality, representing the quality of individual employees, has been paid more attention. University students are the main force of future economic development [3]. And their ideological and political level will have an important influence on the development direction of the whole socialist market economy. To improve the ideological and political level and the overall quality of university students, an important measure is to strengthen the education and evaluate the results scientifically. That can not only scientifically evaluate the evaluation system, but also help students improve their ideological quality, thus providing more comprehensive talents for the development of a socialist market.

2.3 Beneficial to the supervision of improving ideological and political education

Each university should carry out ideological and political education, which is the core activity of education. Ideological and political education can help students establish a correct ideological and political concept, thus improving the quality and level of students. But with the development of society, the ideological and political environment has undergone some changes, so the relative content should be changed correspondingly in the university. In order to examine the timeliness of the ideological and political education, the evaluation of it seems very necessary.

2.4 Beneficial to establish a correct ideological and political concept for students

Ideological and political education in university affects students' ideological and political concept. This requires the concept students received should be advanced, scientific and reasonable. Ideological and political education evaluation system can help universities find its shortage, thus improving the education. Only under such a system of ideological and political education can more excellent students, who are suitable for economic and social development, be cultivated. Therefore, the evaluation system is beneficial for the university to prepare scientific and rational content of ideological and political education, thus helping students to establish a correct, advancing concept.

3 PROBLEMS IN IDEOLOGICAL AND POLITICAL EDUCATION EVALUATION SYSTEM

3.1 Scope of the evaluation system does not keep pace with the times

With the development of society, the scope of ideological and political has been deepening and reforming, with the content increasingly rich. In this dynamic environment of ideological and political, the ideological and political education evaluation system of the university should make some corresponding adjustments without remaining unchanged. In addition, the ideological and political level of university students has a big difference with the past with the development of society. And the education they received and their living environment are also different from the past. However, many universities do not adjust the ideological and political education evaluation system with the development of society and the change in people's thinking. Following the original standard of ideological and political education in universities is not beneficial to form a correct and advanced ideological and political system of students. Moreover, it is also not beneficial to the further development of the evaluation system of the university. Therefore, what problems of ideological and political education should university students know? Which education method and system should be adopted to educate their ideology and politics? What should universities do to keep the ideological and political education evaluation system with the times? These issues of ideological and political education evaluation system are what the university should consider.

3.2 Fuzzy of the nature of ideological and political education

Literally, ideological and political education is to teach people political issues and deepen their understanding of political environment, aiming at making them be loyal to their motherland and people. However, the current situation is that the ideological and political education evaluation of many universities does not only aim at political aspects, also mixed with morality. University students receive not only simply political education, but more about the character and moral ideology knowledge. Actually, this teaching method has distorted the connotation of ideological and political education. Ideological and political education should highlight the political knowledge and patriotic content, rather than morality. If unchanged, the nature of the ideological and political education evaluation system will become blurred, then whether evaluating political or ideological education becomes confused. Such fuzzy nature and content is not beneficial to the development of the ideological and political education of the university.

3.3 Evaluation method cannot be enough scientific and keep pace with the times

In the ideological and political education evaluation system of universities, using which method to operate an objective evaluation more fairly is an important issue. That is what university evaluators should consider. With the development of society, some scientific and convenient evaluation methods are in need for the evaluation of the ideological and political education system. At present, many universities evaluate the system by referring to documents and datum. Although this traditional evaluation method has its advantage, it cannot keep pace with the times where efficiency is stressed. Moreover, inefficient and time-consuming of traditional evaluation method brings a lot of inconvenience to evaluation.

3.4 *Fuzzy of the subject and object problems in evaluation system*

In the ideological and political education evaluation system of universities, the foundation problem is that who is the evaluator and who is the evaluated object. With the development of universities, the division of the functional departments of universities becomes more obvious, but the overlapping of functions still exists in some departments. The evaluators, who conduct ideological and political education evaluation in universities, are not the members of certain evaluation department, but those with the ability universities admit. This has resulted in the confusion of identity of the main evaluation personnel, as well as some unprofessional phenomenon in evaluation. Therefore, universities should determine the evaluation personnel who can really play a role in the evaluation of the system. As for the evaluated object, some people think that the evaluation of the evaluator is to guide the evaluated object to a certain extent. And some others think evaluators are the ones who evaluate the work and thought of the evaluated object in a specific stage, and both are equal. Therefore, the present relationship between the evaluator and the evaluated object still do not have a clear position.

4 COUNTERMEASURES AGAINST PRESENCE PROBLEMS IN IDEOLOGICAL AND POLITICAL EDUCATION EVALUATION SYSTEM

4.1 *Enriching the content of ideological and political education*

Universities are the advanced force in social development. Their ideological and political education, with the development of society, should keep pace with the times, and make exploration and innovation rather than remain unchanged. And development of ideological and political education should be consistent with the thinking of university students in the new era. With the development of the times, students receive more and more new ideas and things, and universities also develop in a complex environment. In order to make the ideological and political education really work and perfect the evaluation system, universities should conduct reform and adjustment of the content. Adding new ideas and concepts of the new era into ideological and political education, then the education can keep pace with the times. In addition, while teaching theoretical knowledge to students in ideological and political education, the students' practical ability should also be emphasized. Combining the theory and practice in ideological and political education, students' practical ability and social competence will be exercised, thus providing more content for the evaluation system.

4.2 *Reforming the ideological and political education evaluation system*

While teaching new ideas and concepts in ideological and political education, the reform and innovation of ideological and political education evaluation system should also be emphasized. And the reform and innovation should focus on the content and methods of evaluation. With the development of society, the content of ideological and political education is changing. Therefore, the evaluation content and standards of evaluation system should also be constantly changed to adapt to the changing environment of ideological and political education. And the changes of content and standards in ideological and political education evaluation system will also promote the changes of evaluation methods. In ideological and political education, using traditional evaluation methods, like referring to documents and datum, no longer meets the needs of modern society. To operate ideological and political education evaluation more scientifically, efficiently and rationally, a specialized study by evaluators is needed to determine more scientific and reasonable evaluation methods. For example, to take full advantage of network technology in the evaluation, specific evaluation procedures developed developer can make it. Linking internet technology with the evaluation closely, the scientific and rational of evaluation will be improved comprehensively, as well as the efficiency of the evaluation. What's more, it can also save labor costs, which can develop the projects of other academic aspects. So, reform of education evaluation system brooks no delay.

4.3 *Differentiating morality and politics*

Fuzzy nature of ideological and political education lies is the in the overlap of political and moral education. Political and moral education has similarities and differences. Adding the element of moral education appropriately while operating political education, students can improve their ethical standards during political and moral education simultaneously. But during the ideological and political education, the emphases of evaluation should prefer political aspects to moral. Thus, the ideological level of university students will get overall supervision, which is beneficial to the development of students, as well as the whole universities and society.

4.4 *Establishing special evaluation department in universities*

Due to the fuzzy of subject and object, the function of evaluation cannot make full play, and the good effectiveness cannot be achieved. In this case, universities should establish special departments that are responsible for the evaluation of the cases of universities, thus improving the work efficiency. Therefore, it is necessary to establish a special evaluation team of ideological and political education. Specialized evaluators have more energy to study ideological and political education and learn advanced evaluation methods, so as to make the most of the function of evaluation.

5 CONCLUSIONS

Social development impels the reform of the content of ideological and political education, as well as the perfection of the evaluation system of ideological and political education. Scientific and rational evaluation system of ideological and political education of universities needs the support of society, as well as the joint efforts and cooperation of evaluators and the evaluated. Currently, the evaluation system of ideological and political education of universities is facing a huge adjustment and reform. In order to better start the reform process, to make the evaluation system play a greater role and to provide talents with sound thought and excellent character, a more broad development platform should be provided for the development of the evaluation system. Then, it can operate more scientific and reasonable evaluation on dynamic ideological and political education of university, thus making greater contributions to the development of students.

ACKNOWLEDGEMENT

The work was sponsored by the fund of university philosophy and social science fund of Jiangsu Provincial Education Department in 2013 for the project "Effectiveness research based on ideological and political education of socialist core values", grant Number: 2013SJB710010

REFERENCES

[1] Yuan Weiguo, *Characteristics and Innovation of Ideological and Political Education Evaluation*, School party building and ideological and political education, 2012 (9):38–39
[2] Dong Xuejun, *Discussion on the evaluation mechanism of ideological and political education*, Journal of Shenyang Normal University: Social Science Edition, 2011 (3):17–19
[3] Tang Huirong, *Reflection and Reconstruction of ideological and political education evaluation system*, Journal of Shanxi College for Youth Administrators, 2007 (4):54–56

Technical analysis of anti-seismic design for construction engineering

Bo Yang
Haikou College of Economics, Haikou, Hainan, China

ABSTRACT: With the development of technology of construction engineering, natural diseases have warned the construction industry that the anti-seismic design must be applied to the construction engineering to ensure the construction quality. In the work, the anti-seismic design technique is used to solve the problems of the modern construction engineering in order to realize the high quality standard of construction engineering. In addition, the corresponding measurement and suggestion are proposed based on the anti-seismic maintenance and management of construction engineering.

Keywords: Construction Engineering; Anti-seismic Design; Technique; Analysis

1 INTRODUCTION

Before the building construction, strict quality control of the building materials is important. In the process of the building construction, it is compulsive for the quality management department to strengthen the force of supervision in the whole process of construction and master the problems possibly occurred in building construction. For recent years, the construction engineering quality in China has presented certain problems especially in the process of earthquake disaster, causing some loss [1]. Consequently, it is necessary to mange in the aspect of anti-seismic design in order to ensure the construction engineering quality and improve the construction engineering system.

2 ANALYSIS OF PROBLEMS OF MODERN BUILDING CONSTRUCTION

Anti-seismic design is necessary for construction design. In order to ensure the minimum loss of construction engineering in natural disasters, design and building techniques of construction should be analyzed. However, the construction bearing capacity in recent natural disasters shows that the construction engineering quality has certain problems and hidden dangers [2]. So, it is necessary to analyze the characteristics and problems of modern construction engineering at first.

2.1 *Analysis of the characteristics of modern construction engineering*

Firstly, complexity and variety are the main characteristics of modern construction engineering. To pursue aesthetic feeling of modern construction, the coordination and integrity of interior design and the concordance with outdoor environment are considered. Besides, the types of construction and requirement of decoration style are different. Secondly, it is difficult for the building construction. Because the requirement of modern building construction is higher and higher, all kinds of new techniques and materials are developed with the development of the industry technology [3]. The technicians have to regularly advance the technical abilities and familiarity with construction materials. Manual operations increase the difficulty of building.

2.2 *Analysis of problems of modern building construction currently*

1) The constructors have a low level of professional technology. Lots of constructors in the market of modern building construction do not have professional techniques, and even some directly work without training, causing inaccurate understanding of the design drawing. Because most companies consider that practice makes perfect and new recruits can follow skilled constructors in operation. However, practice without good base will cause the increase of the substandard products and affect the building construction quality.

2) The supervisors do not make a good job. It is necessary for the supervisors to master the basic knowledge and techniques like the procedure, the quality and the proportion of investment combination of materials. However, most of the supervisors do not investigate each related procedure and make it enter the next procedure, causing many hidden dangers of quality.

3) There are many irresponsible phenomena of companies in modern building construction. At present, many modern building construction companies directly subcontract the work to subcontractors after winning the bid. Some unqualified subcontractors do not have special construction and technical management staffs and give labor contractors full power to handle the whole building construction. The irresponsible attitude will affect the construction quality. E.g. some constructors directly carry out construction without checking the design drawings, thus affecting the building structure.

2.3 Importance of building technology and management in modern construction engineering

The technical management is applied as the leading factor of each modern construction, going through the whole construction process. It is an important step to take measures like plan, organization, control of management to supervise and direct in each procedure of the construction process and ensure teamwork of types of work. So the quality of modern building construction can be ensured. Consequently, it can improve the industry prestige, decrease unnecessary cost by reasonably arranging each procedure and cross operation and increase the quality. The requirement of the new technology and update of equipments promote the technicians to strengthen the technical learning, thus ensuring that the technical level, manage ability, the materials and equipment can follow the development of the times.

3 TECHNICAL ANALYSIS OF ANTI-SEISMIC DESIGN OF BUILDING CONSTRUCTION

It is necessary to analyze the construction condition and environment in detail and then carry on the anti-seismic drawing design based on the anti-seismic design of building construction. Among them, the geological condition should be considered to ensure that the working place is not in the earthquake area. In addition, the location is important. If the active zone is selected, it will cause great influence to the construction process including the maintenance. The construction materials should be examined, and the qualified anti-seismic materials should be selected for construction, thus ensuring the stability of the construction. Otherwise, there will be lots of problems even if the construction is finished. Consequently, to ensure the rationality of anti-seismic design, the following measurements should be done:

Firstly, the location must be rationally selected. The earthquake active zone or the geologic active zone must not be selected.

Secondly, the construction materials must be qualified and examined. The supervision system can be used to deny the unqualified design.

Thirdly, in the construction process, the design drawing should be modified repeatedly to ensure reasonable construction materials, and the construction quality should be examined to ensure the qualified rate of the building construction.

4 QUALITY MANAGEMENT ANALYSIS OF ANTI-SEISMIC DESIGN OF BUILDING CONSTRUCTION

The anti-seismic design refers to the technical layer, the construction process and the maintenance, thus ensuring the construction quality.

4.1 Beforehand preparation measures of anti-seismic design of modern construction project

At first, the division of the post is important. The special personnel are arranged to be engaged in professional jobs according to different specialties and abilities of the personnel. Then the construction link is distributed to the person in charge, making all the constructors understand the content of jobs themselves in order to avoid unclear responsibility. One of the responsibilities of the supervisors is to verify the qualification of the materials. Besides, they should train the personnel in basic quality and technical level, including acquaintance of the building drawing, the construction design drawing and the key techniques of each item. E.g. to make the exterior wall show elevation effect, the types of external tile, the mixing rate of the cement and the construction process should be considered, the supervisors with serious attitude needed, and the whole construction scheme ascertained. In addition, the key quality and safe problems should be taken seriously to reduce construction cost based on better quality. At last, the purchase department should buy qualified materials according to the construction schedule ahead of time in order to avoid affecting the construction scheme.

4.2 Quality assurance of anti-seismic design of modern construction engineering

1) Each construction procedure should be strictly supervised. Firstly, although supervision is carried out by the supervisors, the examination should be done to judge and analyze the whole construction quality to control the quality between the procedures. Secondly, the weak link or key part should be examined more strictly and the selective examination should be conducted to increase the scheme on the premise of ensuring the quality at intervals. Thirdly, when the procedures are connected, the last procedure will be strictly examined by the person in charge of the next procedure and enter the next procedure after ensuring quality. Fourthly, the mandatory measurement is taken to the behavior affecting construction quality seriously, e.g. the

individuals that scrimp work and stint materials will be expelled. Fifthly, a standard for quality control should be established according to the characteristics of the procedures and the requirements of the construction and the employer. The decorators and inspectors should work according to it.

2) The construction requirement of each item should be stipulated. Each item of the construction engineering has a schedule, specific to everyday quantities, and the using time and kind of instruments and materials are calculated to ensure that the construction can be finished on time and by amount. Besides, the materials and equipments on site should be stored in concrete places, thus spurring each decorator on to regulate constructions, reducing the wasting time for finding tools and materials and improving the construction efficiency.

3) The harmless construction materials should be selected because constructions are provided for people. Nowadays many materials in construction market, advertised as environmental and pollution-free materials, have different kinds and prices. So the constructors should responsibly select harmless products, and the polluted materials should be tested and replaced to ensure the health of people if the concentration index is beyond a certain standard.

4) The techniques and quality of construction should be improved with the development of times. The traditional construction technology level cannot satisfy the requirement of times because of the occurrence of new techniques, equipment and materials, so the regular learning should be arranged to constructors. Besides, it is important to be familiar with the new materials because some quality problems can be caused by the unacquaintance with new materials.

4.3 Control measures of construction quality of anti-seismic design in building construction

In the construction process, the construction unit and the contractor often discuss the engineering problems or change the construction scheme. The change of construction should be seriously examined, because a little change in the drawing can be a large change of the engineering and cause quality problems of the engineering. Especially, each step of the construction operation of the contractor should be supervised and controlled, avoiding quality problems caused by shortening the construction period. Any requirement proposed by the construction unit should be analyzed and constructively responded after field testing. The requirement of the contractor should be cooperated, explained and understood. And the problems of the engineering are solved with a positive attitude as soon as possible, thus avoiding the standstill of the construction. In construction process, the prolonged construction period caused by the weather, geology or the construction change can easily bring about the emotional change and tired work of the constructors, thus causing the distraction of the constructors to the construction quality. Then the constructors and the contractor unit should encourage the construction team, be themselves strictly and finish the construction assignment according to the construction requirement. The construction unit should test the finished construction assignment. If it is found that the construction operation was not strictly carried on according to the construction drawings, or the engineering quality problems occurred, the supervisor should inform the construction unit to rework the operation. After that, the supervisor should test again to ensure no quality problems or hidden dangers in this stage, and carry on the next engineering assignment. The quality problems should be assigned to the construction unit by the supervisor in an oral or written form, thus making the construction unit avoid changing the construction drawings by themselves and the operation going against the drawings, and the construction units should sign and confirm it.

5 CONCLUSIONS

The constructions in China have made a bad performance in natural disaster, especially in earthquake disasters in recent years. E.g. the earthquake disaster has caused great damage to Wenchuan in Sichuan Province. So, the anti-seismic design must be applied to modern construction engineering in China, and the characteristics and problems of modern construction engineering should be definite. The construction and management of anti-seismic design should be supervised by the professional supervisions to ensure the improvement of the engineering quality. Consequently, the anti-seismic quality of the construction engineering should be ensured by the way of finding, solving and handling the problems. The management and maintenance of the construction engineering is necessary. The qualified maintenance and the finding of the problems in time can ensure the qualified anti-seismic design of the construction engineering.

REFERENCES

[1] Wang Jianhong, Zhang Jinli, Establishment of Quality Prevention and Control System of Construction Engineering, Gansu Science and Technology, 2011(20): 162–164.

[2] Yang Zhongcai, How to Establish Whole Process Quality Control System in Process of Construction Project, Development Guide to Building Materials, 2012(7):133–134.

[3] Jing Mingxun, Discussion of Quality Supervision System and Control Method of Construction Engineering, Fortune World, 2010(22):20–20.

Channel transmission technology of computer network environment

Fei Wang
Hohhot Vocational College, Hohhot, Inner Mongolia, China

ABSTRACT: Computer network technology is one of the major branches generated during the development of computer technology. In order to smooth remote communication, necessary network devices will be applied to achieve the network coverage. Through analyzing the existing computer network technology, the work unscrambled channel transmission in the network environment, and analyzed the methods of improving the network transmission efficiency to achieve efficient transmission technology.

Keywords: computer network; channel transmission technology; analysis; network environment

1 INTRODUCTION

As the theoretical basis of the internet, computer network achieves the globalization of computer network platform through the integration of host, switch, router and terminal. During accessing the network, acquiring and exchanging information can be achieved by browsing the internet. Technically, in order to realize efficient communication in computer network, analysis of its software and hardware system is required to integrate them. In a modern computer network, broadband or optical fiber are applied to signal transmission, which is related with channel [1]. With the significance of realizing generalized signal transmission, channel is the medium for this process, largely influencing the stability and efficiency of the signal. In order to further understand signal transmission in the internet, the work will analyze the channel transmission technology in the network environment.

2 BRIEF ANALYSIS OF THE NETWORK ENVIRONMENT

The formation of network is mainly based on the development of computer technology. The combination of software and hardware technology is well reflected in the computer network. The related network devices such as switch, host and router are all based on the hardware to achieve the network platform. And in the level of software, by means of various network protocols, computer network achieves direct access interval of software, in order to exchange information in the network by different domain names and analysis servers [2]. Therefore, the analysis on computer network environment can be done in different aspects to understand it accordingly.

2.1 *Hardware environment of computer network and analysis on its working principle*

The basic identification code of computer is binary, which is relatively easy to identify. It is a simple identification code in computer language, but for better identification in the interaction of human and computer, the transformation or more advanced process is required for binary data. For example in the network environment, communication between people is based on characters, requiring analysis on the corresponding binary identification codes [3]. Based on the above principle of information transmission, the biggest hardware device in computer network is the host. Actually, the computer network is a virtual world, and the host is the basic facility for this world. For better resource matching and network layout, the facility for allocating resources should be introduced based on the host. That is why the switch is invented. By means of resource matching, the switch allocates resources properly for the host. In a vast area, several switches can be applied for allocating resources of the host. Meanwhile, in order to share the resources, a router is introduced to allocate resources of the switch. It can establish a small local area network (LAN) to achieve information interaction between different computer terminals. Therefore, the hardware environment of a computer network can be concluded as follows: host—switch—router—terminal. This relation can ensure the interaction and exchange of information under the environment of the whole network.

2.2 *Software environment of computer network and analysis on its working principle*

Network protocol is the most important in the network environment. During visiting the network resources, for targeted visit and reasonable resource management, protocol of network visit is required as a basic management regulation. Only the corresponding login is allowed to visit, achieving the resource sharing. The process of accessing information is achieved by requiring the input of Uniform Resource Locator (URL) and domain name. For example, in order to use a search engine, it is required to login the domain name of the search engine to access to the related resources. Moreover, hierarchical processing is also required when using the computer network. It is actually a kind of classified management for different data. Because during the transmission of network information, the passing data comes from different layers, the single layer management is not conductive to the transmission efficiency, hindering the information interaction efficiency. In order to improve the efficiency of network information interaction, the hierarchy of computer network is introduced. Generally the 7-layered structure is applied, which is hierarchical processing on the physical layer, data link layer, network layer and analysis layer. Targeted analysis on data is processed to improve the efficiency of network information interaction.

3 INFLUENCE OF CHANNEL ON SIGNAL TRANSMISSION

The network information transmission also requires signal principle, which refers to information source, transmission channel and information receiver. As for the internet, information management is unified, and information transmission is not a onefold channel. The internet is a complex interactive network with multi-link, where different receivers can also be transmitters and vise versa. Therefore, such kind of bidirectional communication demands higher for the channel. As for the carrier of network information transmission, transmission via radio wave is selected. As the carrier, the signal will highly increase the information transmission rate to improve the user experience of computer network. Then, what influence will signal transmission of internet have on the channel?

3.1 *Property of channel medium largely determines transmission rate of the signal*

The channel is actually a kind of medium to transport signal. It plays an important role in the completed signal transmission system. As the transmission rate of signal varies a lot according to different channels, the computer network experienced the following stages during its development:

The first stage is dial-up. In the early stage of network development, limited by network information and technology, dial-up is applied. It actually utilizes the home telephone to transport network resources through telephone lines. However, adjustment on dialer circus is required in this method to connect to the internet. As telephone lines are used as the channel of this dial-up method, unstable network signal and slow transport rate are quite common, which are obvious drawbacks existing in the dial-up method.

The second stage is broadband. With the expansion of network resources and innovation of technology, exclusive channels instead of telephone lines are applied for information transmission, highly improving the network speed. From dial-up to broadband, the transport rate of computer network has progressed a lot. Broadband also optimizes the utilization of network resources. The host got the chance to show its performance and the application and development of the internet were promoted all because of broadband.

The third stage is optical fiber. The introduction of this technology truly realized the modernization of computer network. As the speed of light is the fastest in the nature, optical fiber can maximize network transportation and highly expand network development. As for the user experience, it has been well achieved with the high speed of network.

In a word, in different stages, computer network applied different channel medium to improve surfing speed, with its development being promoted to truly realize global network system.

3.2 *Property of channel medium highly influences the accuracy of signal transmission*

The property of channel medium is already analyzed in a previous chapter. As different medium result in different transmission efficiency, user experience partially depends on medium property. Moreover, requirement on accuracy is also strict to achieve better inner environment of computer network. The process of transmission is usually influenced by noises, and the whole system refers to decoding. Therefore, strict requirement on medium property is necessary to improve the efficiency of information transmission. For example, using optical fiber as the transport medium will achieve high signal accuracy. As the transmission principle is total reflection, transmission loss of optical fiber is negligible. In addition, total reflection hinders the participation of noises, reducing the influence of noise on signal transmission to improve decoding. This property contributes a positive effect on transmission accuracy. Therefore, as for the transport medium, the property of channel medium highly influences the accuracy of signal transmission.

4 CHANNEL TRANSMISSION TECHNIQUE BASED ON COMPUTER NETWORK ENVIRONMENT

Requirements on channel transmission technique are very high under a computer network environment. The high utilization ratio of channel owes to the closed loop multi communication applied by computer network. In the process of channel transmission, two aspects have to be ensured. The first one is transmission efficiency. With the development of technology, people's requirements are increasing with their needs.

The network transmission efficiency plays an important role because high transmission efficiency can meet people's requirement on computer network. Secondly, absolute accuracy should be ensured. As the network transmission is based on accuracy, one mistake will influence the users a lot, resulting in great loss for them. Further strengthening the channel transmission technology is required to achieve the accuracy of computer network. Applying optical fiber channel to computer network is the mainstream now. There are two points making optical fiber outweigh the other technologies. Firstly, optical fiber guarantees high accuracy of the signal. Its total reflection principle has little influence on transmission loss, and can also reduce the signal loss to a certain extent. Thus great saving is achieved in the integration and utilization of resources. Total reflection also hinders the participation of noises, reducing the influence of noise on signal transmission to improve decoding, which contributes positive effect on transmission accuracy. On the other hand, hierarchical processing is applied to computer network. It is actually a kind of classified management for different data. Because during the transmission of network information, the passing data comes from different layers, the single layer management is not conductive to the transmission efficiency, hindering the information interaction efficiency. Therefore, in designing a network channel, the classified structure of network should be taken into consideration to meet its transmission property. The interaction and high efficiency of network should be ensured in information exchange and transmission to lay a solid foundation for achieving high performance channel transmission of computer network.

5 CONCLUSIONS

By analyzing the property of channel transmission and its influences on the internet, it is concluded that in considering the medium property of signal under the environment of computer network, transporting by optical fiber will contribute to network development. Because of its property, optical fiber cannot only increase transmission speed, but also maximize transmission accuracy. The development of channel transmission technology drives the improvement of other industries. By achieving multi channels for multi communication in the internet, the traditional network mode also benefits a lot. In a word, for the channel transmission technology under the environment of computer network, not only transmission speed but also transmission accuracy should be ensured. Only with both of them can the technique be applied to information transmission to achieve the collaborative development of computer network technology.

REFERENCES

[1] Zhao Huiling. *Evolution of Core Technology of Network Exchange*. Telecommunications Technology, 2009(01):63–65
[2] Deng Wuhua, Dou Zheng, Yang Xiaodong, Guo Dongmei. *Achieving TCP/IP Protocol Based on Switching Structure of Software Radio Network*. Applied Science and Technology, 2006(10):15–18
[3] Wang Bing, Chen Xiaojuan. *Development of Network Switching Technology*. Science & Technology Ecnony Market, 2007(10):2–3

Influence of digital technology on urban public art and its interactivity—a case study of cities in Jiangxi

Jing Li
School of Art and Design, Jiangxi Science and Technology of Normal University, Nanchang, Jiangxi, China

ABSTRACT: Based on the impact of technology on the art, this paper makes a concrete analysis about the influence of digital technology on urban public art and its interactivity, combined with the current facts of urban public art of Jiangxi Province, in order to give an outlook for the development of the urban public art theory and its interactive of Jiangxi Province.

Keywords: Digital technology; urban public art; interactivity; Jiangxi

1 INTRODUCTION

Urban public art is art which faces and serves the city public, and generally it is displayed in city public spaces with crowded people, such as city streets, parks, squares, airports, railway stations, subway and other places. Generally speaking, urban public art is presented mainly in sculptures and murals; however, with the improvement of modern society and technology, digital technology affects our lives more and more. At the same time, the impact of digital technology on the urban public art is great undoubtedly [1]. The urban public art is no longer presented only in sculptures and murals, more and more works of art created with electronic technology show up, using digital media as a medium.

Using digital technology for the creation of urban public art is that the digital technology is involved in the creation of artworks, which also means increasing expression forms for artworks. In today's digital era, discussing the expression forms of urban public art also means discussing how the digital technology affects the urban public art as a new medium carrier [2]. And this is involved in the relationship between technology and art, as well as the influence of the science and technology when creating works of art.

Traditional urban public art is to decorate urban environment and is appreciated by the public, which is a one-way relationship with the public, while the digitized urban public art can interact with the public, including playing the role of traditional public art. We can see an interactive design concept which is applied in the whole creative process of works.

2 IMPACT OF SCIENCE AND TECHNOLOGY ON ART

"Beauty is an emotional show of rationality," said Hegel. It exactly describes the impact of science and technology on art. Art is to show the emotion, which is emotional, while science is rational, calm and objective [3]. Art emphasizes aesthetic creation, while science stresses the understanding of the objective world. Science pursues truth, while art pursues beauty. Although there are many differences between science and art, they will meet each other now and then when we pursuit truth and beauty.

Art had a close relationship with natural science during the Renaissance. There were a lot of people who were not only artists, but also scientists; Leonardo da Vinci is a typical representative. This is closely related to the development of natural science during the Renaissance. The impact of science to art is directly reflected in the application of perspective, which makes the three-dimensional widely applied in painting. That is very important for the development of Western art in the later period.

The Industrial Revolution which began in the 18th century has a significant impact on the world pattern, including art. After the 19th century, science impacted more on art. Artists in the Neoclassicism, Romantics, Baroque and Rococo eras during and after the Renaissance have paid attention to the use of perspective, and get the most out of them. Vivid and realistic three-dimensional scenes are presented on a flat surface; this kind of realistic technique is singularly impressive. However, after entering the 19th century, with the invention and use of the camera, people can easily copy real life scenarios, just by pressing the camera button. There is no need to take a long time to describe and sketches before. The invention of photographic technique makes paintings performance back on the plane from the three-dimensional space. In the early 20th century, the machine aesthetic basically dominated all contemporary and modern movement. People usually think that Futurism, Cubism, Constructivism, Bauhaus, part Dada and even Surrealism are concerned with the machine aesthetic. This machine aesthetic showed in art schools before World War II is called "classic machine aesthetic", and it is considered as a core element of "classic modernism". Since the mid-20th century, with the advent and popularity of the computer, a new art form is created—Multimedia Art. Today, in the 21st century, computers, mobile phones and other electronic products appear more and more in our lives; meanwhile, digital technology has completely penetrated into the artistic creation. The relationship between technology and art becomes more and more close.

3 INFLUENCE OF DIGITAL TECHNOLOGY ON THE URBAN PUBLIC ART AND ITS INTERACTIVITY

3.1 *Influence of digital technology on the urban public art*

Urban Public Art was born in the West, and China gradually pays attention to urban public art since the reform and opening up, which is represented by the appearance of the mural in the capital airport in 1979. That impacts greatly. With the city construction, the demand for urban public art has increased as well. In the late 20th century, the domestic urban public art is presented mainly in sculptures, and then murals. There are historical stories, heroes, myths and legends, and scenes, etc., in terms of the expressive subject matter. Concrete and abstract are both exit in the technique of expression. These urban public artworks are often placed at a fixed location for the public to enjoy.

The impact of digital technology on urban sculpture is obvious. How is an urban sculpture created without using digital technology? Usually artists make a sculpture artwork with small clay, and then magnify it after the scheme is determined. It can be made of gypsum, bronze or other materials. While an urban sculpture applied digital technology can be output directly to work, using a three-dimensional engraving machine? What is a three-dimensional engraving machine? It is a device that combines computer technology, laser technology with machining technology. Through a variety of technology, the device can carve out works with three-dimensional effect with a variety of materials such as wood, metal, PVC, glass, plastic, silicone and plexus glass. Three-dimensional carving machine is widely used, even replaces hand-carving techniques, and laid the foundation for the rapid production of large quantities of products.

Take Buddha sculpture making for example, which is a kind of urban sculpture. If they use artificial carving, it is hard for them to grasp its physical proportions, and clothing, etc., even according to the plate in the past, for the ignorance of the prevailing style characteristics of original statues at that time. While if they use three-dimensional carving technology to produce the Buddha, the same statues with exactly the same proportion, clothing and others can be carved out as long as they choose a good sample. They can even batch produce. Generally speaking, there is a step called turnover formwork in the process of sculpture carving, in this part if the plaster is not turned over well which is often happened, they have to re-create or repair the sample mode. In other cases for example, some sculptures is complicated in pattern of manifestation, so it is troublesome to the chosen materials. However, using three-dimensional carving techniques to make sculptures can avoid these problems, and they can set materials, choose design samples going without interruption from the very beginning.

Thus it can be seen that digital technology makes it more convenient for the way to show urban public art, more diverse to present urban public art, more optional materials can be chosen, and the performance of the content becomes more rich as well. That is to say, digital technology makes it more diversified and stronger in expressive force, no matter in terms of material choice, technique of expression or expressive subject matter, comparing to the traditional urban public art. Therefore, digital technology will bring urban public art into a new era!

3.2 *Influence of digital technology on the urban public art and its interactivity*

The influence of digital technology to the urban public art lies in the art itself. It makes it more diverse in the form of urban public art; on the other hand, the most important thing is that digital urban public art pays more attention to the audience, which is obvious different from traditional public art. In short, the digital urban public art contains a strong interaction design concept, interactivity is in the first consideration when creating, which gets it closer to the audience with art.

Traditional urban public art is mainly controlled by government leaders and artists from production to be completed, and the works is shown on a fixed location in the city after completed. The audiences just appreciate it then. The digital urban public art put the viewer in the first place in the whole process from creation to completion, which is different. It focuses on the interaction of public art and the audience. Only when the audience experience and interact with the work, the work of art has been complete finally, not when the work is presented. Thus, the audience has become a part of the work, and it is an integral part. Evidently this has been raised a level in the design concept. The urban public art under the background of digitization not only integrate with the urban architecture and the landscape environment, but also communicate and interact with people on city life and cultural. In recent years, both at home and abroad, the urban public art develops in the direction of digitization, and many outstanding works appear, for example, the PepsiCo on the bus station in London launched a new digital billboard, which makes people waiting for the bus experience a "thrilling and fun" time. This digital billboard on the bus station uses the high-tech digital technology; very realistic images are showed in a rectangular transparent screen. Passengers can see blasting, UFO, bombing robot, the tiger running to the crowd and other scary scenes, feeling a false alarm. There is also a camera in the digital billboards, so that passengers can self-timer, take photos with virtual robots on the billboards screen, and interact. That is very interesting. This kind of dynamic advertising promotes the company image, effectively using the passengers' waiting time; meantime, it brings passengers a lot of cheers and laughter, making people feel good about the company and the city. Another example is the Winter Olympics in Sochi this year, there is a spectacle—a large dynamic face sculpture. First the viewers in the Winter Olympics need to scan their faces through a "3D photo booth", and then digitized high-tech brakes begin to move, forming a large scanning three-dimensional model of the face eventually. The participants' faces can be displayed within 20 seconds, and also can be seen by the participant themself. Their three-dimensional faces will be shown in the recorded 20 seconds video. This digital public art is a highlight of the Winter Olympics in Sochi undoubtedly, allowing viewers to experience a star feeling. Another example is the 4D movie shown in the Thailand Pavilion in the 2010 Shanghai World Expo, more advanced than 3D technology. Viewers can experience vivid feeling not only in the visual sense, but also in the sense of touch and smell. When it rains in the film, really raindrops hit the audience, and when the flowers appear in the film, the real fragrance of flowers can be smelled. This effect is really unprecedented, pleasantly surprised!

Thus it can be seen that digital technology has brought the audience deep communication and interaction with urban public art. The audience can experience beauty and joy while interacting with the artwork. That makes the audience closer to the art. Art is no longer a piece of art placed in the museum, lofty untouchable, but integrated into the public life, so that the public can feel the charm of art in their everyday life.

4 DISCUSSION ON THE JIANGXI URBAN PUBLIC ART AND ITS INTERACTIVITY UNDER DIGITIZATION BACKGROUND

In the 1980s, Jiangxi urban public art is mainly represented by monumental sculptures with red culture, such as Museum of August 1 Nanchang Uprising and city sculptures in Jinggangshan Sculpture Park. These sculptures are created to show the difficulties of the war during the Red Army Period, and to eulogize Red Army soldiers for their bravery. They set an example of heroic dedicated spirit for us. Since the 1990s, Jiangxi urban public art develops in the decorative direction, emphasizing expression of the city's history and culture. For example, the cultural relief walls of "Jiujiang Soul" on the Jiangxi Jiujiang Bund depict the Jiujiang's history and the flood scenario then. Besides, there is purely decorative urban public art, such as the lantern sculptures of "the Qingshan Lake Pearl" in Nanchang, playing the role of beautifying the city and decorative urban environment.

With the popularity and promotion of digital technology home and abroad, digital urban public art is gradually increasing. Jiangxi urban public art develops in the digital direction, but there are not many digital public artworks with quality and interactive concept. Currently in our country, interactive urban public art presents mainly in the media and touch screen. The audience interact in ways of taking photos with the artwork, inputting their own head images into some devices and then images change with the change of sound, more interesting than the traditional urban public art. With the implementation and depth of the interactivity concept, the thought of people-oriented appears as well, so that everyone can participate in artistic creation, realizing their artistic dreams. Jiangxi urban public art has accepted and used digital technology gradually, and I believe that with the progress of time, more and more digital public artworks will be presented in Jiangxi, and be interacted by the public, allowing the public to experience the art's beauty, to feel the culture and the charm of Jiangxi's city.

5 CONCLUSIONS

Today with the electronics fully penetrating into our lives, it is an inevitable trend to use digital technology to make the urban public art. Urban public art under digital background cleverly combines truth in science and beauty in the art. The audience marvel at advanced science and technology, experience the beauty of artworks, feel the city's culture and concept while interacting. Not only does digital urban public art decorate the city, but also combines the public with technology, art and city together, bringing us a visual feast and spiritual sublimation.

ACKNOWLEDGEMENTS

This work is one of the achievements of Jiangxi provincial arts and sciences projects—"Research and Design on the Influence of the Urban Public Art and Its Interactivity under the Digital Background in Cities of Jiangxi", with granted number YG2013025.

REFERENCES

[1] OU YANG Hua, Function of Urban Public Arts in Portraying City's Culture and Character, Journal of Central South University of Forestry & Technology (Social Sciences), Vol. 2, No. 3, 2008, pp. 124–126.
[2] DONG Qi, DAI Xiao-ling, Urban Public Art Planning: A New Study Area, Journal of Shenzhen University (Humanities & Social Sciences), Vol. 28, No. 3, 2011, pp. 147–152.
[3] LV Jing, WANG Huai-liang, Concrete Embodiment of Public Culture of City Public Art, Journal of Jilin Teachers Institute of Engineering and Technology, Vol. 30, No. 3, 2014, pp. 56–67.

Future transformation and development of physical education in colleges and universities

Li Wang
Nanjing Technical College of Special Education, Nanjing, Jiangsu, China

ABSTRACT: At present, the level of physical training in colleges and physical quality of college students are getting lower and lower, indicating a decreasing trend of the overall health of college students. Therefore, the reform and transformation of physical education in colleges and universities becomes an inevitable result. Playing an important role in the national fitness program, physical training in colleges and universities is an important indicator to study the overall physical quality of national people. The work analyzed current problems in the transformation of physical education and further discussed the future developing plan of physical education, hoping to present more positive implications for the reform and transformation of colleges.

Keywords: colleges and universities; transformation of physical education; future developing direction

1 INTRODUCTION

Since the 21st century, the Chinese economy and society have been making continuous improvement, leading to a rising living standard for Chinese people. People's need has increased from solving the problem of food and clothing to meeting their spiritual demands. As a result, varied service industries emerged and have been developed in recent years. While people focus on satisfying their spiritual demands, they should consider health as the foundation. As the proverb goes, a healthy body is the capital of revolution [1]. Therefore, in order to meet the demands to improve people's health status, the Chinese government promotes a national fitness program to raise their awareness of physical health, encouraging them to actively participate in physical exercises. The national fitness program has improved the overall health status of the whole nation. Physical education in colleges and universities plays an important role in the national fitness program. Since college students are the major force of socialist modernization, their physical quality is the basic guarantee to the future development of Chinese socialist modernization. However, current physical education in colleges and universities does not meet the requirements of social development in China [2]. Therefore, it is dispensable to carry out reform and transformation for college physical education, which is also an important aspect of the whole social reform and transformation.

2 CURRENT STATUS OF PHYSICAL EDUCATION IN COLLEGES

2.1 *Low attention to physical education*

Exam-oriented education plays the main role in the education system of China; thus physical education does not obtain much attention from it. For a long time, it has been called for to promote students' comprehensive development of moral, intellectual, and physical, aesthetic, labor-round abilities. However, physical education and quality-oriented education are totally neglected under the influence of exam-oriented education. Students go to school in the purpose of achieving a high grade, giving little care to other aspects and resulting in a serious decline in students' physical quality. It is found in the national population census report that students' physical quality is declining and that many college students fail in their physical fitness tests. This phenomenon further proves that colleges and universities pay little attention to physical education and overlook students' physical quality, resulting in continuous decline of the overall physical quality of college students. In a word, current problems of physical education show the necessity of physical quality transformation, and they also form a driving force of physical quality transformation in colleges [3].

2.2 Weak concept of physical education transformation

With the development of society, the transformation and reform of physical education continues to expand in colleges and universities and is constantly mentioned by people since government and society began to emphasize physical education transformation in colleges. However, under the major social context of developing economy, the goal of colleges is to cultivate talents suitable and capable of developing social economy. Colleges don't make deep analysis and exploitation about the concept of physical education transformation, leading to students' misunderstanding of the necessity and schools' overlooking of physical education transformation. As a result, weak concept of physical education transformation becomes a severe obstacle for physical education transformation in colleges and universities. Therefore, in order to better accelerate physical education transformation, colleges and universities should strive to promote physical education transformation deeply into people's mind.

2.3 Wrong positioning of college physical education

According to statistics, most colleges and universities have biases to the positioning of physical education, specifically in the following aspects: scientific development of physical education, education direction, curriculum, teaching mode selection and teaching methods. Varied reasons contribute to those biases and have different influence on the transformation of physical education. Most colleges have vague knowledge and no clear definitions about the above aspects. Correct positioning of physical education is vital to the development of overall physical career in colleges. It determines whether college physical education develops toward the right direction, whether college education plays its full function and whether students can benefit from physical education. A correct positioning of college physical education is beneficial to solve those problems and thus ensure a smooth transformation of physical education in colleges.

2.4 Little emphasis on physical education from college students

Students are the subject and also the direct beneficiary of physical education transformation in colleges and universities. The decline of students' physical quality has become a concern of the entire society, so the purpose of physical education transformation is in fact to improve students' physical quality. Only on this basis can students make a contribution to the national fitness program and help to improve the physical quality of the whole nation. However, college students don't pay enough attention to physical education, and don't make enough effort in physical sports classes. Instead, most students just care about the final score of sports courses, ignoring the actual content and significance of those physical sports courses. The main reasons for this situation includes: firstly, students suffer great pressure from professional learning and employment, and it's difficult for them to spend time and energy in physical sports classes; secondly, students show a negative attitude towards physical education, failing to realize the importance of physical education; finally, score-oriented teaching mode does not help to inspire students' interest in physical sports courses.

2.5 Urgent need to improve teachers' qualities

Some physical teachers in college have problems on the concept of physical education. In class, some teachers always guide students to do what the teacher want them to do, while others just let students to do whatever they want to do. Those concepts can hardly achieve the purpose to encourage students to participate in physical exercises. Moreover, some teachers are unable to guide students to take physical exercises due to the limitation of their personal experiences and teaching levels. More specifically, they don't have rich and extensive knowledge to guide students and train them from multiple aspects. As a result, it becomes difficult to develop the potential of students on physical ability. In addition, several classes usually combine together to have the same physical sports class in colleges, resulting in the teaching content lack of good pertinence. Consequently, students can only react to teachers' guidance and cannot choose a proper training project according to their own interests. As a result, the enthusiasm and initiative of students to physical activities will be greatly affected, leading to physical training hardly to meet the standard and students' physical quality gradually decline.

3 MEASURES TO SOLVE THE PROBLEMS FOR PHYSICAL EDUCATION IN COLLEGES

The transformation of physical education in colleges and universities should be targeted to improve the physical quality of college students and to strengthen the positioning of physical education. Based on the guiding ideologies summed up from the transformation, governments and colleges should formulate corresponding measures such as policies and regulations to reform the teaching content of physical education and to govern the reform at the same time. The implementation of those measures has a very important influence on the transformation of physical education in colleges and universities.

3.1 Correct positioning of physical education in colleges and universities

Physical education in colleges and universities is not paid much attention mainly because colleges have no appropriate positioning for physical education, making physical education at the edge of education. Colleges should develop an appropriate positioning for physical education in order to achieve a comprehensive and proper transformation. Thus, colleges should develop clear stipulations to explain the importance of education, and to define corresponding teaching mode, teaching method and teaching content. Moreover, students' performance in physical sports classes should be considered as an important indicator to evaluate students' achievement in school. Therefore, physical education will gradually develop as one of the key disciplines, manifesting the importance of physical education.

3.2 Strengthening the investment of physical education

Any enterprise develops on the basis of sound capital support, so the urgent transformation of physical education needs the support of massive funds to ensure physical education reforming measures implements smoothly. Therefore, colleges should pay more attention to the development of physical education, changing their ideas and strengthening capital investment on physical education. In addition, they should make efforts to improve infrastructure construction and to prepare for the development of physical education. For example, colleges should purchase complete sports equipment, maintain sports facilities, increase capital investment on teacher cultivation and improve teachers' welfare.

3.3 Reforming teaching content of physical education

To reform teaching content of physical education is also an important measure of physical education transformation. Traditional teaching content of physical education just repeats some simple knowledge without any innovation content, overlooking the basic knowledge of health education and physical exercises. Those teaching content only weakens students' initiative and enthusiasm to take physical exercises and hardly initiates students' learning interest, which is not conducive to build a healthy body for students. Therefore, the reform of teaching content for physical education is inevitable. And it is also indispensable to enhance the interest and practicability of teaching materials, broaden students' knowledge, strengthen the education of basic theory knowledge to students and improve their theoretical level, finally training students to make better use of theory to guide practice in their life.

3.4 Reforming teaching method of physical education

Traditional teaching methods, based on teacher-oriented teaching or freedom-oriented teachings, are no longer suitable for the requirements of developing college physical education in modern time. Those traditional teaching methods cannot enhance students' interest in physical sports learning, and cannot achieve the purpose of enhancing students' physical exercises. Therefore, colleges should develop some teaching methods according to the practical characteristics and conditions of both the school and their students when carrying out reform of teaching methods. It is very important for colleges to analyze specific issues and organize group study for students with similar interest, thus to stimulate their learning interests and to inspire their potential in physical sports. In the end, colleges should summarize the experience through those reforms for the purpose of improving the level of physical quality, finalizing a detailed reform of teaching methods for physical education transformation.

3.5 Cultivating a well-trained and qualified teacher team

Teachers are the direct source for students to obtain physical knowledge, and they also play a vital role in the transformation of physical education. In fact, they are the main force of education reform. At present, the overall quality of physical teachers in colleges is relatively low, which is not conducive to the transformation of physical education. Therefore, colleges must be equipped with a well-trained and qualified teacher team so as to ensure successful transformation and a better future for physical education in colleges. In order to cultivate teachers with high qualities, colleges should: firstly, invest capital to recruit teachers of high qualities so as to improve the overall qualities of teachings from the first step; secondly, assess physical teachers on a regular basis, announce the assessment results publicly and record the results in documentation; finally, establish a profound measures of rewards and punishments in order to improve teachers' working attitude and initiative.

3.6 Learning advanced western experience of transformation

As a developing country, China is not well developed in many aspects, and there is still a certain gap with developed countries. China can learn much advanced experience from developed countries, which can help China to avoid many problems in the process of reforms. Therefore, colleges and universities should make good use of related lessons and experience from colleges and universities in developed countries. Besides, they should also develop suitable plans and policies to serve Chinese physical education according practical development requirement, actively promoting the combination of domestic colleges and universities with the society and the world.

4 DEVELOPMENT DIRECTION OF PHYSICAL EDUCATION IN COLLEGES

4.1 Developing traditional physical sports with ethnic characteristics

China consists of 56 ethnics, and each ethnic has its own unique customs and competitive sports. In recent years, the concept of ethic equality has deepened in people's life with the development of society. Therefore, people gradually realize the great significance of learning competitive sports of minority people to the development of the whole nation.

Chinese traditional ethnic sports should be included as an important part of physical education in colleges. Ethnic physical sports contain the wisdom and spirit of Chinese people, which is of great significance to Chinese traditional culture and ethnic communications.

First of all, based on the need of developing personality, ethnic traditional physical sports combine the advantages of each ethnic, and it helps to achieve the purpose of spreading Chinese traditional culture system and enhancing quality-oriented education through the teaching of ethnic traditional physical sports. Secondly, learning the theory knowledge of ethnic traditional physical sports enables students to learn the historical, political and military background of that ethnic. Improving students' culture quality and scientific and cultural knowledge plays a positive guiding role in implementing a comprehensive fitness strategy. Finally, learning the core spirit and idea of traditional physical sports can help to improve the level of students' morality, thus contributing to the construction of the moral standards for the whole society.

4.2 Eliminating uneven physical education level in colleges to avoid "localized" education

Uneven economy developing level of each region in China contributes to different economic conditions and social conditions for varied colleges, resulting in big differences of students' thinking mode. Therefore, colleges should adopt a localization strategy to carry out specific analysis for physical education form in different region. Anyway, the ultimate goal of physical education in China is to improve students' physical quality. Only when students understand this goal, can they participate in physical education in a better way. Therefore, colleges and universities should carry out analysis of physical education based their own characteristics and adopt appropriate methods and ideas to create more vitality to the reform and transformation of physical education.

5 CONCLUSIONS

The ultimate goal of Chinese physical education in colleges is to improve students' physical quality, but the current physical education system can hardly achieve this goal. Therefore, the transformation of physical education in colleges is indispensable in order to create more favorable conditions to achieve this goal. Colleges should make careful analysis and adopt specific measures during physical education reforms so as to get twice the results with half the effort. The reform of physical education in colleges has great significance to the physical quality of the whole society. Therefore, colleges and universities must adopt feasible plans to steadily promote the transformation of physical education, developing toward a glorious future of physical education.

REFERENCES

[1] Chen Xiaorong, Zhu Baocheng. Current Status of Chinese Physical Education in Colleges and Its Future Development. Journal of Shanghai Physical Education Institute, 2010(1): 13–21
[2] Liu Zhihong. Current Status of Chinese Physical Education in Colleges and the Strategies for Its Future Development. Hua Zhang, 2011(7): 32–44
[3] Tan Xianglie. Future Development Trend and Idea of Chinese Physical Education in Colleges. Journal of Changsha Railway University, 2013(1): 70–74

Problems and policies of financial products innovation

Yuhua Shen
Guiyang University, Guiyang, Guizhou, China

ABSTRACT: As the financial reform in China is moving forward, the competition in financial industry is becoming increasingly fierce. Business operators are faced with the problems on how to choose innovation strategies for financial products to pursue profit maximization. Based on the current development of financial industry in China, this study, according to the theory of finance and industrial economies, analyzed the problems of insufficient market segmentation and single service object in financial innovation and proposed some effective advice, conducting diversified businesses based on the market environment.

Keywords: Financial Products Innovation; Market Environment; Financial Industry

1 INTRODUCTION

Financial products innovation refers to the process of producing products for market requirements by using precise mathematical tools in capital raising and investment as well as derivative financial business. Financial products innovation needs to follow not only the special rules of financial industry, but also relevant theories of industrial economies, especially industrial relationship [1]. Firstly, this study analyzed basic principles of financial products innovation to explicit the overall train of thought of financial products innovation; based on this analysis, this study pointed out some problems of financial products innovation. Finally, countermeasures for the optimization of financial products innovation are proposed according to the current situation of the financial industry of China.

2 PRINCIPLES OF FINANCIAL PRODUCTS INNOVATION

This study first analyzed the general selection of financial products innovation to provide guidance for further analysis. The intent of financial products innovation is to help enterprises to achieve profit maximization in a constant changing market [2]. The constant change of market environment is a direct premise for the financial products innovation.

Firstly, financial products innovation should adapt to market environment. Enterprises should continually adjust their product design and promotion program based on the changes of segment market. Financial products need to meet different investment and financing requirements of customers. With the continuous development of market economy, the financing market tends to be more refined and stratified. Since individual investor demands higher capital gains and property income while different investors of different incomes have distinct risk preference, financial institutions have to emphasize the enhancement of the market segmentation level of financial products innovation so as to provide more diversified financial products for financial market.

Secondly, controlling financial risks is an important goal of financial products innovation. The risk control of financial products mainly demonstrates capital-raising process. As the sale of financial products is virtually a capital-raising process, many financial institutions should pursue the goal of gaining profits while controlling interests and enhancing anti-risk capabilities in designing the financial products.

Lastly, improving settlement efficiency is another problem that current financial enterprises need to face. The improvement of settlement efficiency is mainly to expand off balance sheet businesses. With the gradual opening up of the financial market of China, on-balance sheet businesses are encountered with problems like fiercer competition and atrophic profit space. Therefore, many banks have been expanding their off balance sheet businesses as a new source of profit growth. Off balance sheet businesses mainly involve capital settlement as well as expenses and receipts, and credit operations are not included. Comparatively speaking, the profit of singe credit operation is low so enterprises should rely on settlement operations to improve their profitability.

3 PROBLEMS OF CURRENT FINANCIAL INNOVATION PRODUCTS

Nowadays, financial products innovation of China still needs to be improved. Due to the higher risk and complexity of financial products, the financial products innovation should be conducted under the guidance of overall principles of financial innovation according to the market structure of Chinese financial market. Since China underwent a long period of planned economy, the financial reform in China is still moving towards the marketization, and some problems caused by institutional factors as well as business practice still remain.

Firstly, the innovation products produced by financial institutions still have problems on coarse market segmentation in the initiation of products promotion and design. Market segmentation is the study and judgment of market requirements. The inherent requirements of market segmentation is to divide the whole market requirements into individual requirements with distinct characteristics and the whole financing market into different small markets. Currently, investors and investees have different recognitions of capital use and expected revenues, and different individuals have diversified investment channels. Against this backdrop, financial institutions should continuously improve their policies of market segmentation to appropriately divide market segmentation and achieve profit maximization by introducing corresponding products according to the market characteristics.

Secondly, the main businesses of Chinese financial institutions are still on-balance sheet businesses. Many regional institutions do not have deep recognition of modern financial innovation products. The financial innovation is often conducted from top to bottom. The supply chains of financial products are original push supply chains, which have low speed of market information feedback. These problems certainly have some connections to the ongoing de-administration reform of Chinese financial institutions [3]. However, the fact that different levels of Chinese commercial banks have on-balance sheet businesses as their main businesses also leads to the weak initiative of self-dependent innovation.

Finally, the current financial institutions in China are faced with the problem of single service object. Financial institutions in China have loaned much of their capital to urban industry and commerce while the capital source of enterprises mainly rely on urban investors, which causes the problem that rural businesses are often neglected by the financial industry in China. With the deepening reform of state-owned forest rights and the circulation of rural collective land, rural businesses will bring more profits to financial institutions in China; meanwhile, it can solve the problem of insufficient credit customers, which is significant for the development of Chinese financial industry. The financial innovation is based on the breakthrough and innovation of on-balance sheet businesses and off-balance-sheet businesses. In addition, the late development of rural financial industry of China offers opportunities for financiers to break through original business model.

In a word, the problems of financial products innovation in China consist in coarse market segments, backward way of business development and excessively single service objects, which are caused by long period planned economy and excessively administrative management of financial industry. Apart from factors like system and mechanism, the late development of financial innovation is also an important reason for the immature financial innovation of China. The resolution of these problems is of vital significance for the development and advancement of Chinese financial industry.

4 STRATEGIES FOR IMPROVING THE LEVEL OF FINANCIAL PRODUCTS INNOVATION

Through the above analysis, it is clear that the general principle of financial products innovation is to correctly divide market demands into different categories according to the market situation; under the requirement of reducing business risks, financial institutions should unceasingly enhance the settlement efficiency of financial businesses and establish a financial and economic system corresponding to marketization and opening up policy. Against this background, the current financial principle innovation of China has the problems of coarse market segment and single service object, without much positivity or initiative to conduct modern financial innovation businesses. All these problems are caused by obstacles of system and mechanism as well as the traditional business model of the financial industry of China. To solve this problem, we need to emphasize market analyzing ability, product design and market service space so as to reform and adjust the financial innovation process of China.

Firstly, enhancing the market analyzing ability is the priority for the reform of the financial industry. In order to continuously improve the ability of financial products innovation, financial institutions should have clear recognition of market fluctuation. The market environment analysis includes two basic aspects, market segment judgment and market structure judgment. Market analysis is the first link of designing innovation products. The goal of market analysis is to recognize the main competitors on the market, the profitability of enterprises and the ability to control market. This process is regarded as the measurement of market power within the scope of industrial economics. The measurement index is industrial concentration index and Lerner index. The Lerner index is the measurement of the ability to fix a price above marginal cost. It concludes customers' response to market prices. The lower Lerner index is, the weaker the ability to control market is, requiring increasing differentiation of products. In other words, the homogenization of products on the market is more serious. If the previous products innovation of an enterprise is commanded by other enterprises, the relevant technology will be generalized and the enterprise will have no monopoly advantage over such a technology and the ability of product launch. The judgment of Lerner index is to provide an evaluation standard for the management of enterprises. If the Lerner index tends to be smaller, enterprises should know that the differentiation advantage is being dispelled. Therefore, they need to redesign their products and promote new products.

Secondly, it is important to separate innovation and promotion of products from on-balance sheet businesses. As current financial institutions in China still rely on on-balance sheet businesses, the establishment of a new financial innovation department can help banks gradually get rid of intrinsic business models and make much progress in financial products innovation. China has administrative management on financial institutions, so there is a relationship of administrative subordination between upper financial institutions and lower ones. The operating decisions are given to the lower financial institutions in the form of administrative instructions. Therefore, the changes of market cannot be detected and recognized by the managers of financial institutions in a short time. For the management of enterprises, this top-down decision-making model is subsidiary to push-type product supply chain. The core enterprises are in the main position in decision-making, so the customers' demands often cannot be cared and recognized, finally leading to the loss of ability to occupy the market. To establish departments independent financial products design does not mean to establish a parallel departments in financial institutions, which cannot solve relevant problems. The managers of enterprises need to establish a new business and management model, which connects the traditional business and management model to modern business and management models. De-administration is a core while flattening management is a direct result of this process. To obtain this goal, enterprises should regard the department of financial products design as an independent department and give it some independent research and decision-making rights. The relevant departments ought to be responsible for the market research while other departments should provide information in time.

Thirdly, it is another important strategy to use market-oriented means to conduct the promotion of financial products innovation. The promotion of products contains two aspects, the introduction of products and its relationship with enterprises. The first aspect follows the rule of supply and demand while the second aspect directly demonstrates the objective requirements of industrial relationship. The reform of market-oriented promotion is indispensable in order to solve the current problems of enterprises, which have to face the influence and challenges of the existing market power when they try to enter a new market. For financial institutions, an advantageous obstructive measure is to limit price and use non market-oriented methods. Compared with western countries, the regulations on Chinese financial industry are still rigid and have more restrictive conditions. The state-owned commercial banks with strong capital advantage can weaken the competitive willingness of outsider enterprises by limiting price for a certain period. They can even achieve this goal by bringing wrong recognition to outsider enterprises as to believe that the state-owned commercial banks are low cost competitors. The restrictive measures on overcapacity in industrial organization theory cannot function well for small-scale financial institutions.

Lastly, it is an important sally port for modern financial innovation to broaden the sphere of business and march to the rural areas. Compared with urban financial industry, the rural counterparts lag behind more with few service varieties and low capital flowability. For enterprises, the external pressure brought by the promotion of financial innovation products is small. Even though enterprises fail to promote products in relevant fields, not much loss will be caused. Compared with the financial innovation in urban areas, the development of rural finance can be test fields of financial innovation, which is significant for the increasing competition among financial enterprises. The financial businesses in rural areas can integrate the off balance sheet business and the on-balance sheet business as well as the credit business and settlement business. Since the rural finance needs to be improved, the comprehensive businesses can give direction to financial institutions to research and development financial innovation products that can meet the requirements of rural investors.

5 CONCLUSIONS

Financial products innovation is an important part and key link of China's financial innovation. The financial innovation can bring more profits to enterprises. The current problems of China's financial innovation derive from administrative management model of financial industry. Facing the market, using more flexible ways to conduct relevant businesses and broadening the sphere of business to rural areas will help solve relevant problems.

REFERENCES

[1] Jiang Yuexiang, Jiang Ruibo. *Regional Financial Innovation: Efficiency Evaluation, Environment Effects and Disparity Analysis*. Journal of Zhejiang University (Humanities and Social Science), 2013(4):52–65.
[2] Gu Shengzu, Wang Min. *Strategic Considerations of Financial Support for Technological Innovation in Agriculture*. Forum on Science and Technology in China, 2012(8):28–31, 48.
[3] Chang Zhongze. *Reflection on U.S. Financial Crisis – Angle of Financial Innovation Tianjin Social Sciences*, 2010(6):74–79.

… *Industrial, Mechanical and Manufacturing Science – Zheng (Ed.)*
© 2015 Taylor & Francis Group, London, ISBN: 978-1-138-02656-8

Analysis on visual field and skill training in table tennis teaching

Hao Li
Xi'an Physical Education University, Xi'an, Shaanxi, China

ABSTRACT: Table tennis teaching, including lots of skill training, is an important part of physical teaching. However, it has many problems to be improved in many ways such as improvement of visual field and skill training, etc. In the work, in order to improve physical teaching quality, it is innovated that construction suggestions are proposed in table tennis teaching through process and meaning analysis of visual field and skill training.

Keywords: Table Tennis Teaching; Visual Field Training; Skill Training; Process Analysis

1 INTRODUCTION

Table tennis is the national game in China, occupying a very important position in the physical field. As a kind of popular physical competition and fitness sport, it brings a lot of fun to the aged and children. Since 1960, the team of table tennis in China has obtained a lot of honor in the world games and even won all the table tennis championships of the whole game. Table tennis, popularized for these years, can prompt cooperation of limbs and increase reaction ability in daily life. In the teaching process, most people take exercise according to basic rules, paying not enough attention to PE [1]. However, table tennis, popular with students, is taken as an important physical lesson in many universities. The students can be directed and taught some skills of table tennis to develop the reputation of the national game.

2 CURRENT SITUATION OF TABLE TENNIS TEACHING

Some teachers are traditional and teach according to the original PE principle without establishing the thought of lifelong education. So many students consider that only the weak students need physical exercise and training, causing the inactivity of many students in PE lessons and little interest in table tennis. In addition, the students cannot have a thorough understanding of the knowledge and learn enough skills with few table tennis lessons [2]. Some teachers also cannot teach the comprehensive knowledge with the limit of class our, causing the decrease of teaching quality. In the teaching process of table tennis, because the learning behavior and process cannot be controlled well, the learning results are different. Many students cannot master some motion or skills of kind of ball. The training time and amount of exercise are limited by the large amount of students [3]. So the students are not familiar with some detail, causing not enough interest of table tennis. Consequently, in the teaching process, the students should be taught in accordance with their aptitude and knowledge of various fields should be taught and connected separately. Then the students can understand the principle of table tennis, master real skills and get more confidence and happiness in sports.

3 ANALYSIS ON VISUAL FIELD TRAINING IN TABLE TENNIS TEACHING

3.1 *Analysis on the meaning of visual field training in table tennis teaching*

With a small volume, table tennis requires athletes with attention, stability and judgment. Firstly, in nervous competition process, the athletes should control psychology themselves well and deal with the pressure of the opponents and outside disturbance, making attention training necessary. The visual field training can make the athletes effectively focus and easily devote themselves to the competition in competitions and other situations. Meanwhile, the visual field training can effectively exercise stability and judgment. It can make athletes judge the position and direction of the ball more accurately and strike back better. Also it can improve the stability of the athletes and make them calmly analyze the situation and strike back better in an unexpected situation. So the visual field training has a significant meaning in table tennis teaching. And the training and learning of it can bring much fun in sports of table tennis.

3.2 *Analysis on the process of visual field training in table tennis teaching*

The usual visual field training methods in table tennis teaching, including the focus expansion training method and the focus moving training method, are different in process, but providing help for the table tennis teaching. Firstly, the former is as follows: the eyes are focused in the center of the table tennis, decreasing the time and frequency of blink; the position and movement of the head are kept still and the eyes rotational quickly, and the above exercise is taken by the way of up-down, left-right, first-quarter and last-quarter. Some reversal exercise can be taken with certain proficiency. Each movement should be kept above ten seconds, thus expanding the horizon, focusing, activating brain and improving reaction. The longtime insistence can increase the advantage and achievability, thus making people happy and healthy while having fun. Secondly, the later is as follows: the sight line should follow the movement of table tennis; in the moving process the eyesight should be focused at a point quickly and duly stay still; the frequency of blink should be decreased. Insisting focusing in the moving process for a long time can make the point and the line clear, improve the stability and make themselves quickly judge and strike back in a short time. The visual field training is a preparing and assistant training, but occupying a very important position in table tennis and determining the result of the competition to some content. So the visual field training should be concerned to improve the teaching quality and motor skills in process of table tennis teaching.

4 ANALYSIS ON SKILL TRAINING OF TABLE TENNIS

4.1 Analysis on point training method

Point training can be divided into straight line to point training method and diagonal line to point training method, and they can transform and complement each other. When one hits by straight line, the other will hit back by straight line or diagonal line, or transform the way deliberately and randomly hit back without a fixed impact point. The students should have the acquired reflex trained constantly by combining straight line to point training method with diagonal line to point training method. Then they can deal with the situation that the impact points are different. Although the training method takes time, it can take effect by combining with other skills and cooperating with partners, and the advantage of point training method can be found.

4.2 Three deep and two drop training method

Three deep and two drop training methods add more block factors, control the force of service at random and transform the impact points into unfixed points. Meanwhile the time when table tennis is in the air is different, making the movement of table tennis transform at random. This method can make the students master front-back footwork and find the range and rule of return, thus increasing the integral conception of the students in the table tennis sports by combining with skills of short court, middle court and back court. So the ball with all impact points and angles can be dealt with and the judgment error can be reduced in the competition. In the original training process, the students should closely cooperate with confirmed partner and have the service speed and frequency controlled, improving the teaching quality and the skills and sports level of table tennis through the constant training.

4.3 Single or dead line training method

Single line training method is used to train the students according to fixed line and direction, and it focuses on the weakness and disadvantage of the students. If the student takes reaction slowly to the left attack, then exercise will be taken based on the left service line, thus complementing one another and making all the skills develop in balance. Dead line training method is the expansion of single line training method, making the training line change from the single to moving direction. Normally, the training line moves from left to right. This training has all the lines applied in one training process by combining the stability of the single line with the activity of the multiple lines. Actually, the two training methods are used together. In the original training process, exercise is taken according to the single line training method. After the training purpose and procedure are understood, the training is expanded by adding difficulty. The students can easily accept this method and find achievability and interest in the training process. Training by this method in proper sequence can motivate the activity of the students and improve the efficiency of the table tennis teaching.

4.4 Combination method of multiple skills

In table tennis teaching process, the teachers should have all the skills and abilities trained comprehensively according to the actual situation of the students. If a student is bad at the skills of the right half court, the teacher will direct him alone and make him combine all the skills within the range of the right half court. In the daily training, the tactical awareness of the students should be trained, and the students should consider the training as the competition, thus achieving a better result. In the special direction of some part, the multiple skills of serving, receiving, chopping, speeding, hitting and blocking should be integrated into one. In the situation of combining multiple skills, moving back and forth, moving left and right, weight lifting and dropping and splitting step can keep the students moving constantly, make them reduce the error and deal with the movement striking back in the training process. The students should review and consolidate the daily skill training according to the requirement of the teacher, so that the skills can be applied to the real competition better. Insisting on one thing by focusing energy is not easy, however, if the student insists on learning the skills comprehensively, the skill problems of table tennis will be solved and the health condition will be improved.

4.5 *Mental quality training*

In physical teaching, the more important is mental quality training except the strength training. The influence of mental quality to sports is direct. When the strengths are comparative, mental quality will determine the competition. Table tennis requires high concentration and attention, and trained students are adaptable to the surrounding noise, keeping good mentality and concentration in a confused environment. Consequently, the students should estimate and image the attack of the opponent according to the training situation themselves and the understanding of table tennis, thus improving the skills of the students, cooperating manipulation with step and striking back quickly. Besides, the students should imply that they will devote every effort to get the score, encourage themselves, regulate the mental condition and concentrate attrition in the competition of each round. In addition, they should recall the training skills and understand the tactics of the opponent to adjust the step and the corresponding strategy. Actually, watching the opponent is also the key to improve. In fact, the achievement of physical teaching is useful in daily life. This mental quality training can make the students get something useful in learning and more happiness in sports. The serious attitude is the way for success. So mental quality training is very important in table tennis teaching, providing table tennis with base and service.

5 CONCLUSIONS

The table tennis lessons are set for the purpose of strengthening the body of students, motivating the interest to sports, increasing activity and improving health. Table tennis is more and more popular, although there are some problems in the teaching process. In the work, the problems are analyzed and the methods of visual field and skill training are proposed. E.g. the visual field training method contains the focus expansion training method and moving training method and the skill training method contains the point training method, three deep and two drop training method, single or dead line training method, a combination method of multiple skills and mental quality training method. With the effort of the education department, the teachers and the students, the teaching process of table tennis will be more and more abundant and more and more students will be interested in table tennis and change the attitude to sports and table tennis. Meanwhile, the teachers will become more and more and the table tennis teaching quality will be increased, making contributions to the improvement of the national constitution and the development of the comprehensive talents.

REFERENCES

[1] Zeng Yang, Educational Reform of Table Tennis Selective Course in colleges in China, Time Education, 2012 (1).
[2] Lin Shuhua, How to Increase Teaching Effect of Table Tennis Selective Course in Common Colleges, Science and Wealth, 2011 (3).
[3] Ou Xiaoli, Development and Strategy of Table Tennis Selective Course in Common Colleges in Heilongjiang Province, Journal of Harbin Institute of Physical Education, 2010 (2).

Native language transfer on English listening and speaking skills

Hua Wei

Jiangxi College of Foreign Studies, Nanchang, Jiangxi

ABSTRACT: The study of native language transfer is important to English learning, especially for English listening and speaking skills. Due to the deep-rooted impact of Chinese as the native language on English learning and using, the correct reference to rational thinking of Chinese is conducive to English listening and speaking skills; while habitual thinking of Chinese being copied indiscriminately into English learning will cause Chinglish (Chinese-style English). This work analyzed the impact of native language transfer on English listening and speaking skills training from similarities and differences of cultural background, pronunciation, grammar, syntax and other aspects, so as to promote positive transfer of native language.

Keywords: Native language transfer; English learning; Listening and speaking skills

1 INTRODUCTION

As a kind of social consciousness, language emerges in concerning with certain social existence. Different languages possess both differences and something in common. Long-term living in a particular cultural context as well as learning and using a certain language will inevitably prompt habits adapted to this language [1]. And these habits will impact second language acquisition positively or negatively.

2 NATIVE LANGUAGE TRANSFER THEORY

The study of language transfer especially native language transfer is an important content of second language acquisition. It refers to the impact of existing language knowledge to target language acquisition. Such effects can play a positive role of promotion, called positive transfer; they can also play a negative role of interference, called negative transfer.

The acquisition of second language is different from native language. Also known as first language, native language refers to the first language people have access to when learning language after born [2]. At this time, people have a blank brain without any interference and live in a certain cultural atmosphere, making it easier to learn and master native language. However, it is different for second language acquisition. On the one hand, people have formed a large number of deep-rooted native language concepts in mind as well as certain language thinking patterns and habits. On the other hand, different cultural atmosphere and heritage will generate difficulties in second language acquisition.

3 POSITIVE TRANSFER OF NATIVE LANGUAGE IN ENGLISH LISTENING AND SPEAKING SKILLS TRAINING

English listening and speaking skills are important parts of English learning, closely related to other English knowledge. As languages, English and Chinese both are kinds of social consciousness and reflections of social existence. Native language transfer plays a catalytic role in learning similar knowledge of English and Chinese.

3.1 *Help on understanding of grammar and vocabulary*

Chinese as the native language makes it easier for people to understand its expression and accept its meaning [3]. Therefore, Chinese thinking can be used to understanding complex grammar and vocabulary of English. In addition, lots of Chinese knowledge can be used to memorize English words to help students master them. For example, Chinese and English can be combined to memorize the word morose (depressed). Pronunciation /mə/means no in our Shandong dialect, which turns morose into no rose. How can people be not depressed without roses on Valentine's Day?

Some basic grammars of English and Chinese are same or similar. For example, English and Chinese both have SVO structure such as I study English (我学英语). It provides convenience to promote English learning and positive transfer of native language.

3.2 *Help on voice and reading*

Some parts and methods of pronunciation in English and Chinese are same or similar. So it can be taken advantage of to promote English listening and speaking skills development.

In reading and writing, English and Chinese have some same stylistics, such as narration, exposition, argumentation, and so on. Take argumentation for example, the three composing elements of it are thesis, argument and argumentation in both English and Chinese. The already mastered knowledge might as well be used in reading and writing to help in learning English.

4 NEGATIVE TRANSFER OF NATIVE LANGUAGE IN ENGLISH LISTENING AND SPEAKING SKILLS TRAINING

4.1 *Impact on voice*

Belonging to different language systems, Chinese and English have lots of differences in pronunciation, apart from the similarities. The pronunciations of English and Chinese are similar but not identical. It means that similar English pronunciation can be found in Chinese, but not the exact same pronunciation.

The pronunciation parts of single vowels and consonants are different in English and Chinese. So it is difficult for Chinese students to fully grasp the correct pronunciation of English. For example, the three consonants / b /, / d /, / g / are quite different. In English, these three consonants can be located anywhere in a word, the pronunciations being voiced sound not expiratory; while as phonetic alphabets in Chinese, they are voiceless not expiratory. Because of this difference, learners will make mistakes when they are located at the end of a word.

In Chinese, consonants are always separated by vowels. As a result, English learners often insert a vowel in the middle of consonant tonal and add a vowel at the end of words ending with a consonant. For example, clean and glass are often pronounced as /kəli:n/ and /gəla:s/; put and book are often pronounced as /'putə/ and /'bukə/.

Any languages are with sociality. Different regions have different cultures, producing different dialects. The pronunciation of English and Chinese are different originally and articulation can be much harder in regions of dialects. For example, sort and thought are pronounced the same, so are sink and think. Some people even pronounce though as zhou, articulation of a Chinese character. Another example, many dialects in southern China have no sound of n but l.Thus no I'm not is misread as low I'm not; on Sunday night as on Sunday light.

From the ideographic, Chinese is tonal language while English is intonation language. Chinese distinguishes meaning through four tones of yin ping, rising tone, third tone and falling tone, while English through four intonations of rising tone, falling tone, lifting tone and falling rising tone. Therefore, without considering this difference in learning spoken English, our students speak English bluntly and make native speakers of English puzzled.

4.2 *Impact on vocabulary forms*

Being basic and the most active element of language, vocabulary should be the start of learning any kind of language. As basic vocabulary of Chinese, characters are all monosyllabic and often combine with each other to form words to express meanings; while English is mostly polysyllabic. For example, the word teacher consists of two characters in Chinese but only one vocabulary in English.

Chinese uses less prefixes and more suffixes in word change, meanwhile most of the suffixes indicate people engaged in particular industries, such as accountants, hairdressers, consultants, and so on. However, English has both abundant prefixes and suffixes. Take employee and employer for example, different suffixes being added to employ expresses different meanings. While in Chinese, these two meanings need to be presented in two different words which are worker and boss. This is an inspiration to English learners that Chinese learning experiences cannot be copied indiscriminately. For example, the word university cannot be big school or large school literally translated from Chinese.

4.3 *Impact on semantics*

Only mastering word information, English learning beginners often take the words too literally, apply Chinese knowledge mechanically and thus make mistakes. For example, envisage and visualize both can be translated as assumption in Chinese, but they have different meanings in English and cannot replace each other. Envisage is not an imagination of visual effect but a forecast for future, without picture; while visualize refers to a specific scene outlined in mind, an imagination of visual effect and can be explained by the word imagine.

On third personal pronouns, she, he and it refer to women, men and nonhuman things in both languages. Their distinctions are evident in written Chinese, but they pronounce the same in oral expression of Chinese; while in English, they have respectively three pronunciations. Getting used to the same punctuation, Chinese students easily mistake identities in English oral expression, confusing the usage of she, he and it.

The same verb in Chinese can express different meanings in different occasions, while in English, different words must be used. For example, read books, watch TV and look all have the verb meaning of look in Chinese but with different verbs in English. Another example, open the door, turn on the TV and term begins all have the verb meaning of open in Chinese but with different verbs in English.

These differences in semantics often lead to English learners making mistakes. For example, saying open the TV will let native speakers of English stare dumbfounded.

4.4 Impact on grammar

Chinese has no change of possessive case, person and number, while English has strict requirements. For example, the expression that I am / you are / he is a teacher and we are / they are teachers in English actually has only one verb and one object in Chinese. English learning beginners are easier to ignore these consistency requirements, making errors such as I is and you is.

Chinese and English differ a lot in expressing tenses. English expresses tenses mainly through verb form changes, while Chinese through adding adverbs as adverbial or empty words as complement. For example, English expressions can only be he visited me (yesterday) or he will visit me (tomorrow); while Chinese can express both tenses by adding time adverbial yesterday or tomorrow to the original sentence without changing verb forms.

There are some related words often used in Chinese, such as because…so…, although…but…, and so on. They are not actually one-to-one correspondence with English. English often uses only one related word at a time. For example, the sentence although Hainan is small, but it's a beautiful island is wrongly stated; the correct statement is although Hainan is small, it's a beautiful island or Hainan is small, but it's a beautiful island, with one related word although or but. At this point, habitual usage of related words in Chinese will make things go wrong in English.

4.5 Impact on sessions under different contexts

Languages are all related with certain environments and cultures. Our long-term native language culture is different from English culture. Without consideration of cultural issues, troubles will occur in communication in English. For example, did you eat is Chinese people's most common greeting, using it in exchanges with foreigners will cause misunderstandings that they are invited to dinner. Another example, Chinese people often ask what are you doing when meeting; while in English culture, it will make people uncomfortable, feeling being interrogated. Chinese and Western cultures differences must be paid attention to in exchange process. Chinese culture background cannot be applied mechanically to English communication.

5 CONCLUSIONS

Native language transfer is an important factor affecting English listening and speaking skills training. It may play both a positive role of promotion and a negative role of interference. Educators and learners should make efforts based on the similarities and differences between English and Chinese, to promote positive transfer and reduce negative transfer of native language.

REFERENCES

[1] Sun Xiaoyun, How to Use Native Language Transfer Strategy Effectively in English Teaching, Yu Shu Wai Xuexi: Junior Edition (mid), 2014 (02):78.
[2] Zuo Lifang, On Positive Transfer of Native Language in Second Language Acquisition, Social Scientists, 2007 (s1):202–203.
[3] Gao Yang, Study on Language Transfer in Second Language Acquisition, The Socialism Institute, 2013 (02):300–301.
[4] Wang Lei, Chen Jun, Study on Native Language Transfer in Second Language Acquisition, Journal of Xi'an Institute of Foreign Affairs, 2007 (04):73–78.

Ideological resources of humanistic care of ideological and political education

Xin Wang
Xi'an Physical Education University, Xi'an, Shaanxi, China

ABSTRACT: Scientific development perspective pays more attention to people oriented humanistic care. The Marxist humanistic care theory is the essence of humanistic care that people should be respected, satisfied and focused. The work focused on how to accurately recognize the various aspects of ideological resources at present. This work obtains the relations with all aspects of humanistic care of ideological and political education through the analysis of ideological resources of humanistic care, hoping to enlighten people's understanding and accepting of the education theory.

Keywords: humanistic care; ideological and political education

1 INTRODUCTION

With the development of economy, social environment and life conditions, people's needs have increased from basic food and clothing problems to spiritual needs. For better development, the society has to focus on meeting the spiritual needs, as well as conducting ideological and political education. It is the main route to solve the contradictory in socialist construction, and is also the main content of socialist construction. Guiding socialist construction of ideology and policies matches people-oriented scientific development proposed by former general secretary Hu Jintao, which is also the main content pointed out in the report of the Seventeenth National Congress. The new ideology focusing on humanistic care injects new ideas and life into socialist construction, as a symbol event of social development in this new period [1]. In order to successfully and harmoniously promote progress and development of humanistic care, the government has to thoroughly study the ideological resources, and continually innovate to meet the needs of social development.

way', the new idea of promoting economic development was proposed for the first time, carrying on the concern of humanism to society in the period of socialist construction. That ideological and political education is developed from education of humanistic care and psychological counseling, which reflects the party's consideration on people, society's care on people, and the williness to revise original serious ideological and political education. The transformation of ideological and political education, focusing on humanistic care and people oriented, is a representative word and policy of party and society's care on people, leading to a closer relationship between the party and people [2].

Ideological and political education of humanistic care is mainly concerning about people. The education is a care for people's living conditions and development, an education ideology which cannot exist without *human*, a respect and inclusion for all-round development of people, a development and extension of ideological and political education content in the new period. Directions for the ideological and political education are also provided.

2 HUMANISTIC CARE CONTENT OF IDEOLOGICAL AND POLITICAL EDUCATION

According to the report in the Seventeenth National Congress 'building a harmonious socialist society, and developing socialist spiritual civilization, ideological and political work should be strengthened and improved, focusing on humanistic care and psychological counseling, handling human relationship in right

3 IDEOLOGICAL RESOURCES OF HUMANISTIC CARE OF IDEOLOGICAL AND POLITICAL EDUCATION

According to the humanistic care and scientific development perspective of general secretary Hu Jintao pointed out in the report of the Seventeenth National Congress, ideological resources of humanistic care should be various, and conform to the summary of social development trend. The ideological resources mainly include Marxist theory, Chinese traditional culture, modern western theory of ideological and political education, and objective needs of development and progress of the economic society [3].

3.1 Research on humanistic care theory from Marxist theory

Corresponding to its core value, humanistic care is the most important problem studied in Marxist philosophy. Therefore, it should learn from Marxists by theory integration.

Marxists focused on the people's freedom and development all the time. Marxists pointed out that humanistic care theory had a lot to do with life conditions. In that period, the people living at the bottom of capitalist countries were so poor that their food and clothing problems could not be solved, which have great effects on advocating the socialist society of Marx and Engels. Meanwhile they thought everyone was born equally, and had rights to pursue happiness and freedom. Therefore, Marxists believed that people's material needs should be satisfied. Once people had enough materials, they could spare more energy to work, create and make history, which were the most basic needs below the spiritual needs, guaranteeing people's living and development and motivating the development of humanistic care of Marxists.

After people's needs are satisfied, improving people's spiritual life is the main purpose of Marxist humanistic care theory. Once people's basic materials are guaranteed, their desire for spiritual needs will become stronger and stronger. However, surveys revealed that people are mainly concerned about their dignity and freedom. As a fundamental difference from other animals, human's desires for respect and living in a free society are essence power existing in their hearts. The desire for dignity and freedom will promote people's desire for social development. But the reason of social contradictions is people's grievance and complaint on unsatisfied needs in their lives. Therefore, society should meet people's material needs and spiritual needs to ensure people a balanced life and minimized social conflicts, to make contributions to the construction of harmonious society. After society developing to a certain stage, Marxist humanistic care theory, which is the corollary of social development, appears to meet material needs and embodies a theory concerning about people's spiritual needs; as the corollary of social development, the theory is an important step in the course of social development, embodying institutional reforms and system reform of superstructure determined by economic base.

In the 21st century, the development of socialism modernization requires more assistance from the development of socialist spiritual civilization to promote the construction of harmonious society, which is the corollary of social development. However, with the developing society, changing social environment and improved life conditions, contradictions of social development are improved constantly, and the differences between ideology and economic development become the most serious barrier of the whole country development. Therefore, the party and country should guide and educate people with ideological and political education to rich their corresponding knowledge. In recent years, with the developing society, the proposed ideological and political education of humanistic care combines humanistic care and reality of our country together to change the traditional ideological and political education. Based on people's fundamental interests, their initiatives and voluntary of ideological and political education are improved in the education mode of people oriented principle, humanistic care and psychological counseling. As the intent and purpose, people's interests should be the aim of Communist Party of China. Therefore, Marxist is the important ideological resource of humanistic care of ideological and political education.

3.2 Influence of humanism spirit in Chinese culture with a long history

In order to develop the humanistic care theory of ideological and political education with Chinese characteristics, we need thorough study on the important thought and theory of humanistic spirit in Chinese history and culture. It will be integrated with the development of modern society for innovation and improvement, making contributions to the construction of ideological and political education of China's socialism society.

Rising during the *spring and autumn and warring stage period,* humanism in Chinese traditional culture has a long history through the whole development of Chinese traditional history and culture. The core idea of humanism spirit in Chinese traditional culture is Confucian kindheartedness. As the representative of Confucian, Confucius advocated *love*, meaning well-intentioned and harmony interaction between each other. He also emphasized that never imposed your beliefs on anyone else.

Confucius is the founder of the Confucian humanism. His descendants developed humanism on the basis of his thought of *love*. Mencius is one of his outstanding successors, who suggested the thought of *Love All People and World* stating the relationship between relatives, ordinary people and all things on the world. He thought that love would keep good things last longer, as well as a good relationship between each other.

Although Mencius's views supported the lords, his thoughts also met the requirements of human-centric ideological and political education in modern society. After that, Dong Zhongshu introduced some innovation and improvement on kindheartedness. His theory played an important role in the development of Confucianism. Meanwhile, it provided a good sample for us to learn from the humanistic thought of the ancients.

Kindheartedness meets the needs of the socialist modernization construction so well that the government should preserve it as essence in Chinese traditional culture and integrate it with ideological and political education in modern times at the same time. Only in this way can they make contributions to the social development.

3.3 *Reasonable absorption of western humanism*

The tradition of western humanist is the original human care theory. It is very necessary to learn from western humanism as well as Lu Xun's theory of *take doctrine* during the innovation and development process of humanistic care. In addition, more innovations based on China's basic national conditions are required for suitable humanistic consideration.

Humanism has a great influence on the development of the whole world. Its core values include affirmation of humanity and realization of human values, requiring human individuality liberation, freedom and equality. Humanism spirit has shown full respect for human dignity and freedom by promoting human rights and opposing theocracy and ignorance. It fully reflects the high development and consciousness of people-oriented ideology. It is not only a revolution in the history of human thought, but also a great outbreak to create history and refresh history at the same time, which can be a great reference for ideological and political education.

Firstly, humanistic care theory of people-oriented ideology in western humanism plays an important role in the corresponding ideological and political education. It is the theoretical basis for the conformation, development and innovation, as well as important idea resources for socialist society development. Secondly, liberation of personality is advocated in the ideology, corresponding to the ideological and theoretical of humanistic care. Both of them emphasize human autonomy, and confirm the positive effect of individual liberation on ideological and political education. Last but not least, the ideology promotes scientific, opposes ignorance and motivates people to contribute to the development of society with their talents, providing the basis for the integration of scientific spirit into humane care at the same time. To help others get rid of the state of ignorance and superstition, promoting their comprehensive development is one of the ultimate goals of humanistic care in ideological and political education. Moreover, the relevant significance of western humanism to humanistic care in ideological and political education involves many aspects, such as homage to people's status, personality differences, and satisfaction of the diverse needs of people. It also helps people solve more practical problems to stimulate their creativity and enthusiasm. By carving human creation of life into people's minds, the nation's cohesion will be improved, breaking the superstition of god dominance. The world will be much better when the development of society and human are combined.

3.4 *Objective needs of economic and social development*

Humanistic care of ideological and political education is the product of social development when it comes to a certain level. Social development is promoted by economy, and it demands more for the work of ideological and political education. On the other hand, continuous progress of ideological and political education provides more highly qualified professionals to the economic development of the society. Both economic development and ideological and political education are absolutely essential throughout social development.

China has come to the key stage of the socialist transformation now, when any direction of development in economy or society will have an enormous impact on the whole economic society. It seems that a stable social environment is a guarantee at this critical time, where enough development space can be ensured for the economy. Moreover, ideological and political education is an important method to resolve social conflicts and to build a socialist spiritual civilization. Humanistic care theory of ideological and political education changes the method from education-oriented to human care-oriented, making people more willing to participate in ideological and political education. And then it will play a very important role in the development of the ideological and political education and the stability of society.

4 CONCLUSIONS

The human care theory of ideological and political education is a milestone of the social development process, which embodies both the nation's attention to the people and the expectation for social stability. Ideas of humanistic care have generated from many different resources. To build socialism, integrating Chinese traditional humanism with western humanistic spirit and merging these resources based on Marxist theory are required, with the consideration of people's interests. With continuous development and innovation, a new theory suitable for the development of socialist society with Chinese characteristics will be established. Definitely, it will have a significant impact on the overall economic development of the society.

REFERENCES

[1] Yongxin Liu. *The theory origin of humanistic care of ideological and political education.* Journal of Hubei Adult Education Institute. 2010(3): 22–25

[2] Dongli Wang. *Thought Resources of humanistic care of ideological and political education.* Zhejiang Academic Journal. 2010(3): 9–12

[3] Yonghong Han. *A general overview of humanistic care of ideological and political education.* Journal of Changchun University of Technology. 2011(2): 20–24

Visualization of numerical methods for ordinary differential equations based on computer network

Linlin Li
School of Mathematical Science Inner Mongolia University, Hohhot, Inner Mongolia, China
School of Statistics and Mathematics, Inner Mongolia University of Finance and Economics, Hohhot, Inner Mongolia, China

ABSTRACT: Study and application of higher mathematics have been more and more popular in recent years. Higher mathematics, involving lots of experiment content, is closely related to other courses. In the field of equations, the mathematical solution is difficult because of some unknown functions. So, many courses make great progress by the application of ordinary differential equations. In the work, the visualization solutions based on computer network are discussed according to the analysis of numerical methods for ordinary differential equations.

Keywords: Computer Network, Ordinary Differential Equations, Numerical Methods, Visualization, Analysis

1 INTRODUCTION

Reform and development of education has made influence on higher mathematical education in colleges. Application of higher mathematical theory is more and more important because the course boundary is more and more blurred. Especially, differential equations of higher mathematics are widely used in physics, and the solutions are obtained by the combination of the two methods. Besides, ordinary differential equations are used in some professional fields. E.g. ordinary differential equations are applied to solve constants and variables in signal system. Consequently, ordinary differential equations can be combined with other courses to solve problems [1]. In the work, visualization analysis is used to solve the visualization problems of Numerical Methods for ordinary differential equations based on the computer network.

2 ANALYSIS OF ORDINARY DIFFERENTIAL EQUATIONS

Equation, as the most important concept and method, can be used to solve the unknown factors and the actual problems. In elementary mathematics, there are many equations, such as unary quadratic equations, high-order and multivariate equations, etc. These equations are used to solve problems according to the direct relation of the unknown and known data. However, the special equations, namely ordinary differential equations, are used to solve problems in the actual production. It is confirmed that many concepts and theorems are established in ideal state in elementary mathematics and natural sciences [2]. So, higher natural sciences focus on numerical and scientific research in non-ideal state, thus enlarging the range and content of research.

2.1 *Context of ordinary differential equations*

Theories and concepts of elementary functions and natural sciences, including the concept of equation of elementary functions, are established in ideal state. However, actual production and research involve many problems in non-ideal state. E.g. in motion of substance the absolute uniform motion is impossible to exist, that is, the motion of substance including the variation cannot be directly analyzed by using the method in ideal state. In addition, the operation track of the aircraft and the environmental factors in the research of aerospace are the key points to be considered in actual production [3]. These problems cannot be solved by using elementary functions and equations because the existing equations cannot be applied to these problems. So, ordinary differential equations are used to analyze the existing data and solve the unknown.

2.2 *Solving thought of ordinary differential equations*

In the research, many equations cannot be solved by using the mode of elementary equations, so in order to achieve the solution of the unknown, the mode of ordinary differential equations should be used. Normally, ordinary differential equations are widely used in the research field of physics to solve the unknown functional relation according to the derivation of the known functional relation. Besides, the solving thought of ordinary differential equations is different from the traditional elementary equations, because the content to be solved is not some numbers but the functional relations in these problems.

Analysis of ordinary differential equations is different with traditional elementary equations, but the solving thought is similar. And the thought is to analyze the unknown relation by the known and solve the problems by the relation of several equations. The unknown functional relations, instead of the separate variables, are solved.

2.3 Solving method of ordinary elementary equations

The knowledge of differential and derivative is used in the solving process of ordinary elementary equations, that is, the solving method is widely changed in the solving process. So, the equations, expressed as the relation of the unknown function and the variable, can be called ordinary elementary equations.

3 NUMERICAL METHODS FOR ORDINARY DIFFERENTIAL EQUATIONS

In general, ordinary differential equations can be solved by two methods: analytic methods and numerical methods. Analytic methods are not adaptable to solve the actual problems, because many uncertain factors have influence on the difficulty of solution. So, numerical methods are usually used to solve the actual ordinary differential equations by computer. The actual case is analyzed by using numerical methods.

E.g. ordinary differential equations can be expressed as:

(1) $y' = kx$, where k is constant
(2) $(y - 2xy)\,dx + x^2\,dy = 0$
(3) $mv'(t) = mg - kv(t)$

These elementary ordinary differential equations can be solved by using numerical methods, thus achieving a simple solution process. If analytic methods are used, the equations will be unsolved and the unknown relation will be uncertain, causing the difficulty of solution. So, numerical methods can ensure the efficiency and accuracy of solution to a large extent.

4 TECHNOLOGY APPLICATION OF COMPUTER NETWORK

With the development of computer technology, technology application of computer network is more and more popular, making visualization technology of computer network the main stream.

4.1 Application of computer network

Computer network, widely used in the internet field, has changed the traditional life in information interaction. In order to accelerate the communication of people, computer networks use the hierarchical design method to ensure that the data of different forms can interact according to the division of layers. So, the hardware of computer network should be specially designed to ensure the accuracy of information interaction in high speed transmission. Computer network can be applied to many fields of production and life.

Firstly, computer network, as the main path of communication, become the basic architecture of internet. Currently, network has been more and more popular and become the part of people's life. In addition, many companies based on internet have occurred, thus making people engaged in internet getting more and more.

Secondly, computer network can realize the information browse and interaction. In general, long-distance information interaction can be realized by internet, thus achieving the long-distance interaction of people. The occurrence of computer network makes communication of people more convenient.

At last, the design of software and hardware has laid the foundation for computer network in order to intensify the application of computer network. Among them, the software uses the hierarchical design method so that data can be divided according to the hierarchical structure to ensure the accuracy of information transmission; the hardware uses the hierarchical pattern of the main engine, the exchange board and the router, ensuring reasonable application of network resource. Besides, computer network security, as the main application problems of computer network, should be strengthened. At present, the problems can be solved by some measures based on software and hardware. So computer network security is very important with the development of computer network.

4.2 Combination of computer network technology

Design of web page is the most important in the application of computer network, and web has been a visual form of internet. Web page is the most popular application of visualization. In addition, computer network technology, combined with other fields, has many cases. E.g., web page is used as the entrance to ensure the experience of visualization and effect of other fields in the application of network visualization. Consequently, computer network can be used in many fields and can make complicated contents of natural science easy to understand. Computer network technology can play a role by combining with other fields.

5 VISUALIZATION ANALYSIS OF NUMERICAL METHODS OF ORDINARY DIFFERENTIAL EQUATIONS BASED ON COMPUTER NETWORK

The above analysis shows that the combination application of computer network technology is widely used especially in the solving process of ordinary differential equations, making computer network visualized. Numerical methods should be used to make the solution of ordinary differential equations simple and convenient. However, analytic methods cannot visualize computer network, and there will be many problems in the solving process. So, numerical methods are more suitable for computer network to realize visualization.

Firstly, for the visualization research of computer network, the computer logic language is used to compile the codes to ensure distinct appearance of the data and the visualization in the solving process of ordinary differential equations. E.g. the input box is established for the input of some constants and variables, and the data are computed by using functions and codes, thus making the solving process more convenient and quick.

Secondly, the internal call and use of functions is convenient. Some frequently used methods are designed for the database by using the principle of program and function, thus making them more convenient to use. Some constants and variables are used to make the functions more effective. In general, combination of computer technology is to make the original solving methods more convenient and quick, so the consumer should be able to enter the constants and variables and efficiently distinguish the constants and variables in the computing process. The call functions can be used to achieve the results as long as the data are entered by the consumer.

At last, the visualization should be realized by the correct call of input box and input functions on basis of computer network technology. The codes should be compiled correctly in order to realize visualization. E.g. in the solving process the frequently used visual software can be used to make the computing process clear and visualized.

Consequently, numerical methods of ordinary differential equations based on computer network technology are established on basis of computer programming and visualization technology to make the solution of ordinary differential equations simple and quick and make the solving process visualized, thus increasing the interest of solving ordinary differential equations.

6 CONCLUSIONS

In the work, the definition and solving methods of ordinary differential equations are analyzed and found to be different with those of elementary equations. In many research fields of physics, ordinary differential equations are widely used to solve the unknown relations. In addition, with the development of computer network technology, visualization of solving process of traditional ordinary differential equations has been one of research projects in many fields currently. Visualization can make the solving process transparent and easy to understand and analyze, thus realizing reducing the solving difficulty of complicated differential equations. Consequently, ordinary differential equations are widely used in the research of many fields. Visualization of numerical methods based on computer network can make the solution of ordinary differential equations simple, convenient and visualized.

REFERENCES

[1] Shi Hongjuan, Gan Qiuge, Gong Youhui, Gan qiuwei, Reasons and Solutions of Slowness of Local Area Network, Journal of Yangling Vocational and Technical College, 2006(02): 50–52.
[2] Zhu Jianguo, Application and Development of Computer Network, Computer Knowledge and Technology, 2010(03): 560–562.
[3] Tang Shaoze, Development of Computer and Network Technology, Science and Technology Economic Market, 2007(03): 208–208.

Study on quality control model of CPA firms under the current situation

Hongguang Xiang
Hunan University of Finance and Economics, Changsha, Hunan, China

ABSTRACT: Nowadays, CPA firms' accounting and audit quality are challenged by greater diversity and uncertainty in financial contents of enterprises and institutions, as well as anthropogenic negative energy existing in actual practices. In order to confront this challenge, a CPA firm should make adjustments in 3 aspects, i.e., inherent internal control, internal and external related accounting law and regulation, and integrated management model. This work discussed CAP firms' quality control model in the current situation to improve their accounting and audit quality.

Keywords: Quality Control Model; Auditing; Accounting Legislations; Integrated Management

1 INTRODUCTION

Many managers agree that reforming in accordance with requirements of the time is the best way to survive for an organization or institution under the Current Situation; as an organization responsible for certain social functions, each CPA firm shall make necessary changes, too. The basic function of a CPA firm is to organize and assign accountants to an organization/institution to do effective and precise financial accounting and supervision works for a specified economic entity, and to produce accurate and systematic financial statements, so to propose scientific and reasonable decision making suggestions to the organization/institution. However, because of changes in forms of organizations/institutions' financial affairs and CPA firms' practices, some so-called "usual practices" which are illegal and morally indefensible are increasingly impairing the national accounting and audit quality [1]. It is inevitable for Chinese managers to find solutions of this problem, and to prevent potential financial risks.

2 INFLUENCES OF THE CURRENT SITUATION ON QUALITY CONTROL MODEL OF CPA FIRMS

In view of the difference of enterprises in accounting and audit quality and great impacts of certain entities' financial status on the national economy and the smooth operation of domestic market, the Chinese Institute of Certified Public Accountants (CICPA) has made a statement calling for more attentions to accounting and audit quality improvements, standardization of CPA practices, and powerful guarantee for enterprises' smooth operations [2]. For CPA firms, quality defects may be originated from 2 aspects: On one hand, some small scale firms, which are congenitally deficient and limited by poor employment and organization, have to struggle to manage heavy annual workloads as well as serious competition and low employee loyalty. In this case, employees of a small CPA firm are much easier to be tempted by money as well as power in organization level within the firm, and may bring serious consequences to the firm, because of more livelihood problems. In this case, the firm may degrade its accounting and audit quality more or less for long term stability and profitability; and sometimes, it may go in this way so far, exceeding the extent legally permitted. It is similarly worldwide. For example, in the ENRON Scandal revealed in 2001, the United States Department of Justice obtained a criminal indictment against Anderson (the entire firm) for obstruction of justice, mainly because of its illegal destruction actions during the judicial investigation. Anderson is the first accounting firm brought to justice in America [3].

On the other hand, poor internal management is another negative factor of a CPA firm's accounting and audit quality. In 2013, the Ministry of Finance executed a nationwide investigation on CPA firms, 23396 enterprises and institutions and 1398 CPA firms were involved. The investigation results indicated that most CPA firms obeyed the "Accounting Law of PRC", the "Law of the PRC on Certified Public Accountants" and other related laws, carried out strict accounting procedures and rules, and provided basically reasonable and lawful audit and supervision service. The investigation also revealed some intractable problems in CPA firms' internal control, financial accounting, tax payment and other aspects, including false financial accounting, unstandardized profit adjustments and disorganized internal management etc., to which attentions should be paid by relevant authorities. These problems may directly degrade a CPA firm's accounting quality; more seriously, they will disturb an enterprise/institution's financial statements, disorder its account items, make the enterprise/institution act in violation of laws and regulations, even confuse the enterprise/institution about real status of its overall financial operation and result in complete rupture of the enterprise/institution's capital chain. Not only would a CPA firm's quality control impact enterprises and institutions the firm is working for, but it would also play the Butterfly Effect to the national economy. A slight negligence will cause an undesirable consequence.

In one word, the current economic situation requires immediate changes in the quality control model of CPA firms. It is important to make timely adoptions and adjustments for maintenance of credibility and reliability of the accounting industry.

3 IMPROVEMENT OF QUALITY CONTROL MODEL OF CPA FIRMS

3.1 *Strengthening internal control and improving systematization*

Internal control is more practically significant than external control for each organization, company or enterprise, and it is an important way of longitudinal development for each individual enterprise/institution, company member or employee. Consisting of environmental factor control, maximum/minimum risk estimation, necessary internal inspection and self-evaluation, decision making related data/information processing, and final decision making etc., internal control should be implemented effectively in each step of production and operation to improve the overall operational efficiency and to achieve strategic objectives. It is noteworthy that CPA firms are independent audit service suppliers which could effectively execute necessary financial accounting and auditing supervision for other enterprises and institutions, and they play a key role in all society sectors' internal quality control, from exterior to interior. So, the internal control of CPA firms is much more important. Responsible for a special social position, CPA firms shall build up a robust internal control system, and provide reasonable risk assessment/analysis and crisis management service to other organizations, so to achieve sustainable development of them.

A good work atmosphere is necessary for a CPA firm to ensure all accountants not only obey relevant moral rules, practicing standards, laws and regulations, but also cooperate with each other friendly. This kind of atmosphere could ease employees' nervous tension and eliminate negative effects of anxious emotions, also could raise work enthusiasm and improve work efficiency, from another point of view.

Within the firm's internal organization, cooperation among employees could enlarge each person's work scope and improve his/her ability, raise the sense of unity, and build trust with each other. These positive effects could reduce employee illegalities and irregularities to a certain extent, and benefit the quality control of the whole firm. In addition, CICPA asks accountants strengthen the execution of auditing standards further, implement the risk-based auditing concept, maintain professional skepticism, stress the importance of audit repeatedly, and ask for sufficient and real audit evidences legitimately, during financial statement auditing processes.

This kind of improvement takes both internal management control and professional service requirements into consideration reasonably; it could not only ensure effective coordination and integration, but also create a brand new institutional situation to the management of CPA firms. Robust and distinct systems could guarantee accountants' practicing levels and work quality, and accountants need these systems in work orientation, problem solution, and self-improvement, to carry out effect, normalized and high quality practices. The management of a CPA firm would become a duck in water with robust and distinct systems.

3.2 *Strengthening accounting legislations' enforcement and supervision*

Although, most CPA firms obey the "Accounting Law of PRC", the "Law of the PRC on Certified Public Accountants" and other related practicing standards, some typical misconducts still exist, e.g. private coffers of some enterprises, false invoices or other accounting documents, casual profit appropriation changes asked by operators, false financial accounting for tax evasion/fraud/evasion, non-execution of 2-line tax-nontax income separate accounting in some public institutions (resulting in misappropriation of dedicated funds and IOU notes in accounting) and poor accounting infrastructures (including unqualified accounting personals and accounting books, and ill accounting management) in some small enterprises, etc. Actually, all above mentioned misconducts are disobedient to the "Accounting Law of PRC", so they are illegal actions. But because of insufficient investigations of CPA firms, industrial and commercial administrations, and audit offices, most of them could get away. In this case, the accounting industry suffers an immoral behavioral momentum and a serious vicious circle, and enterprises/institutions suffer a financial bubble. If things go on like this, a potential crisis will be created in operation of CPA firms and enterprises. Once an enterprise or CPA firm is investigated for legal liabilities, many other enterprises/institutions/CPA firms would be implicated, and another round of supervision and law enforcement will rise like a storm.

Nowadays, these "usual practices" are accepted as rightful by some accountants who disregard accounting laws and regulations, as well as professional ethics, for personal interests. Besides rich experiences of the accountants, another reason why illegal accounting actions could escape punishment is relaxed and weak management of supervisory authorities, which connive with an unhealthy tendency in the accounting industry. In a nationwide CPA firm investigation, financial administrations have inflicted punishments on many CPA firms by warning, fining, confiscation of illegal gains, and suspension of businesses etc., for illegal or inappropriate accounting; and even by revocation of business licenses on a CPA firm from Bohai, Tianjin, and a CPA firm from Buxin, Liaoning. Some certificated accountants were also punished by fining, suspension of practices and revocation of certificate in the investigation. Investigations executed occasionally are not enough. In order to make relevant laws and regulations go into full force and provide legal guarantee for healthy development of the accounting industry, the country shall build up regular and standardized supervision procedures, and CPA firms shall cooperate with supervisor administrations in law enforcement and supervision.

3.3 Integrated management required by the current situation

CPA firms of different business scales are challenged by management problems of different characteristics. First of all, no matter how many employees a CPA firm has, integrated management shall be emphasized. As a professional organization, a CPA firm shall have industrialized and standardized operation, streamlined management and efficient management, so it shall build up uniform behavior norms, work norms and work principles to unify employees' subjective awareness, so to unify their behavioral habits and work procedures effectively.

In addition, integrated management is the trend of function conversion of companies, public institutions and authorities, and is a kind of management method which could take effect rapidly under the current situation of high data/information integration and interdisciplinary collaboration. Not only could it guarantee reasonability and smoothness of works, but it could also guarantee a development characterized by personality and branding for an organization, in the integrated economic environment, in a certain period of time. It is an opportunity to improve management levels for large enterprises, as well as CPA firms. CPA firms shall focus on standardization in implementation of the integrated management, including further studying on relevant laws and regulations, uniform evaluation on accountants' performances, and even uniform administration on each employee's behaviors.

Not only could standardization perfectly enable a CPA firm's management but it could also play an important role in maintenance of its professional specialty.

Integrated management is advantageous in urging managers to make long term planning for the CPA firm, actively or passively, and to improve the existing model of management (including business management, HR management and so on) further, and advantageous in improving the member loyalty, employee enthusiasm and attention to details. All these aspects are key factors to build up an advanced and high efficiency management system. Moreover, integrated management will facilitate the firm's expansion and branch establishment. It is easier for a CPA firm under integrated management to grow into an elite group with huge longitudinal and horizontal development spaces, to follow the trend of integration in management of various sectors, and to increase the value of the firm.

4 CONCLUSIONS

As professional organizations responsible for accurate auditing and supervision on companies' financial status and providing reliable data/information evidences for the social economic construction, CPA firms bear heavier legal and social duties than common companies or enterprises. In order to correct the unhealthy tendency existing at present, the CICPA requires CPA firms to improve their accounting and auditing quality by management model improvements and adjustments. Therefore, CPA firms shall strengthen their internal management, complete audit jobs better, improve employees' legal awareness and enrich their legal knowledge cooperating with external authorities, get tougher with malfeasances and illegal actions, implement a trend following integrated management model, and do their best to play the role of auditing and supervision.

REFERENCES

[1] Li Xiaoxi. Thinking on Accounting Standard Setting. Journal of Beijing Institute of Business: 23–26.
[2] Xinmin Dai, and Na Zhou. The Influencing Factors of Audit Quality in China—Basing on the New Institution Economics. Anhui University of Technology: 67–73.
[3] Qinghui Ye. Rethinking the Judgments of Accounting's Materiality. Xiamen University, 2013(17): 45–50.

Author index

An, X.L. 161
Ao, L.X. 151

Bai, F. 117
Balayan, A.E. 71
Bian, P. 183
Butyrin, M.V. 63

Cai, X.Y. 5
Cao, G.X. 135
Cao, P. 223
Chai, X. 45
Chen, H. 191
Chen, J. 205
Chen, Y. 79
Chi, H. 25
Cui, X. 17

Deng, G.C. 147
Dong, Z.H. 83
Du, B. 259
Du, M. 75
Duan, Q.M. 135

Feng, L.B. 135
Fu, X.Y. 103

Gan, J. 125
Gao, Z. 187
Gong, B. 25
Gong, S. 183
Gong, Y.L. 29
Gu, H.W. 217
Guan, C.Z. 93
Guo, B. 87
Guo, M. 107

Han, S. 17
Hao, J. 39
He, X. 39
He, Z. 183
Hu, G. 59
Huang, L. 75
Huang, W. 33
Huang, X. 49

Iosifovich, S.D. 67

Jia, Y.H. 93
Jiang, L. 141
Jiang, X. 267
Jin, J.W. 99

Jin, Y. 121
Ju, S. 251

Kang, Z.H. 1
Konovalov, A.S. 63
Konovalova, E.Yu. 71

Lai, Y. 175, 179
Li, C. 9
Li, H. 295
Li, J. 141, 283
Li, L. 307
Li, N. 201, 235
Li, Q. 1
Li, S. 117
Li, X. 25
Li, Y. 1
Lin, J. 135
Liu, F. 49
Liu, H.R. 29
Liu, J.D. 99
Liu, K. 45
Liu, Q. 183
Liu, S. 59, 169, 187
Liu, Y. 175, 179
Luo, Z.G. 175, 179

Ma, D.X. 175, 179
Ma, H. 247
Ma, M.J. 53
Ma, R. 157
Ma, S. 87
Ma, X. 113

Nie, Z. 187
Ning, Y. 195

Peng, B. 117
Peng, X. 113
Prohorovna, K.R. 67
Protasov, E.S. 71

Qian, Y. 75, 79, 271
Qin, J. 141
Qu, A. 239
Qu, J. 141

Ren, Y. 187

Saksonov, M.N. 63
Shen, Y. 291
Shi, N. 235

Shi, Y. 141
Song, D.Y. 33
Song, J. 141
Song, Z. 141
Sorooshian, S. 13
Stom, D.I. 63, 71
Sun, H. 255

Tang, C. 79
Tang, W.C. 107
Tao, J.C. 147
Tian, F. 25
Tolstoy, M.Yu. 71
Tong, W. 211
Tyutyunin, V.V. 63, 71

Wan, Z.P. 107
Wang, C. 187
Wang, F. 279
Wang, H. 227
Wang, J. 131
Wang, L. 287
Wang, M. 165
Wang, S.J. 29
Wang, S.Y. 21
Wang, W. 165
Wang, X. 303
Wang, Y.F. 1
Wang, Ying 187
Wang, Yuan 187
Wang, Z. 87
Wei, H. 299
Wu, G. 113
Wu, J.F. 103
Wu, J.Y. 217
Wu, S. 205, 211

Xiang, H. 311
Xie, H. 125
Xu, L. 5
Xu, Q. 121
Xu, Y. 169
Xu, Y.C. 147
Xue, L.Y. 33
Xue, X. 141

Yan, X. 49
Yan, Z.G. 107
Yang, B. 275
Yang, F. 169
Yang, G. 157
Yang, X. 75, 79

Yang, X.F. 99
Ye, X.Z. 99
Yi, J. 125
Yi, K.J. 33
Yi, X.F. 135
Yu, D.M. 99
Yu, T. 165
Yu, Z. 113
Yuan, H. 217
Yue, D.Q. 53
Yurevich, K.A. 67

Zang, H. 25
Zhan, W. 243

Zhang, C. 9
Zhang, D. 187
Zhang, F. 161
Zhang, K. 157
Zhang, L. 21
Zhang, R. 201
Zhang, S. 263
Zhang, T. 141
Zhang, W. 141
Zhang, Y. 29
Zhang, Z. 231
Zhao, B. 53
Zhao, C. 45
Zhao, Z. 125

Zheng, W. 157
Zheng, X. 107
Zhou, H. 5
Zhou, M.J. 147
Zhou, X.B. 151
Zhu, X. 191
Zhu, Y. 165
Zhu, Y.Z. 103
Zhuo, H.C. 5
Zong, Z.L. 83
Zou, H. 191